Student Solutions Manual
to accompany

Charles P. McKeague

Elementary Algebra

EDITION 5

Patricia K. Bezona

Valdosta State University

SAUNDERS COLLEGE PUBLISHING
Harcourt Brace College Publishers

**Fort Worth   Philadelphia   San Diego   New York
Orlando   Austin   San Antonio   Toronto
Montreal   London   Sydney   Tokyo**

Bezona: Student Solutions Manual to accompany Elementary
Algebra, 5e. McKeague.

ISBN 0-03-009782-7

890123 021 98765

PREFACE

This Student Solutions Manual accompanies **Elementary Algebra, Fifth Edition,** by Charles P. McKeague. I believe this manual will help students who need more guidance with the mathematics problems than is usually available through class time or individual assistance.

The manual contains solutions to every other odd-numbered problem, all the review exercises and all the chapter text exercises. The solutions should be followed, step-by-step, to solve any problem that is confusing to you. The solutions may be a guide on homework assignments.

I would like to thank Charles P. McKeague for giving me this opportunity and for all his guidance; Marc Sherman, my editor, for his patience and understanding; and Suzanne Garlow for the keyboarding, proofreading, and final formatting of the manual. A special thanks to Theo Reed, Ann Sturgeon, my husband Ron, and my daughter Tammy for their constance source of encouragement during this project.

Patricia K. Bezona

TABLE OF CONTENTS

SOLUTIONS TO SELECTED PROBLEMS

CHAPTER 1

SECTION 1.1

1. $x + 5 = 14$ Remember **sum** means add and **is** means equal

5. $3y \leq y + 6$ Remember **product** means to multiply and **sum** means to **add**

9. $3^2 = 3 \cdot 3 = 9$ Base 3, exponent 2

13. $2^3 = 2 \cdot 2 \cdot 2 = 8$ Base 2, exponent 3

17. $2^4 = 2 \cdot 2 \cdot 2 \cdot 2 = 16$ Base 2, exponent 4

21. $11^2 = 11 \cdot 11 = 121$ Base 11, exponent 2

25. $2(3 + 5) = 2(8) = 16$ Twice the sum of 3 and 5 is 16

29. $(5 + 2) \cdot 6 = 7 \cdot 6 = 42$ Add inside parentheses first, then multiply

33. $5(4 + 2) = 5(6) = 30$ Add inside parentheses first, then multiply

37. $(8 + 2)(5 + 3) = 10(8) = 80$ Add inside parentheses first, then multiply

41. $5 + 2(3 \cdot 4 - 1) + 8 = 5 + 2(12 - 1) + 8$ Multiply inside parentheses first

        $= 5 + 2(11) + 8$ Then, subtract inside parentheses

        $= 5 + 22 + 8$ Multiply

        $= 35$ Add left to right

45. $4 + 8 \div 4 - 2 = 4 + 2 - 2$ First, divide

        $= 6 - 2$ Then, add

        $= 4$ Finally, subtract

49. $3 \cdot 8 + 10 \div 2 + 4 \cdot 2 = 24 + 5 + 8$ Multiply and divide left to right

        $= 37$ Then, add

53. $5^2 - 3^2 = 25 - 9$ Simplify numbers with exponents

        $= 16$ Subtract

57. $4^2 + 5^2 = 16 + 25$ Simplify numbers with exponents

        $= 41$ Add

61. $2 \cdot 10^3 + 3 \cdot 10^2 + 4 \cdot 10 + 5$

        $= 2 \cdot 1000 + 3 \cdot 100 + 4 \cdot 10 + 5$ Simplify numbers with exponents

        $= 2000 + 300 + 40 + 5$ Multiply from left to right

        $= 2,345$ Then, add

65. $4[7 + 3(2 \cdot 9 - 8)]$

    $= 4[7 + 3(18 - 8)]$

    $= 4[7 + 3(10)]$ Simplify within the innermost grouping symbols, first

    $= 4[7 + 30]$ Multiply inside the parentheses

    $= 4[37]$ Add

    $= 148$ Multiply

69. $3(4 \cdot 5 - 12) + 6(7 \cdot 6 - 40)$

    $= 3(20 - 12) + 6(42 - 40)$ Inside parentheses, first

    $= 3(8) + 6(2)$ Then, subtract inside parentheses

    $= 24 + 12$ Add

    $= 36$

73. $5^2 + 3^4 \div 9^2 + 6^2$

    $= 25 + 81 \div 81 + 36$ Simplify numbers with exponents

    $= 25 + 1 + 36$ Divide $81 \div 81$

    $= 62$ Add

77. $3(50) - 14$ Tripled means **three** times, and loses means to **subtract**

81. There are two servings per package with 210 calories per serving so $2 \cdot 210 = 420$ calories.

85. If 2,000 calories have 65 g fat and 2,500 calories have 80 g fat then 3,000 calories which is 500 calories more will have 15 g more fat. So, 3,000 calories have 95 g of fat.

89. 10 is the next even number in this sequence.

93. $2 + 0 = 2$, $2 + 2 = 4$, $4 + 2 = 6$, $6 + 4 = 10$;   10 is the next number in the sequence.

SECTION 1.2

Answers to the odd numbers 1 - 5 (See in the back of the textbook.)

9. $\frac{3}{4} = \frac{3 \cdot 6}{4 \cdot 6} = \frac{18}{24}$ Property 1, multiply the numerator and denominator by 6, since $4 \cdot 6 = 24$.

13. $\frac{5}{8} = \frac{5 \cdot 3}{8 \cdot 3} = \frac{15}{24}$ Property 1, multiply the numerator and denominator by 3, since $8 \cdot 3 = 24$.

17. $\frac{11}{30} = \frac{11 \cdot 2}{30 \cdot 2} = \frac{22}{60}$ Property 1, multiply the numerator and denominator by 2, since $30 \cdot 2 = 60$.

21. opposite $-\frac{3}{4}$, reciprocal $\frac{4}{3}$, absolute value $\frac{3}{4}$

25. opposite 3, reciprocal $-\frac{1}{3}$, absolute value 3

29. opposite $-x$, reciprocal $\frac{1}{x}$, absolute value $|x|$ (The distance between x and 0 on the number line.)

33. $-3 > -7$ Use the greater than sign because -3 is to the right of -7 on the number line.

37. $7 > -|-7|$ Use the greater than sign because $-|-7| = -(7) = -7$ which is to the left of 7 on the number line.

41. $-\frac{3}{2} < -\frac{3}{4}$ Use the less than sign because $-\frac{3}{2}$ is to the left of $-\frac{3}{4}$ on the number line.

45. $\left|5 \cdot 2^3 - 2 \cdot 3^2\right| = |5 \cdot 8 - 2 \cdot 9|$ First, simplify each number with an exponent

$= |40 - 18|$ Then, multiply left to right

$= |22|$ Subtract

$= 22$ Absolute value

49. $10 - |7 - 2(5 - 3)| = 10 - |7 - 2(2)|$ Simplify within the innermost grouping symbols first

$= 10 - |7 - 4|$ Then, multiply

$= 10 - |3|$ Subtract

$= 10 - 3$ Absolute value

$= 7$ Subtract

53. $\frac{2}{3} \cdot \frac{4}{5} = \frac{2 \cdot 4}{3 \cdot 5} = \frac{8}{15}$

57. $\frac{1}{4}(5) = \frac{1}{4} \cdot \frac{5}{1} = \frac{1 \cdot 5}{4 \cdot 1} = \frac{5}{4}$

61. $6\left(\frac{1}{6}\right) = \frac{6}{1} \cdot \frac{1}{6} = \frac{6 \cdot 1}{1 \cdot 6} = \frac{6}{6} = 1$

65. $\left(\frac{3}{4}\right)^2 = \frac{3}{4} \cdot \frac{3}{4} = \frac{3 \cdot 3}{4 \cdot 4} = \frac{9}{16}$

69. $\left(\frac{1}{10}\right)^4 = \frac{1}{10} \cdot \frac{1}{10} \cdot \frac{1}{10} \cdot \frac{1}{10} = \frac{1 \cdot 1 \cdot 1 \cdot 1}{10 \cdot 10 \cdot 10 \cdot 10} = \frac{1}{10,000}$

73. $1, \frac{1}{4}, \frac{1}{9}, \frac{1}{16}, \frac{1}{25}$ The sequence is the reciprocals of the squares. Since the next square is 25, the

reciprocal is $\frac{1}{25}$.

77. Perimeter $= 2l + 2w = 2(1.5 \text{ inches}) + 2(0.75 \text{ inches})$

$= 3.0 \text{ inches} + 1.50 \text{ inches} = 4.5 \text{ inches}$

Area $= lw = (1.5 \text{ inches})(0.75 \text{ inches}) = 1.125 \text{ inches}^2$

81. The loss of 8 yards corresponds to -8 on the number line. A gain of six yards is 6 and a loss of 8 yards is

-8, therefore: $6 + (-8) = -2$ (a two yard loss)

85. One hundred feet below the surface is -100. If Steve descends five more feet, it is -105.

89. At 120 lbs on the chart: Handball 544, for 2 hours   1,088

Bicycling for 1 hour   299

1,387 calories

SECTION 1.3

1. $3 + 5 = 8$ Apply the rule for adding positive and negative numbers.

$3 + (-5) = -2$

$-3 + 5 = 2$

$-3 + (-5) = -8$

5. $6 + (-3) = 3$ Subtract the smaller absolute value from the

9. $18 + (-32) = -14$ larger. The answer will have the sign of the number

13. $-30 + 5 = -25$ with the larger absolute value.

17. $-9 + (-10) = -19$ Simply add their absolute values and use the common sign.

21. $5 + (-6) + (-7)$
    $= -1 + (-7)$ Apply the rule for order of operations
    $= -8$ Add left to right

25. $5 + [6 + (-2)] + (-3)$
    $= 5 + (4) + (-3)$ Add inside the brackets first
    $= 9 + (-3)$ then add left to right
    $= 6$

29. $20 + (-6) + [3 + (-9)]$
    $= 20 + (-6) + (-6)$ Add inside the brackets first
    $= 14 + (-6)$ then add left to right
    $= 8$

33. $(-9 + 2) + [5 + (-8)] + (-4)$
    $= (-9 + 2) + [-3] + (-4)$ Add inside the brackets,
    $= (-7) + (-3) + (-4)$ then the parentheses,
    $= -10 + (-4)$ then add left to right
    $= -14$

37. $(-6 + 9) + (-5) + (-4 + 3) + 7 = 3 + (-5) + (-1) + 7$ Add inside the parentheses then add left to right
    $= -2 + (-1) + 7$
    $= -3 + 7$
    $= 4$

41. $9 + 3(-8 + 10) = 9 + 3(2)$ Add inside the parentheses
    $= 9 + 6$ Then multiply (order of operations)
    $= 15$ Finally, add

45. $2(-4 + 7) + 3(-6 + 8) = 2(3) + 3(2)$ Add inside the parentheses
    $= 6 + 6$ Multiply left to right
    $= 12$ Add

49. 10, 15, 20, 25, ...: Each term is found by adding 5 to the term before it. Therefore, the next two terms will be 30 and 35.

53. 6, 0, −6, ...: Each term is found by adding −6 to the term before it. Therefore, the next two terms will be −12 and −18.

57. Yes, since an arithmetic sequence is a sequence of numbers in which each number (after the first number) comes from adding the same amount to the number before it. And each odd number (except the first number) comes from adding two to the number before it.

61. $[-7 + (-5)] + 4 = [-12] + 4$ Add inside the brackets first
    $= -8$ Add

65. $-8 + ? = -5$ The answer is 3
    $-8 + 3 = -5$

69. $-12 + 4$ Below means negative, risen means addition

73. $-\$30 + \$40 = \$10$ Overdrawn means negative, deposit means addition

SECTION 1.4

1. $5 - 8 = 5 + (-8)$ Change subtraction to addition by adding the opposite
$ = -3$ Add

5. $5 - 5 = 5 + (-5)$ Change subtraction to addition by adding the opposite
$ = 0$ Add

9. $-4 - 12 = -4 + (-12)$ Adding the opposite
$ = -16$ Add

13. $-8 - (-1) = -8 + 1$ Adding the opposite
$ = -7$ Add

17. $-4 - (-4) = -4 + 4$ Adding the opposite
$ = 0$ Add

21. $9 - 2 - 3 = 7 - 3$ Subtract left to right
$ = 4$

25. $-22 + 4 - 10 = -18 - 10$ Add left to right
$ = -18 + (-10)$ Adding the opposite
$ = -28$ Add

29. $8 - (2 - 3) - 5 = 8 - [2 + (-3)] - 5$ Adding the opposite
$ = 8 - (-1) - 5$ Add
$ = 8 + 1 - 5$ Adding the opposite
$ = 9 - 5$ Add
$ = 4$ Subtract

33. $5 - (-8 - 6) - 2 = 5 - [-8 + (-6)] - 2$ Adding the opposite
$ = 5 - (-14) - 2$ Add
$ = 5 + 14 - 2$ Adding the opposite
$ = 19 - 2$ Add
$ = 17$ Subtract

37. $-(3 - 10) - (6 - 3) = -[3 + (-10)] - [6 + (-3)]$ Adding the opposite
$ = -(-7) - 3$ Add inside parentheses
$ = 7 + (-3)$ Adding the opposite
$ = 4$ Add

41. $5 - [(2 - 3) - 4] = 5 - [2 + (-3) + (-4)]$ Adding the opposite
$ = 5 - [-1 + (-4)]$ Adding left to right
$ = 5 - [-5]$ Add
$ = 5 + 5$ Adding the opposite
$ = 10$ Add

45. $2 \cdot 8 - 3 \cdot 5 = 16 - 15$ Multiply left to right
$ = 16 + (-15)$ Adding the opposite
$ = 1$ Add

49. $5 \cdot 9 - 2 \cdot 3 - 6 \cdot 2 = 45 - 6 - 12$ Multiply left to right

$= 45 + (-6) + (-12)$ Adding the opposite

$= 39 + (-12)$ Add left to right

$= 27$ Add

53. $2 \cdot 3^2 - 5 \cdot 2^2 = 2 \cdot 9 - 5 \cdot 4$ Simplify exponents first

$= 18 - 20$ Multiply

$= 18 + (-20)$ Adding the opposite

$= -2$ Add

57. $-7 - 4 = -7 + (-4)$ Adding the opposite

$= -11$ Add

61. $-5 - (-7) = -5 + 7$ Adding the opposite

$= 2$ Add

65. $8 - 5 = 3$ Difference means subtraction

69. $8 - (-5) = 8 + 5$ Adding the oppposite

$= 13$ Add

73. $8 - ? = 10$ Remember: $8 + 2 = 10$ therefore,

$8 + 2 = 8 - (-2) = 10$

The answer is -2

77. $-35 + 15 - 20 = -35 + 15 + (-20)$ Adding the opposite

$= -20 + (-20)$ Adding left to right

$= -40$ Add

81. 4500, 3950, 3400, 2850, 2300... This is an arithmetic sequence because each number but the first number comes from adding -550 to the number before it.

85. 439 feet (In the table, look at the column lableled "Time in Seconds". Find 3 and go horizontally across to "Distance Traveled in Feet" to find 439.)

89. 5 months old $= \;\; 149$

2 months old $= -\underline{68}$

81 pounds

93. A right angle is $90°$.

$90° - 55° = 35°$ (value of x)

SECTION 1.5

1. $3 + 2 = 2 + 3$ Commutative property of addition

5. $4 + x = x + 4$ Commutative property of addition

9. $(3 + 1) + 2 = (1 + 3) + 2$ Commutative and associative

$= 1 + (3 + 2)$ properties of addition

13. $3(x + 2) = 3(2 + x)$ Commutative property of addition

17. $4(xy) = 4(yx)$ Commutative property of multiplication

21. $3(x + 2) = 3(x) + 3(2) = 3x + 6$

25. $3(0) = 0$

29. $10(1) = 10$

33. $(x+2)+7 = x+(2+7)$ Associative property of addition

$\quad\quad\quad\quad = x+9$ Add

37. $9(6y) = (9 \cdot 6)y$ Associative property of multiplication

$\quad\quad\quad = 54y$ Multiply

41. $\frac{1}{3}(3x) = \left(\frac{1}{3} \cdot 3\right)x$ Associative property of multiplication

$\quad\quad\quad = 1x$ Multiply

$\quad\quad\quad = x$ Multiply

45. $\frac{3}{4}\left(\frac{4}{3}x\right) = \left(\frac{3}{4} \cdot \frac{4}{3}\right)x$ Associative property of multiplication

$\quad\quad\quad = 1x$ Multiply

$\quad\quad\quad = x$ Multiply

49. $8(x+2) = 8(x) + 8(2)$ Distributive property

$\quad\quad\quad = 8x + 16$ Multiply

53. $4(y+1) = 4(y) + 4(1)$ Distributive property

$\quad\quad\quad = 4y + 4$ Multiply

57. $2(3a+7) = 2(3a) + 2(7)$ Distributive property

$\quad\quad\quad = 6a + 14$ Associative property $(2 \cdot 3)a$ and multiply

61. $\frac{1}{2}(3x-6) = \frac{1}{2}(3x) - \frac{1}{2}(6)$ Distributive property

$\quad\quad\quad = \frac{3}{2}x - 3$ Remember, $-\frac{1}{2}(6) = -\frac{1}{2}\left(\frac{6}{1}\right) = -\frac{6}{2} = -3$

65. $3(x+y) = 3x + 3y$ Distributive property

69. $6(2x+3y) = 6(2x) + 6(3y)$ Distributive property

$\quad\quad\quad = 12x + 18y$ Associative property and multiply

73. $\frac{1}{2}(6x+4y) = \frac{1}{2}(6x) + \frac{1}{2}(4y)$ Distributive property

$\quad\quad\quad = 3x + 2y$ Remember: $\frac{1}{2} \cdot \frac{6}{1} = 3$ and $\frac{1}{2} \cdot \frac{4}{1} = 2$

77. $2(3x+5) + 2 = 2(3x) + 2(5) + 2$ Distributive property

$\quad\quad\quad = 6x + 10 + 2$ Multiply

$\quad\quad\quad = 6x + 12$ Add

81. No, because a man would not put on his shoes and then his socks.

85. Answers will vary, examples $16 \div 2 \neq 2 \div 16$, $8 \div 4 \neq 4 \div 8$.

89. $P = 2l + 2w$

$\quad P = 2(l+w)$ Distributive property

SECTION 1.6

1. $7(-6) = -42$ Different signs give a negative answer.

5. $-8(2) = -16$ Different signs give a negative answer.

9. $-11(-11) = 121$ Like signs give a positive answer.

13. $-3(-4)(-5) = 12(-5)$ Multiply left to right

$\quad\quad\quad = -60$ Multiply

17. $(-7)^2 = (-7)(-7)$ Definition of exponents

$\quad\quad\quad = 49$ Multiply

21. $-2(2-5) = -2(-3)$ Subtract

$= 6$ Multiply

25. $(4-7)(6-9) = (-3)(-3)$ Subtract

$= 9$ Multiply

29. $-3(-6) + 4(-1) = 18 + (-4)$ Multiply

$= 14$ Add

33. $4(-3)^2 + 5(-6)^2 = 4(9) + 5(36)$ Definition of exponents

$= 36 + 180$ Multiply

$= 216$ Add

37. $6 - 4(8-2) = 6 - 4(6)$ Simplify inside parentheses

$= 6 - 24$ Multiply

$= -18$ Subtract

41. $-4(3-8) - 6(2-5) = -4(-5) - 6(-3)$ Simplify inside parentheses

$= 20 + 18$ Multiply

$= 38$ Add

45. $7 - 3[2(-4-4) - 3(-1-1)]$

$= 7 - 3[2(-8) - 3(-2)]$ Simplify inside parentheses

$= 7 - 3[-16 + 6]$ Multiply inside brackets

$= 7 - 3[-10]$ Add

$= 7 + 30$ Multiply

$= 37$ Add

49. $-\frac{2}{3} \cdot \frac{5}{7} = -\frac{2 \cdot 5}{3 \cdot 7}$ Different signs give a negative answer

$= -\frac{10}{21}$ Multiply

53. $-\frac{3}{4}\left(-\frac{4}{3}\right) = \frac{3 \cdot 4}{4 \cdot 3}$ Same signs give a positive answer

$= \frac{12}{12}$ Multiply

$= 1$ Reduce

57. $\left(-\frac{2}{3}\right)^3 = \left(-\frac{2}{3}\right)\left(-\frac{2}{3}\right)\left(-\frac{2}{3}\right)$ Definition of exponents

$= \frac{4}{9}\left(-\frac{2}{3}\right)$ Multiply left to right

$= -\frac{8}{27}$ Multiply

61. $-7(-6x) = [-7(-6)]x$ Associative property of multiplication

$= 42x$ Multiply

65. $-4\left(-\frac{1}{4}x\right) = \left[-4\left(-\frac{1}{4}\right)\right]x$ Associative property of multiplication

$= 1x$ Multiply

$= x$ Multiply

69. $-\frac{1}{2}(3x-6) = -\frac{1}{2}(3x) - \frac{1}{2}(-6)$ Distributive property

$= -\frac{3}{2}x + \frac{6}{2}$ Multiply

$= -\frac{3}{2}x + 3$ Reduce

73. $-5(3x+4) - 10 = -5(3x) + (-5)(4) - 10$ Distributive property

$= -15x - 20 - 10$ Multiply

$= -15x - 30$ Subtract

77. $2(-4x) = [2(-4)]x$ Associative property of multiplication

$= -8x$ Multiply

81. 1, 2, 4, 8... (Starting with 1, each number is obtained from the previous number by

multiplying by 2 each time.)

85. $1, \frac{1}{2}, \frac{1}{4}, \frac{1}{8}...$ (For each new number, multiply previous number by $\frac{1}{2}$.)

89. $3, -6, 12, -24...$ (For each new number, multiply the previous number by -2.)

93.

5:00	6:00	7:00	8:00	9:00
25°F	19°F	13°F	7°F	1°F

At 9:00 the temperature is 1°F.

SECTION 1.7

1. $\frac{8}{-4} = -2$ Unlike signs give a negative answer

5. $\frac{-7}{21} = -\frac{1}{3}$ Reduce to lowest terms by dividing the numerator and the denominator by 7

9. $\frac{-6}{-42} = \frac{1}{7}$ Reduce to lowest terms by dividing the numerator and the denominator by -6

13. $-3 + 12 = 9$ Add

17. $-3(12) = -36$ Unlike signs give a negative number

21. $\frac{4}{5} \div \frac{3}{4} = \frac{4}{5} \cdot \frac{4}{3}$ Rewrite as multiplication by the reciprocal

$= \frac{16}{15}$ Multiply

25. $\frac{10}{13} \div \left(-\frac{5}{4}\right) = \frac{10}{13}\left(-\frac{4}{5}\right)$ Rewrite as multiplication by the reciprocal

$= -\frac{40}{65}$ Multiply

$= -\frac{8}{13}$ Reduce to lowest terms by dividing numerator and denominator by 5

29. $-\frac{3}{4} \div \left(-\frac{3}{4}\right) = -\frac{3}{4}\left(-\frac{4}{3}\right)$ Rewrite as multiplication by the reciprocal

$= \frac{12}{12}$ Multiply

$= 1$ Divide

33. $\frac{-5(-5)}{-15} = \frac{25}{-15}$ Multiply

$= -\frac{5}{3}$ Reduce to lowest terms by dividing numerator and denominator by 5

37. $\frac{27}{4-13} = \frac{27}{-9}$ Simplify denominator by subtracting

$= -\frac{3}{1}$ Reduce to lowest terms by dividing numerator and denominator by 9

$= -3$

41. $\dfrac{-3+9}{2(5)-10} = \dfrac{6}{10-10}$ Simplify numerator and denominator separately

      $= \dfrac{6}{0}$ Undefined

45. $\dfrac{27-2(-4)}{-3(5)} = \dfrac{27+8}{-15}$ Simplify numerator and denominator separately

      $= -\dfrac{35}{15}$ Add

      $= -\dfrac{7}{3}$ Reduce to lowest terms by dividing numerator and denominator by 5

49. $\dfrac{5^2-2^2}{-5+2} = \dfrac{25-4}{-5+2}$ Definition of exponents

      $= \dfrac{21}{-3}$ Simplify numerator and denominator separately

      $= -7$ Divide 21 by -3

53. $\dfrac{(5+3)^2}{-5^2-3^2} = \dfrac{8^2}{-25-9}$ Remember $-5^2 = -(5)^2 = -(5 \cdot 5) = -25$

      $= \dfrac{64}{-34}$ Simplify exponents and subtract

      $= -\dfrac{32}{17}$ Reduce to lowest terms by dividing numerator and denominator by 2

57. $\dfrac{-4 \cdot 3^2 - 5 \cdot 2^2}{-8(7)} = \dfrac{-4 \cdot 9 - 5 \cdot 4}{-56}$ Definition of exponents

      $= \dfrac{-36-20}{-56}$ Multiply

      $= \dfrac{-56}{-56}$ Subtract

      $= 1$ Reduce to lowest terms by dividing numerator and denominator by -56

61. $\dfrac{7-[(2-3)-4]}{-1-2-3} = \dfrac{7-[(-1)-4]}{-6}$ Simplify numerator and denominator separately

      $= \dfrac{7-[-5]}{-6}$ Subtract

      $= \dfrac{7+5}{-6}$ Subtract

      $= \dfrac{12}{-6}$ Add

      $= -2$ Divide 12 by -6

65. $\dfrac{3(-5-3)+4(7-9)}{5(-2)+3(-4)} = \dfrac{3(-8)+4(-2)}{-10+(-12)}$ Simplify numerator and denominator separately

      $= \dfrac{-24+(-8)}{-22}$ Multiply and add

      $= \dfrac{-32}{-22}$ Add

      $= \dfrac{16}{11}$ Reduce to lowest terms by dividing numerator and denominator by -2

69. $\dfrac{(2-9)-(5-7)}{10-15} = \dfrac{-7-(-2)}{-5}$ Simplify numerator and denominator separately

      $= \dfrac{-7+2}{-5}$ Subtract

      $= \dfrac{-5}{-5}$ Add

      $= 1$ Reduce

73. $\dfrac{(-3+7)-(-4+2)}{-1-5} = \dfrac{4-(-2)}{-6}$ Simplify numerator and denominator separately

      $= \dfrac{4+2}{-6}$ Subtract

      $= \dfrac{6}{-6}$ Add

      $= -1$ Reduce

77. $\frac{?}{-5} = 2$ Answer is -10

$\frac{-10}{-5} = 2$

81. $\frac{-20}{4} - 3 = -5 - 3 = -8$ Remember, quotient means to divide and decrease means to subtract

85. Noom   1 pm   2 pm   3 pm   4 pm

   75°F   ?   ?   ?   61°F

First, we must find the difference in the temperatures.

$$75° - 61° = 14°$$

Then, we divide by the number of sections, which is 4.

$$\frac{14}{4} = \frac{7}{2} = 3.5$$

Therefore, the temperature changed (dropped) 3.5°F each hour.

SECTION 1.8

1. Whole numbers $= \{0, 1\}$

5. Real numbers $= \left\{ -3, -2.5, 0, 1, \frac{3}{2}, \sqrt{15} \right\}$ (All)

9. π is the only irrational number

13. False   (A number is either rational <u>or</u> irrational)

17. True   (The set of integers is a subset of the set of rational numbers)

21. Prime   (37 only has one and itself as factors)

25. $144 = 12 \cdot 12$ (Other sets of factors may be used)

$= 4 \cdot 3 \cdot 4 \cdot 3$

$= 2 \cdot 2 \cdot 3 \cdot 2 \cdot 2 \cdot 3$ Factor to primes

$= 2^4 \cdot 3^2$

29. $105 = 3 \cdot 35$ (Other sets of factors may be used)

$= 3 \cdot 5 \cdot 7$

33. $385 = 5 \cdot 77$ (Other sets of factors may be used)

$= 5 \cdot 7 \cdot 11$

37. $420 = 10 \cdot 42$ (Other sets of factors may be used)

$= 2 \cdot 5 \cdot 6 \cdot 7$

$= 2 \cdot 5 \cdot 2 \cdot 3 \cdot 7$ Factor to primes

$= 2^2 \cdot 3 \cdot 5 \cdot 7$

41. $\frac{105}{165} = \frac{3 \cdot 5 \cdot 7}{3 \cdot 5 \cdot 11}$ Factor to primes

$= \frac{7}{11}$ Reduce to lowest terms

45. $\frac{385}{455} = \frac{5 \cdot 7 \cdot 11}{5 \cdot 7 \cdot 13}$ Factor to primes

$= \frac{11}{13}$ Reduce to lowest terms

49. $\frac{205}{369} = \frac{5 \cdot 41}{9 \cdot 41}$ Factor to primes

$= \frac{5}{9}$ Reduce to lowest terms

53. $6^3 = (2 \cdot 3)^3 = 2^3 \cdot 3^3$

57. $3 \cdot 8 + 3 \cdot 7 + 3 \cdot 5 = 24 + 21 + 15 = 60 = 2^2 \cdot 3 \cdot 5$

61. 8, 21, 34   (Answers may vary)

SECTION 1.9

1. $\frac{3}{6} + \frac{1}{6} = \frac{3+1}{6}$ Add numerators; keep the same denominator

   $= \frac{4}{6}$ The sum of 3 and 1 is 4

   $= \frac{2}{3}$ Reduce to lowest terms

5. $-\frac{1}{4} + \frac{3}{4} = \frac{-1+3}{4}$ Add numerators; keep the same denominator

   $= \frac{2}{4}$ The sum of -1 and 3 is 2

   $= \frac{1}{2}$ Reduce to lowest terms

9. $\frac{1}{4} + \frac{2}{4} + \frac{3}{4} = \frac{1+2+3}{4}$ Add numerators; keep the same denominator

   $= \frac{6}{4}$ The sum of 1, 2 and 3 is 6

   $= \frac{3}{2}$ Reduce to lowest terms

13. $\frac{1}{10} - \frac{3}{10} - \frac{4}{10} = \frac{1-3-4}{10}$ Subtract numerators, keep the same denominator

   $= \frac{-6}{10}$ Subtract terms left to right

   $= -\frac{3}{5}$ Unlike signs give a negative answer, and reduce to lowest terms

17. $\frac{1}{8} + \frac{3}{4} = \frac{1}{8} + \frac{3 \cdot 2}{4 \cdot 2}$ Rewrite each fraction as an equivalent fraction with denominator 8

   $= \frac{1}{8} + \frac{6}{8}$

   $= \frac{7}{8}$ Add numerators; keep the same denominator

21. $\frac{4}{9} + \frac{1}{3} = \frac{4}{9} + \frac{1 \cdot 3}{3 \cdot 3}$ Rewrite each fraction as an equivalent fraction with denominator 9

   $= \frac{4}{9} + \frac{3}{9}$

   $= \frac{7}{9}$ Add numerators; keep the same denominator

25. $-\frac{3}{4} + 1 = -\frac{3}{4} + \frac{1}{1}$

   $= -\frac{3}{4} + \frac{1 \cdot 4}{1 \cdot 4}$ Rewrite each fraction as an equivalent fraction with denominator 4

   $= -\frac{3}{4} + \frac{4}{4}$

   $= \frac{1}{4}$ Add numerators; keep the same denominator

29. $\frac{5}{12} - \left(-\frac{3}{8}\right) = \frac{5}{12} + \frac{3}{8}$ Rewrite each fraction as an equivalent fraction with denominator 24

   $= \frac{5 \cdot 2}{12 \cdot 2} + \frac{3 \cdot 3}{8 \cdot 3}$

   $= \frac{10}{24} + \frac{9}{24}$

   $= \frac{19}{24}$ Add numerators; keep the same denominator

33. $\frac{17}{30} + \frac{11}{42} = \frac{17 \cdot 7}{30 \cdot 7} + \frac{11 \cdot 5}{42 \cdot 5}$ Rewrite each fraction as an equivalent fraction with denominator 210

$= \frac{119}{210} + \frac{55}{210}$

$= \frac{174}{210}$ Add numerators; keep the same denominator

$= \frac{29}{35}$ Reduce to lowest terms

37. $\frac{13}{126} - \frac{13}{180} = \frac{13 \cdot 10}{126 \cdot 10} - \frac{13 \cdot 7}{180 \cdot 7}$ Rewrite each fraction as an equivalent fraction with denominator 1260

$= \frac{130}{1260} - \frac{91}{1260}$

$= \frac{39}{1260}$ Subtract numerators; keep the same denominator

$= \frac{13}{420}$ Reduce to lowest terms

41. $\frac{1}{2} + \frac{1}{3} + \frac{1}{4} + \frac{1}{6}$ (LCD = 12)

We then change to equivalent fractions and add as usual:

$= \frac{1 \cdot 6}{2 \cdot 6} + \frac{1 \cdot 4}{3 \cdot 4} + \frac{1 \cdot 3}{4 \cdot 3} + \frac{1 \cdot 2}{6 \cdot 2}$

$= \frac{6}{12} + \frac{4}{12} + \frac{3}{12} + \frac{2}{12}$

$= \frac{15}{12}$

$= \frac{5}{4}$ Reduce to lowest terms

45. $\frac{7}{8} - \frac{1}{4} = \frac{7}{8} - \frac{1 \cdot 2}{4 \cdot 2}$ (LCD = 8)

$= \frac{7}{8} - \frac{2}{8}$

$= \frac{5}{8}$

49. $\frac{1}{3}, 1, \frac{5}{3}, ...$:   Adding $\frac{2}{3}$ to each term produces the next term. The fourth term will be $\frac{5}{3} + \frac{2}{3} = \frac{7}{3}$. This is an arithmetic sequence.

CHAPTER 1 REVIEW

1. $-7 + (-10) = -17$

2. $(-7 + 4) + 5 = (-3) + 5 = 2$

3. $(-3 + 12) + 5 = 9 + 5 = 14$

4. $4 - 9 = 4 + (-9) = -5$

5. $9 - (-3) = 9 + 3 = 12$

6. $-7 - (-9) = -7 + 9 = 2$

7. $(-3)(-7) - 6 = 21 - 6 = 21 + (-6) = 15$

8. $5(-6) + 10 = -30 + 10 = -20$

9. $2(-8 \cdot 3x) = 2(-24x) = -48x$

10. $-25/(-5) = 5$

11. $(-40/8) - 7 = -5 - 7 = -5 + (-7) = -12$

12. $(-45/15) + 9 = -3 + 9 = 6$

13-18 See the graph in the back of the textbook.

19. $|12| = 12$

20. $|-3| = 3$

21. $\left|-\frac{4}{5}\right| = \frac{4}{5}$

22. $\left|-\frac{7}{10}\right| = \frac{7}{10}$

23. $|-1.8| = 1.8$

24. $-|-10| = -(10) = -10$

25. opposite is -6, reciprocal is $\frac{1}{6}$

26. opposite is $-\frac{3}{10}$, reciprocal is $\frac{10}{3}$

27. opposite is 9, reciprocal is $-\frac{1}{9}$

28. opposite is $\frac{12}{5}$, reciprocal is $-\frac{5}{12}$

29. $\left(\frac{2}{5}\right)\left(\frac{3}{7}\right) = \frac{2 \cdot 3}{5 \cdot 7} = \frac{6}{35}$ Like signs give a positive answer

30. $\left(\frac{4}{5}\right)\left(\frac{5}{4}\right) = \frac{20}{20} = 1$ Like signs give a positive answer

31. $(1/2)(-10) = (1/2)(-10/1) = -10/2 = -5$ Different signs give a negative answer

32. $(1/6)(-18) = (1/6)(-18/1) = -18/6 = -3$ Different signs give a negative answer

33. $\left(-\frac{4}{5}\right)\left(\frac{25}{16}\right) = -\frac{4 \cdot 25}{5 \cdot 16}$ Different signs give a negative answer

$ = -\frac{100}{80}$ Multiply

$ = -\frac{5}{4}$ Reduce to lowest terms

34. $-7\left(-\frac{1}{7}\right) = -\frac{7}{1}\left(-\frac{1}{7}\right) = \frac{7}{7} = 1$ Like signs give a positive answer

35. $-9 + 12 = 3$ Subtract the smaller absolute value from the larger. The answer will have

 the sign of the number with the larger absolute value.

36. $-40 + (-5) = -45$ Simply add the absolute values and

37. $-18 + (-20) = -38$ use the common sign.

38. $-20 + 14 = -6$ (Same rule as #35)

39. $-3 + 8 + (-5) = 5 + (-5) = 0$ Add left to right

40. $9 + (-7) + (-8) = 2 + (-8) = -6$ Add left to right

41. $(-5) + (-10) + (-7) = -15 + (-7) = -22$ Add left to right

42. $(-3) + (-12) + (-9) = -15 + (-9) = -24$ Add left to right

43. $-2 + (-8) + [-9 + (-6)] = -2 + (-8) + (-15)$ Simplify inside brackets first

$ = -10 + (-15)$ Add left to right

$ = -25$ Add

44. $-5 + (-7) + [-3 + (-1)] = -5 + (-7) + (-4)$ Simplify inside brackets first

$ = -12 + (-4)$ Add left to right

$ = -16$ Add

45. $(-21) + 40 + (-23) + 5 = 19 + (-23) + 5$ Add left to right

$ = -4 + 5$ Add left to right

$ = 1$ Add

46. $(-31) + 52 + (-39) + 18 = 21 + (-39) + 18$ Add left to right

$ = -18 + 18$ Add left to right

$ = 0$ Add

47. $6 - 9 = 6 + (-9) = -3$ Change subtraction to addition by adding the opposite

48. $8 - 11 = 8 + (-11) = -3$ (Same rule as #47)

49. $14 - (-8) = 14 + 8 = 22$ Adding the opposite

50. $20 - (-9) = 20 + 9 = 29$ Adding the opposite

51. $-12 - (-8) = -12 + 8 = -4$ Adding the opposite

52. $-15 - (-3) = -15 + 3 = -12$ Adding the opposite

53. $4 - 9 - 15 = 4 + (-9) + (-15)$ Adding the opposite

$ = -5 + (-15)$ Add left to right

$ = -20$ Add

54. $10 - 8 - 11 = 10 + (-8) + (-11)$ Adding the opposite
$= 2 + (-11)$ Adding left to right
$= -9$ Add

55. $-14 + 7 - 8 = -14 + 7 + (-8)$ Adding the opposite
$= -7 + (-8)$ Adding left to right
$= -15$ Add

56. $-18 + 10 - 3 = -18 + 10 + (-3)$ Adding the opposite
$= -8 + (-3)$ Adding left to right
$= -11$ Add

57. $5 - (-10 - 2) - 3 = 5 - [-10 + (-2)] - 3$ Simplify in grouping symbols first
$= 5 - (-12) - 3$ Add in grouping first
$= 5 + 12 + (-3)$ Adding the opposite
$= 17 + (-3)$ Add left to right
$= 14$ Add

58. $5 - (-2 - 8) - 1 = 5 - [-2 + (-8)] - 1$ Adding the opposite in grouping symbols first
$= 5 - (-10) + (-1)$ Add in grouping first
$= 5 + 10 + (-1)$ Adding the opposite
$= 15 + (-1)$ Add left to right
$= 14$ Add

59. $6 - [(3 - 4) - 5] = 6 - [3 + (-4) + (-5)]$ Adding the opposite in grouping symbols first
$= 6 - [-1 + (-5)]$ Add left to right
$= 6 - [-6]$ Add
$= 6 + 6$ Adding the opposite
$= 12$ Add

60. $8 - [(7 - 8) - 9] = 8 - [7 + (-8) + (-9)]$ Adding the opposite in grouping symbols first
$= 8 - [-1 + (-9)]$ Add left to right
$= 8 - [-10]$ Add
$= 8 + 10$ Adding the opposite
$= 18$ Add

61. $20 - [-(10 - 3) - 8] - 7 = 20 - [-(7) - 8] - 7$ Subtract inside the parentheses
$= 20 - [-7 + (-8)] - 7$ Adding the opposite in grouping symbols first
$= 20 - [-15] - 7$ Add inside parentheses
$= 35 + (-7)$ Add left to right
$= 28$ Add

62. $20 - [-(8-6) - 10] - 12 = 20 - \{-[8 + (-6)] + (-10)\} + (-12)$ Adding the opposite
$$= 20 - [-(2) + (-10)] + (-12)$$ Adding inside grouping
$$= 20 - (-12) + (-12)$$ Adding inside grouping
$$= 20 + 12 + (-12)$$ Adding the opposite
$$= 32 + (-12)$$ Adding left to right
$$= 20$$ Add

63. $(-5)(6) = -30$ Different signs give a negative answer
64. $(-9)(-7) = 63$ Like signs give a positive answer
65. $4(-3) = -12$ Different signs give a negative answer
66. $(-9)(-5) = 45$ Like signs give a positive answer
67. $-2(3)(4) = (-6)(4) = -24$ Multiply left to right
68. $-3(4)(-2) = (-12)(-2) = 24$ Multiply left to right
69. $(-1)(-3)(-1)(-4) = 3(-1)(-4) = (-3)(-4) = 12$ Multiply left to right
70. $(-2)(-3)(-1)(-5) = 6(-1)(-5) = (-6)(-5) = 30$ Multiply left to right
71. $\frac{12}{-3} = -4$ Unlike signs give a negative answer
72. $\frac{-48}{12} = -4$ Unlike signs give a negative answer
73. $\frac{-9}{36} = -\frac{1}{4}$ Unlike signs give a negative answer
74. $\frac{-63}{-7} = 9$ Like signs give a positive answer

75. $-\frac{8}{9} \div \frac{4}{3} = -\frac{8}{9} \cdot \frac{3}{4}$ Rewrite as multiplication by the reciprocal
$$= -\frac{24}{36}$$ Multiply
$$= -\frac{2}{3}$$ Reduce to lowest terms by dividing numerator and denominator by 12

76. $-\frac{4}{5} \div \frac{8}{15} = -\frac{4}{5} \cdot \frac{15}{8}$ Rewrite as multiplication by the reciprocal
$$= -\frac{60}{40}$$ Multiply
$$= -\frac{3}{2}$$ Reduce to lowest terms by dividing numerator and denominator by 20

77. $4 \cdot 5 + 3 = 20 + 3 = 23$ Order of operations
78. $2 \cdot 7 + 10 = 14 + 10 = 24$ Order of operations
79. $9 \cdot 3 + 4 \cdot 5 = 27 + 20 = 47$ Order of operations
80. $6 \cdot 7 + 7 \cdot 9 = 42 + 63 = 105$ Order of operations

81. $2^3 - 4 \cdot 3^2 + 5^2 = 8 - 4 \cdot 9 + 25$ Definition of exponents
$$= 8 - 36 + 25$$ Multiply
$$= 8 + (-36) + 25$$ Adding the opposite
$$= -28 + 25$$ Add left to right
$$= -3$$ Add

82. $1^3 - 2 \cdot 3^2 + 10^2 = 1 - 2 \cdot 9 + 100$ Definition of exponents

$= 1 - 18 + 100$ Multiply

$= -17 + 100$ Subtract

$= 83$ Add

83. $12 - 3(2 \cdot 5 - 7) + 4 = 12 - 3(10 - 7) + 4$ Multiply inside parentheses

$= 12 - 3(3) + 4$ Subtract inside parentheses

$= 12 - 9 + 4$ Multiply

$= 3 + 4$ Subtract

$= 7$ Add

84. $16 - 2(3 \cdot 3 - 4) + 6 = 16 - 2(9 - 4) + 6$ Multiply inside parentheses

$= 16 - 2(5) + 6$ Subtract inside parentheses

$= 16 - 10 + 6$ Multiply

$= 6 + 6$ Subtract

$= 12$ Add

85. $20 + 8 \div 4 + 2 \cdot 5 = 20 + 2 + 10$ Order of operations

$= 22 + 10$ Add left to right

$= 32$ Add

86. $30 + 6 \div 2 + 4 \cdot 5 = 30 + 3 + 4 \cdot 5$ Order of operations

$= 30 + 3 + 20$ Order of operations

$= 33 + 20$ Add left to right

$= 53$ Add

87. $2(3 - 5) - (2 - 8) = 2(-2) - (-6)$ Subtract inside parentheses

$= -4 - (-6)$ Multiply

$= -4 + 6$ Adding the opposite

$= 2$ Add

88. $4(5 - 7) - (3 - 4) = 4(-2) - (-1)$ Subtract inside parentheses

$= -8 - (-1)$ Multiply

$= -8 + 1$ Adding the opposite

$= -7$ Add

89. $-4(-5) + 10 = 20 + 10 = 30$ Order of operations

90. $-3(-7) + 15 = 21 + 15 = 36$ Order of operations

91. $(-2)(3) - (4)(-3) - 9 = (-6) - (-12) - 9$ Multiply

$= (-6) + 12 + (-9)$ Adding the opposite

$= 6 + (-9)$ Adding left to right

$= -3$ Add

92. $(-4)(2) - 5(2) - 10 = (-8) - 10 - 10$ Multiply

$= -8 + (-10) + (-10)$ Adding the opposite

$= -18 + (-10)$ Adding left to right

$= -28$ Add

93. $3(4-7)^2 - 5(3-8)^2 = 3[4+(-7)]^2 - 5[3+(-8)]^2$ Adding the opposite inside grouping

$\qquad = 3(-3)^2 - 5(-5)^2$ Adding inside grouping

$\qquad = 3(9) - 5(25)$ Definition of exponents

$\qquad = 27 - 125$ Multiply

$\qquad = 27 + (-125)$ Adding the opposite

$\qquad = -98$ Add

94. $6(6-7)^2 - 9(3-6)^2 = 6(-1)^2 - 9(-3)^2$ Simplifying inside of grouping

$\qquad = 6(1) - 9(9)$ Definition of exponents

$\qquad = 6 + (-81)$ Multiply then add the opposites

$\qquad = -75$ Add

95. $(-5-2)(-3-7) = (-7)(-10) = 70$ Simplify inside of grouping, then multiply

96. $(-3-7)(-2-8) = (-10)(-10) = 100$ Simplify inside of grouping, then multiply

97. $\dfrac{4(-3)}{-6} = \dfrac{-12}{-6}$ Multiply in the numerator

$\qquad = \dfrac{12}{6}$ Like signs give a positive answer

$\qquad = 2$ Reduce to lowest terms

98. $\dfrac{9(-2)}{-2} = \dfrac{-18}{-2} = 9$ Multiply in the numerator, like signs are positive

99. $\dfrac{3^2+5^2}{(3-5)^2} = \dfrac{9+25}{(-2)^2}$ Definition of exponents, simplify in grouping

$\qquad = \dfrac{34}{4}$ Add in the numerator, exponents in denominator

$\qquad = \dfrac{17}{2}$ Reduce to lowest terms

100. $\dfrac{4^2-8^2}{(4-8)^2} = \dfrac{16-64}{(-4)^2}$ Definition of exponents, simplify in grouping

$\qquad = \dfrac{-48}{16}$ Subtract in the numerator, exponents in denominator

$\qquad = -3$ Reduce to lowest terms

101. $\dfrac{15-10}{6-6} = \dfrac{5}{0}$ Undefined

102. $\dfrac{12-3}{8-8} = \dfrac{9}{0}$ Undefined

103. $\dfrac{8(-5)-24}{4(-2)} = \dfrac{-40-24}{-8}$ Unlike signs are negative

$\qquad = \dfrac{-64}{-8}$ Simplify the numerator

$\qquad = 8$ Reduce to lowest terms

104. $\dfrac{9(-6)-10}{2(-8)} = \dfrac{-54-10}{-16}$ Unlike signs are negative

$\qquad = \dfrac{-64}{-16}$ Simplify the numerator

$\qquad = 4$ Reduce to lowest terms

105. $\dfrac{2(-7)+(-11)(-4)}{7-(-3)} = \dfrac{-14+44}{10}$ Multiply in the numerator, add the opposite in the denominator

$\qquad = \dfrac{30}{10}$ Add in the numerator

$\qquad = 3$ Reduce to lowest terms

106. $\dfrac{-3(4-7)-5(7-2)}{-5-2-1} = \dfrac{-3(-3)-5(5)}{-5+(-2)+(-1)}$

Simplify in the parentheses in the numerator, add the opposite in the denominator

$= \dfrac{9-25}{-8}$ Multiply in the numerator and add in the denominator

$= \dfrac{-16}{-8}$ Subtract

$= 2$ Reduce to lowest terms

107. $9(3y) = (9 \cdot 3)y$ Associative property of multiplication

108. $8(1) = 8$ Multiplicative identity

109. $(4+y)+2 = (y+4)+2$ Commutative property of addition

110. $5+(-5) = 0$ Additive inverse

111. $6\left(\dfrac{1}{6}\right) = 1$ Multiplicative inverse

112. $8+0 = 8$ Additive identity

113. $(4+2)+y = (4+y)+2$ Commutative and associative properties of addition

114. $5(w-6) = 5w-30$ Distributive property

115. $7+(5+x) = (7+5)+x$ Associative property of addition

$= 12+x$ Addition of like terms

116. $4(7a) = (4 \cdot 7)a = 28a$ Associative property of multiplication, then multiply like terms

117. $(1/9)(9x) = (1/9) \cdot (9)x$ Associative property of multiplication

$= (1/9) \cdot (9/1)x$ Remember: $9 = \dfrac{9}{1}$

$= 9/9x$ Multiply

$= 1x$ Reduce to lowest terms

$= x$ Multiply

118. $\dfrac{4}{5}\left(\dfrac{5}{4}y\right) = \left(\dfrac{4}{5} \cdot \dfrac{5}{4}\right)y$ Associative property of multiplication

$= \dfrac{20}{20}y$ Multiply

$= 1y$ Reduce to lowest terms

$= y$ Multiply

119. $7(2x+3) = 7(2x) + 7(3)$ Distributive property

$= 14x + 21$ Multiply

120. $3(2a - 4) = 3(2a) - 3(4)$ Distributive property

            $= 6a - 12$ Associative property and multiply

121. $(1/2)(5x - 6) = (1/2)(5x) - (1/2)(6)$ Distributive property

                 $= (1/2)(5x/1) - (1/2)(6/1)$ Remember: $5x = 5x/1$ and $6 = 6/1$

                 $= 5x/2 - 6/2$ Multiply

                 $= (5/2x) - 3$ Reduce to lowest terms

122. $(-1/2)(3x - 6) = (-1/2)(3x) - (-1/2)(6)$ Distributive property

                 $= (-1/2)(3x/1) - (-1/2)(6/1)$ Remember $3x = 3x/1$ and $6 = 6/1$

                 $= (-3x/2) - (-6/2)$ Multiply

                 $= (-3/2x) + 3$ Reduce to lowest terms

123. $-1/3, 0, 5, -4.5, 2/5, -3$ Remember: $\left\{ \dfrac{a}{b} \,\middle|\, a \text{ and } b \text{ are integers } (b \neq 0) \right\}$

124. $0, 5$ Remember: Counting numbers and the number 0

125. $\sqrt{7}, \pi$ Remember: {nonrational numbers; nonrepeating, nonterminating decimals}

126. $0, 5, -3$ Remember: Whole numbers and the opposites of all the counting numbers

127. $90 = 9 \cdot 10$ (Other sets of factors may be used)

      $= 3 \cdot 3 \cdot 2 \cdot 5$ Factor to primes

      $= 2 \cdot 3^2 \cdot 5$

128. $120 = 12 \cdot 10$ (Other sets of factors may be used)

       $= 3 \cdot 4 \cdot 2 \cdot 5$

       $= 3 \cdot 2 \cdot 2 \cdot 2 \cdot 5$ Factor to primes

       $= 2^3 \cdot 3 \cdot 5$

129. $840 = 84 \cdot 10$ (Other sets of factors may be used)

       $= 7 \cdot 12 \cdot 2 \cdot 5$

       $= 7 \cdot 3 \cdot 4 \cdot 2 \cdot 5$

       $= 7 \cdot 3 \cdot 2 \cdot 2 \cdot 2 \cdot 5$ Factor to primes

       $= 2^3 \cdot 3 \cdot 5 \cdot 7$

130. $1024 = 32 \cdot 32$

$\qquad = 4 \cdot 8 \cdot 4 \cdot 8$

$\qquad = 2 \cdot 2 \cdot 2 \cdot 2 \cdot 2 \cdot 2 \cdot 2 \cdot 2 \cdot 2 \cdot 2$ Factor to primes

$\qquad = 2^{10}$

131. $\frac{18}{35} + \frac{13}{42} = \frac{18 \cdot 6}{35 \cdot 6} + \frac{13 \cdot 5}{42 \cdot 5}$ $\qquad$ $35 = 5 \cdot 7$ $\qquad$ LCD $= 2 \cdot 3 \cdot 5 \cdot 7 = 210$

$\qquad = \frac{108}{210} + \frac{65}{210}$ $\qquad$ $42 = 2 \cdot 3 \cdot 7$

$\qquad = \frac{173}{210}$ $\qquad$ Add

132. $\frac{9}{70} + \frac{11}{84} = \frac{9 \cdot 6}{70 \cdot 6} + \frac{11 \cdot 5}{84 \cdot 5}$ $\qquad$ $70 = 2 \cdot 5 \cdot 7$ $\qquad$ LCD $= 2^2 \cdot 3 \cdot 5 \cdot 7 = 420$

$\qquad = \frac{54}{420} + \frac{55}{420}$ $\qquad$ $84 = 2^2 \cdot 3 \cdot 7$

$\qquad = \frac{109}{420}$ $\qquad$ Add

133. 10, 7, 4, 1, . . . For each new number add −3 to the previous number. The next number is −2.

134. 10, -30, 90, -270, 810, . . . For each new number multiply the previous number by −3.

135. 1, 1, 2, 3, 5, 8, . . . For each new number, add the previous two numbers.

136. 4, 6, 8, 10, 12, . . . For each new number add 2 to the previous number.

137. 1, 1/2, 0, -1/2, -1, . . . For each new number add -1/2 to the previous number.

138. 1, -1/2, 1/4, -1/8, 1/16, . . . For each new number multiply the previous number by −1/2.

CHAPTER 1 TEST

1. $x + 3 = 8$

2. $5y = 15$

3. $5^2 + 3(9 - 7) + 3^2$

$\qquad = 25 + 3(9 - 7) + 9$ $\qquad$ Simplify numbers with exponents

$\qquad = 25 + 3(2) + 9$ $\qquad$ Simplify inside parentheses

$\qquad = 25 + 6 + 9$ $\qquad$ Multiply

$\qquad = 40$ $\qquad$ Add

4. $10 - 6 \div 3 + 2^3 = 10 - 6 \div 3 + 8$ $\qquad$ Simplify number with exponents

$\qquad = 10 - 2 + 8$ $\qquad$ Divide

$\qquad = 8 + 8$ $\qquad$ Add or subtract left to right

$\qquad = 16$ $\qquad$ Add

5. Opposite 4, Reciprocal $-\frac{1}{4}$, Absolute value $|-4| = 4$

6. Opposite $-\frac{3}{4}$, Reciprocal $\frac{4}{3}$, Absolute value $\left|\frac{3}{4}\right| = \frac{3}{4}$

7. $3 + (-7) = -4$ Subtract the smaller absolute value from the larger. The answer

                                                                  will have the sign of the number with the larger absolute value.

8. $(-9 + (-6)) + (-3 + 5) = -15 + 2$ Simplify inside the parentheses

                          $= -13$ Add

9. $-4 - 8 = -4 + (-8)$ Adding the opposite

              $= -12$ Add

10. $9 - (7 - 2) - 4 = 9 - (5) - 4$ Order of operations

                      $= 4 - 4$ Subtract left to right

                      $= 0$ Subtract

11. c. Associative property of addition. This problem changes the <u>grouping</u> of the numbers with addition.

12. e. Distributive property. Multiplication of 3 is <u>distributed</u> over $(x + 5)$.

13. d. Associative property of multiplication. The problem changes the <u>grouping</u> of the numbers with multiplication.

14. a. Commutative property of addition. This problem changes the <u>order</u> of $(x + 5)$ and 7.

15. $-3(7) = -21$ Unlike signs give a negative answer

16. $-4(8)(-2) = -32(-2)$ Unlike signs give a negative answer

                $= 64$ Like signs give a positive answer

17. $8\left(-\frac{1}{4}\right) = \frac{8}{1}\left(-\frac{1}{4}\right)$ Unlike signs give a negative number, remember $8 = \frac{8}{1}$

                  $= -\frac{8}{4}$ Multiply

                  $= -2$ Reduce to lowest terms

18. $\left(-\frac{2}{3}\right)^3 = \left(-\frac{2}{3}\right)\left(-\frac{2}{3}\right)\left(-\frac{2}{3}\right)$ Definition of exponents

$= \frac{4}{9}\left(-\frac{2}{3}\right)$ Like signs give a positive answer

$= -\frac{8}{27}$ Unlike signs give a negative answer

19. $-3(-4) - 8 = 12 - 8$ Order of operations, like signs give a positive answer

$= 4$ Subtract

20. $5(-6)^2 - 3(-2)^3 = 5(36) - 3(-8)$ Simplify numbers with exponents

$= 180 + 24$ Multiply

$= 204$ Add

21. $7 - 3(2 - 8) = 7 - 3(-6)$ Simplify within parentheses

$= 7 + 18$ Multiply, (order of operations)

$= 25$ Add

22. $4 - 2[-3(-1 + 5) + 4(-3)]$

$= 4 - 2[-3(4) + 4(-3)]$ Simplify innermost symbols

$= 4 - 2[-12 + (-12)]$ Multiply in brackets

$= 4 - 2[-24]$ Add in brackets

$= 4 + 48$ Multiply, (order of operations)

$= 52$ Add

23. $\frac{4(-5) - 2(7)}{-10 - 7} = \frac{-20 - 14}{-10 - 7}$ Multiply

$= \frac{-34}{-17}$ Add

$= 2$ Reduce to lowest terms

24. $\frac{2(-3-1) + 4(-5+2)}{-3(2) - 4} = \frac{2(-4) + 4(-3)}{-3(2) - 4}$ Simplify parentheses

$= \frac{-8 + (-12)}{-6 - 4}$ Multiply

$= \frac{-20}{-10}$ Add

$= 2$ Reduce to lowest terms

25. $3 + (5 + 2x) = (3 + 5) + 2x$ Associative property

$= 8 + 2x$ Add

26. $-2(-5x) = (-2 \cdot -5)x$ Associative property

$= 10x$ Multiply

27. $2(3x + 5) = 2(3x) + 2(5)$ Distribute 2 over $(3x + 5)$

$= 6x + 10$ Multiply

28. $(-1/2)(4x - 2) = (-1/2)(4x) + (-1/2)2(-2)$ Distribute $-\frac{1}{2}$ over $(4x - 2)$

            $= (-4/2x) + 1$ Multiply

            $= -2x + 1$ Reduce to lowest terms

29. 1, -8   Remember: Whole numbers and the opposites of all counting numbers.

30. 1, 1.5, 3/4, -8   Remember: $\left\{ \dfrac{a}{b} \middle|\, a \text{ and } b \text{ are integers } (b \neq 0) \right\}$

31. $\sqrt{2}$ Remember: (non-rational numbers; non-repeating, non-terminating decimals)

32. All of them. Remember: Counting numbers, whole numbers, integers, rational numbers

     and irrational numbers.

33. $592 = 2 \cdot 296$ (Other sets of factors may be used)

       $= 2 \cdot 2 \cdot 148$

       $= 2 \cdot 2 \cdot 2 \cdot 74$

       $= 2 \cdot 2 \cdot 2 \cdot 2 \cdot 37$ Factor to primes

       $= 2^4 \cdot 37$

34. $1{,}340 = 2 \cdot 670$ (Other sets of factors may be used)

         $= 2 \cdot 2 \cdot 335$

         $= 2 \cdot 2 \cdot 5 \cdot 67$ Factor to primes

         $= 2^2 \cdot 5 \cdot 67$

35. $\dfrac{5}{15} + \dfrac{11}{42}$

     $= \dfrac{5 \cdot 14}{15 \cdot 14} + \dfrac{11 \cdot 5}{42 \cdot 5}$ $15 = 3 \cdot 5$ $\text{LCD} = 2 \cdot 3 \cdot 5 \cdot 7 = 210$

     $= \dfrac{70}{210} + \dfrac{55}{210}$ $42 = 2 \cdot 3 \cdot 7$

     $= \dfrac{125}{210}$ Add

     $= \dfrac{25}{42}$ Divide numerator and denominator by 5 to reduce to lowest terms

36. $\dfrac{7}{12} + \dfrac{23}{30} + \dfrac{11}{45} = \dfrac{7 \cdot 15}{12 \cdot 15} + \dfrac{23 \cdot 6}{30 \cdot 6} + \dfrac{11 \cdot 4}{45 \cdot 4}$ $12 = 2 \cdot 2 \cdot 3$ $30 = 2 \cdot 3 \cdot 5$ $45 = 3 \cdot 3 \cdot 5$

                                                 $\text{LCD} = 2^2 \cdot 3^2 \cdot 5 = 180$

         $= \dfrac{105}{180} + \dfrac{138}{180} + \dfrac{44}{180}$

         $= \dfrac{287}{180}$ Add

37. $8 + (-3) = 5$

38. $-24 - 2 = -24 + (-2) = -26$

39. $-5(-4) = 20$ Like signs are positive

40. $\dfrac{-24}{-2} = 12$ Like signs are positive

41. $-8,\ -3,\ 2,\ 7,\ 12,\ldots$ Add 5 to the previous number

42. $8,\ -4,\ 2,\ -1,\ \frac{1}{2},\ldots$ Multiply the previous number by $-\frac{1}{2}$

CHAPTER 2

SECTION 2.1

1. $3x - 6x = (3 - 6)x$ Distributive property

          $= -3x$ Addition

5. $7x + 3x + 2x = (7 + 3 + 2)x$ Distributive property

          $= 12x$ Addition

9. $4x - 3 + 2x = 4x + 2x - 3$ Commutative property

          $= (4x + 2x) - 3$ Associative property

          $= (4 + 2)x - 3$ Distributive property

          $= 6x - 3$ Addition

13. $2x - 3 + 3x - 2 = 2x + 3x - 3 - 2$ Commutative property

          $= (2x + 3x) + (-3 - 2)$ Associative property

          $= (2 + 3)x + (-3 - 2)$ Distributive property

          $= 5x - 5$ Addition

17. $-4x + 8 - 5x - 10 = -4x - 5x + 8 - 10$ Commutative property

          $= (-4x - 5x) + (8 - 10)$ Associative property

          $= (-4 - 5)x + (8 - 10)$ Distributive property

          $= -9x - 2$ Addition

21. $5(2x - 1) + 4 = 5(2x) - 5(1) + 4$ Distributive property

          $= 10x - 5 + 4$ Multiplication

          $= 10x - 1$ Addition

25. $-3(2x - 1) + 5 = -3(2x) - (-3)(1) + 5$ Distributive property

          $= -6x + 3 + 5$ Multiplication

          $= -6x + 8$ Addition

29. $6 - 4(x - 5) = 6 - 4(x) - (-4)(5)$ Distributive property

              $= 6 - 4x + 20$ Multiplication

              $= -4x + 6 + 20$ Commutative property

              $= -4x + 26$ Addition

33. $-6 + 2(2 - 3x) + 1$

    $= -6 + 4 - 6x + 1$ Distributive property

    $= -6x - 6 + 4 + 1$ Commutative property

    $= -6x - 1$ Addition

37. $8(2a + 4) - (6a - 1)$

    Remember: $-(6a - 1) = -1(6a - 1) = -1(6a) - (-1)(1) = -6a + 1$

    $8(2a + 4) - (6a - 1) = 16a + 32 - 6a + 1$ Distributive property

                      $= 16a - 6a + 32 + 1$ Commutative property

                      $= 10a + 33$ Add similar terms

41. $4(2y - 8) - (y + 7)$

    Remember: $-(y + 7) = -1(y + 7) = -1(y) + (-1)(7) = -y - 7$

    $4(2y - 8) - (y + 7) = 8y - 32 - y - 7$ Distributive property

                    $= 8y - y - 32 - 7$ Commutative property

                    $= 7y - 39$ Add similar terms

45. When $x = 2$, $3x - 1 = 3(2) - 1$ Substitute for the x value

                    $= 6 - 1$ Multiply

                    $= 5$ Subtract

49. When $x = 2$, $x^2 - 8x + 16 = (2)^2 - 8(2) + 16$ Substitute for the x value

                         $= 4 - 16 + 16$ Exponents, multiply

                         $= -12 + 16$ Subtract

                         $= 4$ Addition

53. When $x = -5$, $7x - 4 - x - 3 = 7(-5) - 4 - (-5) - 3$ Substitute for the x value

                          $= -35 - 4 + 5 - 3$ Multiply

                          $= -39 + 5 - 3$ Addition

                          $= -34 - 3$ Addition

                          $= -37$ Addition

57. When $x = -3$ and $y = 5$, $\quad x^2 - 2xy + y^2 = (-3)^2 - 2(-3)(5) + (5)^2 \quad$ Substitution

$$= 9 - 2(-3)(5) + 25 \qquad \text{Exponents}$$

$$= 9 + 30 + 25 \qquad \text{Multiply}$$

$$= 39 + 25 \qquad \text{Addition}$$

$$= 64 \qquad \text{Addition}$$

61. When $x = -3$ and $y = 5$, $\quad x^2 + 6xy + 9y^2 = (-3)^2 + 6(-3)(5) + 9(5)^2 \quad$ Substitution

$$= 9 + 6(-3)(5) + 9(25) \qquad \text{Exponents}$$

$$= 9 + (-90) + 225 \qquad \text{Multiply}$$

$$= -81 + 225 \qquad \text{Addition}$$

$$= 144 \qquad \text{Addition}$$

65. When $x = \frac{1}{2}$, $\quad 12x - 3 = 12\left(\frac{1}{2}\right) - 3 \qquad\qquad\qquad$ Substitution

$$= 6 - 3 \qquad\qquad\qquad\qquad \text{Subtraction}$$

$$= 3$$

69. When $x = \frac{3}{2}$, $\quad 12x - 3 = 12\left(\frac{3}{2}\right) - 3 \qquad\qquad\qquad$ Substitution

$$= 18 - 3 \qquad\qquad\qquad\qquad \text{Subtraction}$$

$$= 15$$

73. $2n + 3$ when $n = 1$, $\quad 2(1) + 3 = 5$

$\quad 2n + 3$ when $n = 2$, $\quad 2(2) + 3 = 7$

$\quad 2n + 3$ when $n = 3$, $\quad 2(3) + 3 = 9$

$\quad 2n + 3$ when $n = 4$, $\quad 2(4) + 3 = 11$

77. $n^2 + 1$ when $n = 1$, $\quad (1)^2 + 1 = 2$

$\quad n^2 + 1$ when $n = 2$, $\quad (2)^2 + 1 = 5$

$\quad n^2 + 1$ when $n = 3$, $\quad (3)^2 + 1 = 10$

$\quad n^2 + 1$ when $n = 4$, $\quad (4)^2 + 1 = 17$

81. $n^2 - 2n + 1$ when $n = 1$, $\quad (1)^2 - 2(1) + 1 = 0$

$\quad n^2 - 2n + 1$ when $n = 2$, $\quad (2)^2 - 2(2) + 1 = 1$

$\quad n^2 - 2n + 1$ when $n = 3$, $\quad (3)^2 - 2(3) + 1 = 4$

$\quad n^2 - 2n + 1$ when $n = 4$, $\quad (4)^2 - 2(4) + 1 = 9$

$\quad 0, \ 1, \ 4, \ 9, \ldots$ a sequence of squares

85. $x - 5$ when $x = -2$, $\quad -2 - 5 = -7$

89. $\frac{10}{x}$ when $x = -2$, $\quad \frac{10}{-2} = -5$

93. $-3 - \frac{1}{2} = -\frac{6}{2} - \frac{1}{2} = -\frac{7}{2}$

SECTION 2.2

1. $\quad x - 3 = 8$

   $x - 3 + 3 = 8 + 3$ $\qquad$ Add 3 to both sides

   $\qquad x = 11$ $\qquad$ Simplify both sides

5. $\qquad a + \frac{1}{2} = -\frac{1}{4}$

   $a + \frac{1}{2} + \left(-\frac{1}{2}\right) = -\frac{1}{4} + \left(-\frac{1}{2}\right)$ $\qquad$ Add $-\frac{1}{2}$ to both sides

   $\qquad a = -\frac{1}{4} + \left(-\frac{2}{4}\right)$

   $\qquad a = -\frac{3}{4}$ $\qquad$ Simplify

9. $\qquad y + 11 = -6$

   $y + 11 + (-11) = -6 + (-11)$ $\qquad$ Add -11 to both sides

   $\qquad y = -17$ $\qquad$ Simplify both sides

13. $\quad m - 6 = -10$

   $m - 6 + 6 = -10 + 6$ $\qquad$ Add 6 to both sides

   $\qquad m = -4$ $\qquad$ Simplify both sides

17. $\qquad 5 = a + 4$

   $5 + (-4) = a + 4 + (-4)$ $\qquad$ Add -4 to both sides

   $\qquad 1 = a$ $\qquad$ Simplify both sides

21. $4x + 2 - 3x = 4 + 1$

   $\qquad x + 2 = 5$ $\qquad$ Simplify both sides first

   $x + 2 + (-2) = 5 + (-2)$ $\qquad$ Add -2 to both sides

   $\qquad x = 3$ $\qquad$ Simplify both sides

25. $-3 - 4x + 5x = 18$

   $\qquad -3 + x = 18$ $\qquad$ Simplify the left side

   $-3 + 3 + x = 18 + 3$ $\qquad$ Add **3** to both sides

   $\qquad x = 21$ $\qquad$ Simplify both sides

29. $-2.5 + 4.8 = 8x - 1.2 - 7x$

   $\qquad 2.3 = x - 1.2$ $\qquad$ Simplify both sides first

   $2.3 + 1.2 = x - 1.2 + \mathbf{1.2}$ $\qquad$ Add **1.2** to both sides

   $\qquad 3.5 = x$ $\qquad$ Simplify both sides

33. $15 - 21 = 8x + 3x - 10x$

$\qquad -6 = x$ Simplify both sides

37. $2(x + 3) - x = 4$

$\qquad 2x + 6 - x = 4$ Distributive property

$\qquad\quad x + 6 = 4$ Simplify the left side

$\qquad x + 6 + (-6) = 4 + (-6)$ Add -6 to both sides

$\qquad\qquad x = -2$ Simplify both sides

41. $5(2a + 1) - 9a = 8 - 6$

$\qquad 10a + 5 - 9a = 8 - 6$ Distributive property

$\qquad\quad a + 5 = 2$ Simplify both sides

$\qquad a + 5 + (-5) = 2 + (-5)$ Add -5 to both sides

$\qquad\qquad a = -3$ Simplify both sides

45. $4y - 3(y - 6) + 2 = 8$

$\qquad 4y - 3y + 18 + 2 = 8$ Distributive property

$\qquad\quad y + 20 = 8$ Simplify the left side

$\qquad y + 20 + (-20) = 8 + (-20)$ Add -20 to both sides

$\qquad\qquad y = -12$ Simplify both sides

49. $-3(2m - 9) + 7(m - 4) = 12 - 9$

$\qquad -6m + 27 + 7m - 28 = 12 - 9$ Distributive property

$\qquad\qquad m - 1 = 3$ Simplify both sides

$\qquad\qquad m - 1 + 1 = 3 + 1$ Add 1 to both sides

$\qquad\qquad\qquad m = 4$ Simplify both sides

53. $\qquad\quad 8a = 7a - 5$

$\qquad 8a + (-7a) = 7a + (-7a) - 5$ Add $-7a$ to both sides

$\qquad\qquad a = -5$ Simplify both sides

57. $\qquad\quad 3y + 4 = 2y + 1$

$\qquad 3y + (-2y) + 4 = 2y + (-2y) + 1$ Add $-2y$ to both sides

$\qquad\qquad y + 4 = 1$ Simplify both sides

$\qquad y + 4 + (-4) = 1 + (-4)$ Add -4 to both sides

$\qquad\qquad y = -3$ Simplify both sides

61. $\qquad\quad 4x - 7 = 5x + 1$

$\qquad 4x + (-4x) - 7 = 5x + (-4x) + 1$ Add $-4x$ to both sides

$\qquad\qquad -7 = x + 1$ Simplify both sides

$$-7 + (-1) = x + 1 + (-1)$$ Add -1 to both sides

$$-8 = x$$ Simplify both sides

65. $$8a - 7.1 = 7a + 3.9$$

$$8a + (-7a) - 7.1 = 7a + (-7a) + 3.9$$ Add $(-7a)$ to both sides

$$a - 7.1 = 3.9$$ Simplify both sides

$$a - 7.1 + 7.1 = 3.9 + 7.1$$ Add 7.1 to both sides

$$a = 11.0$$ Simplify both sides

$$a = 11$$

69. $\frac{1}{5}(5x) = \left(\frac{1}{5} \cdot \frac{5}{1}\right)x$ Associative property

$$= 1x$$ Reciprocals

$$= x$$

73. $-2\left(-\frac{1}{2}x\right) = \left[-\frac{2}{1}\left(-\frac{1}{2}\right)\right]x$ Associative property

$$= 1x$$ Reciprocals

$$= x$$

SECTION 2.3

1. $$5x = 10$$

$$\frac{1}{5}(5x) = \frac{1}{5}(10)$$ Multiply both sides by $\frac{1}{5}$

$$\left[\frac{1}{5}(5)\right]x = \frac{1}{5}(10)$$ Associative property

$$x = 2$$ Simplify: $\frac{1}{5}(10) = \frac{1}{5}\left(\frac{10}{1}\right) = \frac{10}{5} = 2$

5. $$-8x = 4$$

$$-\frac{1}{8}(-8x) = -\frac{1}{8}(4)$$ Multiply both sides by $-\frac{1}{8}$

$$\left[-\frac{1}{8}(-8)\right]x = -\frac{1}{8}(4)$$ Associative property

$$x = -\frac{1}{2}$$ Simplify: $-\frac{1}{8}(4) = -\frac{1}{8}\left(\frac{4}{1}\right) = -\frac{4}{8} = -\frac{1}{2}$

9. $$-3x = -9$$

$$-\frac{1}{3}(-3x) = -\frac{1}{3}(-9)$$ Multiply both sides by $-\frac{1}{3}$

$$x = 3$$ Simplify: $-\frac{1}{3}(-9) = -\frac{1}{3}\left(-\frac{9}{1}\right) = \frac{9}{3} = 3$

13. $$2x = 0$$

$$\frac{1}{2}(2x) = \frac{1}{2}(0)$$ Multiply both sides by $\frac{1}{2}$

$$x = 0$$ Simplify both sides

17. $\frac{x}{3} = 2$

   $\frac{1}{3}x = 2$ Dividing by 3 is equivalent to multiplying by $\frac{1}{3}$

   $3\left(\frac{1}{3}x\right) = 3(2)$ Multiply both sides by **3**

   $x = 6$ Simplify both sides

21. $-\frac{x}{2} = -\frac{3}{4}$

   $-\frac{1}{2}x = -\frac{3}{4}$ Dividing by 2 is equivalent to multiplying by $\frac{1}{2}$

   $-2\left(-\frac{1}{2}\right)x = -2\left(-\frac{3}{4}\right)$ Multiply both sides by **-2**

   $x = \frac{6}{4}$ Simplify: $-2\left(-\frac{3}{4}\right) = -\frac{2}{1}\left(-\frac{3}{4}\right) = \frac{6}{4}$

   $x = \frac{3}{2}$ Reduce to lowest terms

25. $-\frac{3}{5}x = \frac{9}{5}$

   $-\frac{5}{3}\left(-\frac{3}{5}x\right) = -\frac{5}{3}\left(\frac{9}{5}\right)$ Multiply both sides by $-\frac{5}{3}$

   $x = -\frac{45}{15}$ Multiply

   $x = -3$ Reduce to lowest terms

29. $-4x - 2x + 3x = 24$

   $-3x = 24$ Simplify the left side

   $-\frac{1}{3}(-3x) = -\frac{1}{3}(24)$ Multiply both sides by $-\frac{1}{3}$

   $x = -\frac{24}{3}$ Simplify: $-\frac{1}{3}(24) = -\frac{1}{3}\left(\frac{24}{1}\right) = -\frac{24}{3}$

   $x = -8$ Reduce to lowest terms

33. $-3 - 5 = 3x + 5x - 10x$

   $-8 = -2x$ Simplify both sides

   $-\frac{1}{2}(-8) = -\frac{1}{2}(-2x)$ Multiply both sides by $-\frac{1}{2}$

   $4 = x$ Simplify: $-\frac{1}{2}(-8) = -\frac{1}{2}\left(-\frac{8}{1}\right) = \frac{8}{2} = 4$

37. $-x = 4$

   $-1(-x) = -1(4)$ Multiply both sides by -1

   $x = -4$

41. $15 = -a$

   $-1(15) = -1(-a)$ Multiply both sides by -1

   $-15 = a$

45. $3x - 2 = 7$

$3x - 2 + 2 = 7 + 2$ Add **2** to both sides

$3x = 9$ Simplify

$\frac{1}{3}(3x) = \frac{1}{3}(9)$ Multiply both sides by $\frac{1}{3}$

$x = 3$ Simplify: $\frac{1}{3}(9) = \frac{1}{3}\left(\frac{9}{1}\right) = \frac{9}{3} = 3$

49. **Method 1**

$$\frac{1}{8} + \frac{1}{2}x = \frac{1}{4}$$

$\frac{1}{8} + \left(-\frac{1}{8}\right) + \frac{1}{2}x = \frac{1}{4} + \left(-\frac{1}{8}\right)$ Add $-\frac{1}{8}$ to each side

$\frac{1}{2}x = \frac{1}{8}$ $\frac{1}{4} + \left(-\frac{1}{8}\right) = \frac{2}{8} + \left(-\frac{1}{8}\right) = \frac{1}{8}$

$\frac{2}{1}\left(\frac{1}{2}x\right) = \frac{2}{1}\left(\frac{1}{8}\right)$ Multiply each side by $\frac{2}{1}$

$x = \frac{2}{8}$

$x = \frac{1}{4}$ Reduce to lowest terms

Method 2

$8\left(\frac{1}{8} + \frac{1}{2}x\right) = 8\left(\frac{1}{4}\right)$ Multiply each side by the LCD **8**

$8\left(\frac{1}{8}\right) + 8\left(\frac{1}{2}x\right) = 8\left(\frac{1}{4}\right)$ Distributive property on the the left side

$1 + 4x = 2$ Multiply

$1 + (-1) + 4x = 2 + (-1)$ Add **−1** to each side

$4x = 1$

$\frac{1}{4}(4x) = \frac{1}{4}(1)$ Multiply each side by $\frac{1}{4}$

$x = \frac{1}{4}$

53. $2y = -4y + 18$

$2y + 4y = -4y + 4y + 18$ Add **4y** to both sides

$6y = 18$ Simplify

$\frac{1}{6}(6y) = \frac{1}{6}(18)$ Multiply both sides by $\frac{1}{6}$

$y = 3$ Simplify: $\frac{1}{6}(18) = \frac{1}{6}\left(\frac{18}{1}\right) = \frac{18}{6} = 3$

57. $8x + 4 = 2x - 5$

$8x + (-2\mathbf{x}) + 4 = 2x + (-2\mathbf{x}) - 5$ Add $-2\mathbf{x}$ to both sides

$6x + 4 = -5$

$6x + 4 + (-4) = -5 + (-4)$ Add -4 to both sides

$6x = -9$

$\frac{1}{6}(6x) = \frac{1}{6}(-9)$ Multiply both sides by $\frac{1}{6}$

$x = -\frac{3}{2}$ Simplify: $\frac{1}{6}(-9) = \frac{1}{6}\left(-\frac{9}{1}\right) = -\frac{9}{6} = -\frac{3}{2}$

61. $6m - 3 = m + 2$

$6m + (-\mathbf{m}) - 3 = m + (-\mathbf{m}) + 2$ Add $-\mathbf{m}$ to each side

$5m - 3 = 2$

$5m - 3 + \mathbf{3} = 2 + \mathbf{3}$ Add $\mathbf{3}$ to each side

$5m = 5$

$\frac{1}{5}(5m) = \frac{1}{5}(5)$ Multiply both sides by $\frac{1}{5}$

$m = 1$

65. $9y + 2 = 6y - 4$

$9y + (-6\mathbf{y}) + 2 = 6y + (-6\mathbf{y}) - 4$ Add $-6\mathbf{y}$ to each side

$3y + 2 = -4$

$3y + 2 + (-\mathbf{2}) = -4 + (-\mathbf{2})$ Add $-\mathbf{2}$ to each side

$3y = -6$

$\frac{1}{3}(3y) = \frac{1}{3}(-6)$ Multiply both sides by $\frac{1}{3}$

$y = -2$ Simplify: $\frac{1}{3}(-6) = \frac{1}{3}\left(-\frac{6}{1}\right) = -\frac{6}{3} = -2$

69. $5(2x - 8) - 3 = 5(2x) + 5(-8) - 3$ Distributive property

$= 10x - 40 - 3$ Multiply

$= 10x - 43$ Simplify

73. $7 - 3(2y + 1) = 7 + (-3)(2y) + (-3)(1)$ Distributive property

$= 7 - 6y - 3$ Multiply

$= -6y + 4$ Simplify

SECTION 2.4

1. Solve: $2(x + 3) = 12$

   Solution: Our first step is to apply the distributive property to the left side of the equation.

   Step 1a: $2x + 6 = 12$ Distributive property

   Step 2: $2x + 6 + (-6) = 12 + (-6)$ Add **-6** to both sides

                      $2x = 6$ Simplify both sides

   Step 3: $\frac{1}{2}(2x) = \frac{1}{2}(6)$ Multiply both sides by $\frac{1}{2}$

                      $x = 3$

   Step 4: Check: When $x = 3$

           the equation $2(x + 3) = 12$

           becomes $2(3 + 3) \overset{?}{=} 12$

                        $2(6) \overset{?}{=} 12$

                        $12 = 12$ A true statement, our solution checks.

5. Solve: $2(4a + 1) = -6$

   Solution: Distribute the 2 across the sum of $4a + 1$:

   Step 1a: $8a + 2 = -6$ Distributive property

   Step 2: $8a + 2 + (-2) = -6 + (-2)$ Add **-2** to both sides

                      $8a = -8$ Simplify both sides

   Step 3: $\frac{1}{8}(8a) = \frac{1}{8}(-8)$`   Multiply both sides by $\frac{1}{8}$

                      $a = -1$

   Step 4: Check: When $a = -1$

                   the equation $2(4a + 1) = -6$

                   becomes $2[4(-1) + 1] \overset{?}{=} -6$

                                $2(-4 + 1) \overset{?}{=} -6$

                                $2(-3) \overset{?}{=} -6$

                                $-6 = -6$ A true statement, our solution checks.

9. Solve: $-2(3y + 5) = 14$ Solution: We begin by multiplying -2 times the sum of $3y + 5$:

   Step 1a: $-6y - 10 = 14$ Distributive property

   Step 2: $-6y - 10 + 10 = 14 + 10$ Add **10** to both sides

                    $-6y = 24$ Simplify both sides

   *Step 3: $-\frac{1}{6}(-6y) = -\frac{1}{6}(24)$ Multiply both sides by $-\frac{1}{6}$

                    $y = -4$

*Remember to multiply by the same sign as the coefficient so the variable will have a coefficient of positive one when you have completed the problem.

Step 4: Check: When $y = -4$

$$\text{the equation } -2(3y + 5) = 14$$

$$\text{becomes } -2[3(-4) + 5] \overset{?}{=} 14$$

$$-2(-12 + 5) \overset{?}{=} 14$$

$$-2(-7) \overset{?}{=} 14$$

$$14 = 14 \qquad \text{A true statement}$$

13. Solve: $1 = \frac{1}{2}(4x + 2)$

Solution: We begin by multiplying $\frac{1}{2}$ times the sum of $4x + 2$:

Step 1a: $1 = \frac{1}{2}(4x) + \frac{1}{2}(2)$ Distributive property

$1 = 2x + 1$ Multiply

Step 2: $1 + (-1) = 2x + 1 + (-1)$ Add -1 to each side

$0 = 2x$ Simplify both sides

Step 3: $(\frac{1}{2})0 = (\frac{1}{2})2x$ Multiply both sides by $\frac{1}{2}$

$0 = x$ Simplify both sides

Step 4: Check: When $0 = x$

$$\text{the equation } \quad 1 = \frac{1}{2}(4x + 2)$$

$$\text{becomes } \quad 1 \overset{?}{=} \frac{1}{2}(4 \cdot 0 + 2)$$

$$1 \overset{?}{=} \frac{1}{2}(0 + 2)$$

$$1 \overset{?}{=} \frac{1}{2}(2)$$

$$1 = 1 \qquad \text{A true statement}$$

17. Solve: $4(2y + 1) - 7 = 1$

Solution: Distribute the 4 across the sum $2y + 1$

$8y + 4 - 7 = 1$ Distributive property

$8y - 3 = 1$ Combine similar terms

$8y - 3 + 3 = 1 + 3$ Add **3** to both sides

$8y = 4$ Simplify both sides

$(\frac{1}{8})8y = (\frac{1}{8})4$ Multiply both sides by $\frac{1}{8}$

$y = \frac{1}{2}$ $\left(\frac{1}{8}\right)4 = \left(\frac{1}{8}\right)\frac{4}{1} = \frac{4}{8} = \frac{1}{2}$

When $\qquad$ $y = \frac{1}{2}$

the equation $\qquad$ $4(2y+1) - 7 = 1$

becomes $\qquad$ $4\left(2 \cdot \frac{1}{2} + 1\right) - 7 \overset{?}{=} 1$

$$4(1+1) - 7 \overset{?}{=} 1$$

$$4(2) - 7 \overset{?}{=} 1$$

$$8 - 7 \overset{?}{=} 1 \quad \text{therefore} \quad 1 = 1 \qquad \text{A true statement}$$

21. **Method 1** Working with decimals.

$-0.7(2x - 7) = 0.3(11 - 4x)$	Original equation
$-1.4x + 4.9 = 3.3 - 1.2x$	Distributive property
$-1.4x + \mathbf{1.4x} + 4.9 = 3.3 - 1.2x + \mathbf{1.4x}$	Add **1.4x** to both sides
$4.9 = 3.3 + 0.2x$	
$4.9 + (\mathbf{-3.3}) = 3.3 + (\mathbf{-3.3}) + 0.2x$	Add **−3.3** to both sides
$1.6 = 0.2x$	
$\frac{1.6}{\mathbf{0.2}} = \frac{0.2x}{\mathbf{0.2}}$	Divide each side by **0.2**
$8 = x$	

Method 2 Eliminating the decimals in the beginning.

$-0.7(2x - 7) = 0.3(11 - 4x)$	Original equation
$-1.4x + 4.9 = 3.3 - 1.2x$	Distributive property
$\mathbf{10}(-1.4x) + \mathbf{10}(4.9) = \mathbf{10}(3.3) - \mathbf{10}(1.2x)$	Multiply both sides by **10**
$-14x + 49 = 33 - 12x$	
$-14x + \mathbf{14x} + 49 = 33 - 12x + \mathbf{14x}$	Add **14x** to each side
$49 = 33 + 2x$	
$49 + (\mathbf{-33}) = 33 + (\mathbf{-33}) + 2x$	Add **−33** to each side
$16 = 2x$	
$\frac{16}{\mathbf{2}} = \frac{2x}{\mathbf{2}}$	Divide each side by **2**
$8 = x$	

Check: Substituting 8 for x in the original problem, we have

$$-0.7(2 \cdot 8 - 7) \overset{?}{=} 0.3(11 - 4 \cdot 8)$$

$$-0.7(16 - 7) \overset{?}{=} 0.3(11 - 32)$$

$$-0.7(9) \overset{?}{=} 0.3(-21)$$

$$-6.3 = -6.3 \qquad \text{A true statement}$$

25. $\qquad \frac{3}{4}(8x-4)+3=\frac{2}{5}(5x+10)-1$

$\frac{3}{4}(8x)+\frac{3}{4}(-4)+3=\frac{2}{5}(5x)+\frac{2}{5}(10)-1$ Distributive property

$\qquad 6x-3+3=2x+4-1$ Multiply

$\qquad\qquad 6x=2x+3$ Simplify

$\qquad 6x+(-2x)=2x+(-2x)+3$ Add $-2x$ to each side

$\qquad\qquad 4x=3$ Simplify

$\qquad (\frac{1}{4})(4x)=(\frac{1}{4})(3)$ Multiply both sides by $\frac{1}{4}$

$\qquad\qquad x=\frac{3}{4}$ Simplify

29. Solve: $6-5(2a-3)=1$

Solution: Begin by multiplying -5 times the difference of $2a-3$:

$\qquad 6-10a+15=1$ Distributive property

$\qquad -10a+21=1$ Simplify the left side

$-10a+21+(-21)=1+(-21)$ Add -21 to both sides

$\qquad -10a=-20$ Simplify both sides

$-\frac{1}{10}(-10a)=-\frac{1}{10}(-20)$ Multiply both sides by $-\frac{1}{10}$

$\qquad\qquad a=2$

33. Solve: $2(t-3)+3(t-2)=28$

Solution: Begin by applying the distributive property to each parentheses:

$\qquad 2(t-3)+3(t-2)=28$ Original equation

$\qquad 2t-6+3t-6=28$ Distributive property

$\qquad\qquad 5t-12=28$ Simplify the left side

$\qquad 5t-12+12=28+12$ Add 12 to both sides

$\qquad\qquad 5t=40$ Simplify each side

$\qquad \frac{1}{5}(5t)=\frac{1}{5}(40)$ Multiply both sides by $\frac{1}{5}$

$\qquad\qquad t=8$

37. Solve: $2(5x-3)-(2x-4)=5-(6x+1)$

Solution: When we apply the distributive property, we have to be careful with signs. Remember, we can

think of $-(6x+1)$ as $-1(6x+1)$ so that $-(6x+1)=-1(6x+1)=-6x-1$.

$\qquad 2(5x-3)-(2x-4)=5-(6x+1)$ Original equation

$\qquad 10x-6-2x+4=5-6x-1$ Distributive property

$\qquad\qquad 8x-2=-6x+4$ Simplify both sides

$\qquad 8x+6x-2=-6x+6x+4$ Add $6x$ to both sides

$$14x - 2 = 4 \qquad \text{Simplify both sides}$$

$$14x - 2 + 2 = 4 + 2 \qquad \text{Add } 2 \text{ to both sides}$$

$$14x = 6 \qquad \text{Simplify both sides}$$

$$\tfrac{1}{14}(14x) = \tfrac{1}{14}(6) \qquad \text{Multiply both sides by } \tfrac{1}{14}$$

$$x = \tfrac{6}{14} \qquad \text{Simplify}$$

$$x = \tfrac{3}{7} \qquad \text{Reduce to lowest terms}$$

41. $\tfrac{1}{2}(3) = \tfrac{1}{2}\left(\tfrac{3}{1}\right) = \tfrac{3}{2}$

45. $\tfrac{5}{9} \cdot \tfrac{9}{5} = \tfrac{45}{45} = 1$

49. $\tfrac{1}{2}(3x + 6) = \tfrac{1}{2}(3x) + \tfrac{1}{2}(6) = \tfrac{3}{2}x + 3$

SECTION 2.5

1. Substituting $P = 300$ and $w = 50$ into $P = 2l + 2w$, we have

$$300 = 2l + 2(50)$$

$$300 = 2l + 100$$

Now we solve for l

$$200 = 2l \qquad \text{Add -100 to both sides}$$

$$\tfrac{1}{2}(200) = \tfrac{1}{2}(2l) \qquad \text{Multiply by } \tfrac{1}{2}$$

$$100 = l$$

The length is 100 feet.

5. Substituting $x = 0$ into $2x + 3y = 6$, we have

$$2(0) + 3y = 6$$

$$3y = 6$$

$$\tfrac{1}{3}(3y) = \tfrac{1}{3}(6) \qquad \text{Multiply each side by } \tfrac{1}{3}$$

$$y = 2$$

9. Substituting $y = 0$ into $2x - 5y = 20$, we have

$$2x - 5(0) = 20$$

$$2x = 20$$

$$\tfrac{1}{2}(2x) = \tfrac{1}{2}(20) \qquad \text{Multiply each side by } \tfrac{1}{2}$$

$$x = 10$$

13. Substituting $y = 3$ into $y = 2x - 1$, gives us

$$3 = 2x - 1$$

$$3 + 1 = 2x - 1 + 1$$

$$4 = 2x$$

$$\tfrac{1}{2}(4) = \tfrac{1}{2}(2x)$$

$$2 = x$$

17. Solve: $d = rt$ for r

Solution: $\tfrac{1}{t}(d) = r(\tfrac{1}{t})t$ Multiply both sides by $\tfrac{1}{t}$

$$\tfrac{d}{t} = r$$

21. Solve $PV = nRT$ for P

$$\tfrac{PV}{V} = \tfrac{nRT}{V} \qquad \text{Divide both sides by } V$$

$$P = \tfrac{nRT}{V}$$

25. Solve: $x - 3y = -1$ for x

Solution: $x - 3y + \mathbf{3y} = \mathbf{3y} - 1$ Add $\mathbf{3y}$ to both sides

$$x = 3y - 1 \qquad\qquad\qquad\quad \text{Simplify}$$

29. Solve: $2x + 3y = 6$ for y

Solution: $2x + (-2x) + 3y = -2x + 6$ Add $-2x$ to each side

$$3y = -2x + 6$$

$$\tfrac{1}{3}(3y) = \tfrac{1}{3}(-2x + 6) \qquad \text{Multiply both sides by } \tfrac{1}{3}$$

$$y = \tfrac{1}{3}(-2x) + \tfrac{1}{3}(6) \qquad \text{Distributive property}$$

$$y = -\tfrac{2}{3}x + 2 \qquad\qquad\quad \text{Multiplication}$$

33. Solve: $5x - 2y = 3$ for y

Solution: $5x + (-5x) - 2y = -5x + 3$ Add $-5x$ to both sides

$$-2y = -5x + 3$$

$$-\tfrac{1}{2}(-2y) = -\tfrac{1}{2}(-5x + 3) \qquad \text{Multiply both sides by } -\tfrac{1}{2}$$

$$y = -\tfrac{1}{2}(-5x) + \left(-\tfrac{1}{2}\right)(3) \quad \text{Distributive property}$$

$$y = \tfrac{5}{2}x - \tfrac{3}{2} \qquad\qquad\qquad\quad \text{Multiplication}$$

37. Solve: $h = vt + 16t^2$ for v

 Solution: $h + (-\mathbf{16t^2}) = vt + 16t^2 + (-\mathbf{16t^2})$ Add $-16t^2$ to both sides

      $h - 16t^2 = vt$

      $\frac{1}{t}(h - 16t^2) = \frac{1}{t}(vt)$ Multiply both sides by $\frac{1}{t}$

       $\frac{h - 16t^2}{t} = v$

41. **Method 1**   Working with the fractions.

      $\frac{x}{2} + \frac{y}{3} = 1$

    $\frac{x}{2} + \left(-\frac{\mathbf{x}}{\mathbf{2}}\right) + \frac{y}{3} = \left(-\frac{\mathbf{x}}{\mathbf{2}}\right) + 1$ Add $-\frac{x}{2}$ to each side

       $\frac{y}{3} = -\frac{x}{2} + 1$ Simplify

      $\mathbf{3}\left(\frac{y}{3}\right) = \mathbf{3}\left(-\frac{x}{2} + 1\right)$ Multiply each side by 3

       $y = -\frac{3}{2}x + 3$ Distributive property

Method 2   Eliminating the fractions in the beginning.

     $6\left(\frac{x}{2}\right) + 6\left(\frac{y}{3}\right) = 6(1)$ Multiply both sides by LCD $= 6$

      $3x + 2y = 6$

       $2y = -3x + 6$ Add $-3x$ to both sides

       $y = \frac{-3x}{2} + \frac{6}{2}$ Multiply both sides by $\frac{1}{2}$

       $y = -\frac{3}{2}x + 3$ Simplify

45. **Method 1**   Working with the fractions:

     $-\frac{1}{4}x + \frac{1}{8}y = 1$

      $\frac{1}{8}y = \frac{1}{4}x + 1$ Add $\frac{1}{4}x$ to both sides

    $8\left(\frac{1}{8}y\right) = 8\left(\frac{1}{4}x\right) + 8(1)$ Multiply both sides by 8

      $y = 2x + 8$

Method 2   Eliminating the fractions in the beginning.

    $8\left(-\frac{1}{4}x\right) + 8\left(\frac{1}{8}y\right) = 8(1)$ Multiply both sides by LCD $= 8$

      $-2x + y = 8$

       $y = 2x + 8$ Add $2x$ to each side

49. $90° - 45° = 45°$ Complementary angle

   $180° - 45° = 135°$ Supplementary arngle

53. What number is 12% of 2000?

   $N = 0.12 \cdot 2000$

   $N = 240$

57. What percent of 40 is 14?

   $N \cdot 40 = 14$

   $40N = 14$

   $\frac{40N}{40} = \frac{14}{40}$

   $N = \frac{7}{20} = 0.35 = 35\%$

61. 240 is 12% of what number?

   $240 = 0.12 \cdot x$

   $240 = 0.12x$

   $\frac{240}{0.12} = \frac{.012x}{.012}$

   $2000 = x$

65. Let F= 68 in the formula $C = \frac{5}{9}(F - 32)$, and solve for C.

   $C = \frac{5}{9}(68 - 32)$ Substitution

   $C = \frac{5}{9}(36)$

   $C = \frac{180}{9}$ Multiply

   $C = 20$

   $C = 20°$ does agree with the information in Table 1.

69. To find the percent of the total calories of vanilla ice cream which are fat calories, we must answer this question:

   90 is what percent of 150?

   $90 = x \cdot 150$

   $90 = 150x$

   $\frac{90}{150} = \frac{150x}{150}$ Divide each side by **150**

   $0.6 = x$

   $60\% = x$

73. If $C = 44$ and $\pi = \frac{22}{7}$, then

$$C = 2\pi r$$

$$44 = 2\left(\frac{22}{7}\right)r \qquad \text{Substitution}$$

$$44 = \frac{44}{7}r$$

$$\frac{7}{44}(44) = \frac{7}{44}\left(\frac{44}{7}r\right) \qquad \text{Multiply each side by } \frac{7}{44}$$

$$7 \text{ meters} = r$$

77. If $V = 42$, $\quad r = \frac{7}{22}$ and $\pi = \frac{22}{7}$, then

$$V = \pi r^2 h$$

$$42 = \left(\frac{22}{7}\right)\left(\frac{7}{22}\right)^2 h$$

$$42 = \left(\frac{22}{7}\right)\left(\frac{7}{22}\right)\left(\frac{7}{22}\right)h$$

$$42 = \frac{7}{22}h \qquad\qquad \left(\frac{22}{7}\right)\left(\frac{7}{22}\right) = \frac{154}{154} = 1$$

$$\frac{22}{7}(42) = \frac{22}{7}\left(\frac{7}{22}h\right) \qquad \text{Multiply both sides by } \frac{22}{7}$$

$$22(6) = h \qquad\qquad \frac{22}{7}\left(\frac{42}{1}\right) = \frac{22}{1}\left(\frac{6}{1}\right) = 22(6)$$

$$132 \text{ feet} = h$$

81. The sum of 4 and 1 is 5.

85. 2(6+3) Remember: sum means to add and twice means two times.

SECTION 2.6

1. Step 1: **Read** the problem and then mentally <u>list</u> the items that are known and the items that are unknown.

   <u>Known items</u>: The numbers 5 and 13.

   <u>Unknown items</u>: The number in question.

   Step 2: **Assign a variable** to one of the unknown items. Then **translate** the other **information** in the problem to expressions involving the variable.

   Let $x =$ the number asked for in the problem then, "The sum of a number and five" is thirteen translates to $x + 5$.

Step 3: **Reread and write an equation.** The sum of a number and five is thirteen.

$$x + 5 = 13$$

Step 4: **Solve the equation** found in Step 3:

$$x + 5 = 13$$
$$x + 5 + (-5) = 13 + (-5)$$
$$x = 8$$

Step 5: **Write the answer.**

The number is 8

Step 6: **Reread and check.**

The sum of 8 and 5 is thirteen

5. Step 1: **Read and list**

Known items: Five times the sum of a number and 7 is 30

Unknown items: The number in question

Step 2: **Assign a variable and translate information**

Let x = the number and "five times the sum of a number and 7" translates to $5(x + 7)$.

Step 3: **Reread and write an equation.**

Five times the sum of a number and seven is thirty

$$5(x + 7) = 30$$

Step 4: **Solve the equation.**

$$5(x + 7) = 30$$
$$5x + 35 = 30$$
$$5x + 35 + (-35) = 30 + (-35)$$
$$5x = -5$$
$$x = -1$$

Step 5: **Write the answer.**

The number is -1.

Step 6: **Reread and check.**

Five times the sum of -1 and 7 is 30.

Check:
$$5(x+7)=30$$
$$5(-1+7)\overset{?}{=}30$$
$$5(6)\overset{?}{=}30$$
$$30=30 \qquad \text{A true statement}$$

9. Step 1: **Read and list**

   Known items: Two numbers added together then increased by 5 is equal to 25. One number is 4 less than 3 times the other.

Step 2: **Assign a variable and translate information.**

   Let $x =$ the first number. The other is $3x-4$.

Step 3: **Reread and write an equation.** Their sum increased by 5 is 25.

   $$x+(3x-4)+5=25$$

Step 4: **Solve the equation.**

   $$x+(3x-4)+5=25$$
   $$4x-4+5=25$$
   $$4x+1=25$$
   $$4x=24$$
   $$x=6$$

Step 5: **Write the answer.**

   The first number is $x=6$ and the second number is
   $$3x-4=3(6)-4=14$$

Step 6: **Reread and check.** The sum of 6 and 14 increased by 5 is 25.

Check:
$$6+(3\cdot6-4)+5\overset{?}{=}25$$
$$6+14+5\overset{?}{=}25$$
$$25=25 \qquad \text{A true statement}$$

13. Step 1: **Read and list**. Known items: Jack is twice as old as Lacy. In three years the sum of their ages will be 54.

   Unknown items: Jack's age and Lacy's age.

Step 2: **Assign a variable and translate information.**

   Let $x =$ Lacy's age. That makes Jack $2x$ old now. A table can help organize the information in an age problem. Notice how we placed the x in the box that corresponds to Lacy's age.

	Now	In three years
Jack	2x	
Lacy	x	

In three years means to add three to the age now. We use this information to fill in the remaining squares.

	Now	In three years
Jack	2x	2x + 3
Lacy	x	x + 3

Step 3: Reread and write an equation.

$$(2x + 3) + (x + 3) = 54$$

Step 4: Solve the equation.

$$(2x + 3) + (x + 3) = 54$$
$$3x + 6 = 54$$
$$3x = 48$$
$$x = 16$$

Step 5: Write the answer. Lacy is $x = 16$ years old. Jack is $2x = 2(16) = 32$ years old.

Step 6: Reread and check.

If Lacy is 16 and Jack is 32, in three years, Lacy will be 19 and Jack will be 35. The answers check in the original problem.

17. **Step 1: Read and list.** Known items: The figure is a rectangle. The length of the rectangle is 5 inches more than the width. The perimeter is 34 inches. Unknown items: The length and width.

Step 2: Assign a variable and translate information.

Since the length is given in terms of the width, we let $x =$ the width of the rectangle. The length is 5 more than the width so it must be $x + 5$.

Step 3: Reread and write an equation.

Twice the length + twice the width is the perimeter.

$$2(x + 5) + 2x = 34$$

Step 4: Solve the equation.

$2(x + 5) + 2x = 34$	
$2x + 10 + 2x = 34$	Distributive property
$4x + 10 = 34$	Add similar terms
$4x = 24$	Add -10 to each side
$x = 6$	Divide each side by 4

Step 5: **Write the answer**.

The width is $x = 6$ inches.   The length is $x + 5 = 6 + 5 = 11$ inches.

Step 6: **Reread and check**.

If the length is 11 and the width is 6, then the perimeter must be $2(11) + 2(6) = 34$, which checks with the original problem.

21. Step 1: **Read and list**. Known items: The figure is a rectangle. The length of the rectangle is 3 inches less than twice the width. The perimeter is 54 inches. Unknown items: The length and width.

Step 2: **Assign a variable and translate information**.

Since the length is given in terms of the width, we let $x =$ the width of the rectangle. The length is 3 less than twice the width, so it must be $2x - 3$.

Step 3: **Reread and write an equation**.

Twice the length + twice the width is the perimeter.

$$2(2x - 3) + 2x = 54$$

Step 4: **Solve the equation**.

$$2(2x - 3) + 2x = 54$$

$4x - 6 + 2x = 54$	Distributive property
$6x - 6 = 54$	Add similar terms
$6x = 60$	Add 6 to both sides
$x = 10$	Divide each side by 6

Step 5: **Write the answer**.

The width is $x = 10$ inches. The length is $2x - 3 = 2(10) - 3 = 17$ inches.

Step 6: **Reread and check**.

If the length is 17 and the width is 10, then the perimeter must be $2(17) + 2(10) = 54$, which checks with the original problem.

25. Step 1: **Read and list**. <u>Known items</u>: The type of coins, the total value of the coins, and that there are twice as many quarters as dimes. <u>Unknown items</u>: The number of dimes and the number of quarters.

Step 2: **Assign a variable and translate information**.

If we let $x =$ the number of dimes, then $2x =$ the number of quarters. Since the value of each dime is worth 10 cents, the amount of money in dimes is $10x$. Similarly, since each quarter is worth 25 cents, the amoun of money in quarters is $25(2x)$. Here is a table that summarizes the information we have so far:

	Dimes	Quarters
Number	x	$2x$
Value(in cents)	$10x$	$25(2x)$

Step 3: **Reread and write an equation**.

$$10x + 25(2x) = 900$$

Step 4: **Solve the equation**.

$$10x + 25(2x) = 900$$

$$10x + 50x = 900 \qquad \text{Multiply}$$

$$60x = 900 \qquad \text{Simplify the left side}$$

$$x = 15 \qquad \text{Divide each side by 60}$$

Step 5: **Write the answer**.

You have $x = 15$ dimes and $2x = 2(15) = 30$ quarters.

Step 6: **Reread and check**.

15 dimes are worth $10(15) = 150$

30 quarters are worth $25(30) = 750$ cents

The total value is 900 cents. 900 cents = \$9.00.

29. 4 is less than 10

33. $12 < 20$ (less than)

37. $|8 - 3| - |5 - 2| = |5| - |3| = 5 - 3 = 2$

SECTION 2.7

1. Step 1: **Read and list**. Known items: The interest rates, \$2,000 more is invested at 9% than 8% and the total interest earned. Unknown items: The amounts invested in each account.

Step 2: **Assign a variable and translate information**. Let $x =$ the amount invested at 8%. From this, $x + 2000 =$ the amount of money invested at 9%. The interest earned on x dollars invested at 8% is $0.08x$. The interest earned on $x + 2000$ dollars invested at 9% is $0.09(x + 2000)$. Here is a table that summarizes this information.

	Dollars invested at 8%	Dollars invested at 9%
Number of	x	$x + 2000$
Interest on	$0.08x$	$0.09(x + 2000)$

Step 3: **Reread and write an equation**.

Interest earned at 8%		Interest earned at 9%		Total interest earned
$0.08x$	+	$0.09(x + 2000)$	=	860

Step 4: **Solve the equation**.

$$0.08x + 0.09(x + 2000) = 860$$

$0.08x + 0.09x + 180 = 860$	Distributive property
$0.17x + 180 = 860$	Add similar terms
$0.17x = 680$	Add -180 to each side
$x = 4000$	Divide each side by 0.17

Step 5: **Write the answer**.

The amount of money invested at 8% is $x = \$4000$ and at 9% is $x + \$2000 = \$6000$.

Step 6: **Reread and check**.

The interest at 8% is 8% of 4,000 = 0.08(4000) = $320

The interest at 9% is 9% of 6,000 = 0.09(6000) = $540

The total interest is $860

5. Step 1: **Read and list**. Known items: The three interest rates, that there is twice as much money at 9% and three times as much money at 10% and the total interest earned. Unknown items: The amounts invested in each account.

Step 2: **Assign a variable and translate information**. Let $x =$ the amount invested at 8%, $2x =$ the amount invested at 9% and $3x =$ the amount invested at 10%. The interest earned on x dollars at 8% is $0.08x$, on $2x$ dollars at 9% is $0.09(2x)$ and on $3x$ dollars at 10% is $0.10(3x)$. Here is a table that summarizes this information.

	Dollars invested at 8%	Dollars invested at 9%	Dollars invested at 10%
Number of	x	$2x$	$3x$
Interest on	$0.08x$	$0.09(2x)$	$0.10(3x)$

Step 3: **Reread and write an equation**.

Interest earned at 8%	+	Interest earned at 9%	+	Interest earned at 10%	=	Total interest earned
$0.08x$	+	$0.09(2x)$	+	$0.10(3x)$	=	280

Step 4: **Solve the equation**.

$0.08x + 0.09(2x) + 0.10(3x) = 280$	
$0.08x + 0.18x + 0.30x = 280$	Multiply
$0.56x = 280$	Add similar terms
$x = 500$	Divide each side by 0.56

Step 5: **Write the answer**.

The amount of money invested at 8% is $x = \$500$, at 9% is $2x = \$1000$ and at 10% is $3x = \$1500$.

Step 6: **Reread and check**.

The interest at 8% is 8% of 500 = 0.08(500) = $40

The interest at 9% is 9% of 1000 = 0.09(1000) = $90

The interest at 10% is 10% of 1500 and 0.10 (1500) = $150

The total interest is $280

9. **Step 1.** **Read and list.** Known items: The sum of all three angles is 180°, the smallest angle is $\frac{1}{5}$ as large as the largest angle and the third angle is twice the smallest angle. Unknown items: The measure of each angle.

   Step 2. **Assign a variable and translate information.** Let x = the largest angle, $\frac{1}{5}x$ = the smallest angle and

   $2\left(\frac{1}{5}x\right)$ = the middle angle.

   Step 3. **Reread and write an equation.**

Smallest angle	+	Middle angle	+	Largest angle	= 180°
$\frac{1}{5}x$		$2\left(\frac{1}{5}x\right)$		x	180

   Step 4: **Solve the equation.**

   $$\frac{1}{5}x + 2\left(\frac{1}{5}x\right) + x = 180$$
   $$x + 2x + 5x = 900 \qquad \text{Multiply by LCD = 5}$$
   $$8x = 900 \qquad \text{Add similar terms}$$

   Step 5: **Write the answer.**

   Smallest angle $\frac{1}{5}x = 22.5^0$,   middle angle $2\left(\frac{1}{5}x\right) = 2(22.5) = 45^0$,   largest angle $x = 112.5^0$

   Step 6: **Reread and check.**

   The angles must add to 180^0:

   $$22.5 + 45 + 112.5 = 180$$
   $$180 = 180 \quad \text{Our answers check}$$

13. **Step 1:** **Read and list.** Known items: It costs $0.41 for the first minute, $0.32 for each additional minute and the total charges are $5.21. Unknown items: How many additional minutes and how many minutes was the call.

   Step 2: **Assign a variable and translate information.** Let x = each additional minute. To find the length of the call add first minute plus additional minutes.

   Step 3: **Reread and write an equation.**

1st minute	+	Additional minutes total		$5.21
0.41	+	0.32 x	=	5.21

   Step 4: **Solve the equation.**

   $$0.41 + 0.32x = 5.21$$
   $$0.32x = 4.80 \qquad \text{Add } -0.41 \text{ to each side}$$
   $$x = 15 \qquad \text{Divide both sides by 0.32}$$

Step 5: **Write the answer**.

The total time of the call was 16 minutes, because you must add one additional minute, for the first minute, to the 15 additional minutes.

Step 6: **Reread and check**.

$$0.41 + 0.32x = 5.21$$
$$0.41 + 0.32(15) \overset{?}{=} 5.21$$
$$0.41 + 4.80 \overset{?}{=} 5.21$$
$$5.21 = 5.21 \qquad \text{Our answer checks}$$

17. Step 1: **Read and list**. Known items: Adult tickets are $6.00 and children's tickets are $4.50. Stacey sold twice as many adult tickets as children's tickets. The total amount of money is $115.50.

Unknown items: How many of each ticket was sold.

Step 2: **Assign a variable and translate information**.

Let x = the number of children's tickets and $2x$ = the number of adult tickets. The amount of each ticket times how many tickets sold will add to $115.50.

Step 3: **Reread and write an equation**.

Adult tickets	+	Children's tickets	= $115.50
6.00(2x)	+	4.50x	= 115.50

Step 4: **Solve the equation**.

$$6(2x) + 4.5x = 115.50$$
$$12x + 4.5x = 115.50$$
$$16.5x = 115.50$$
$$x = 7$$
$$2x = 14$$

Step 5: **Write the answer**.   Stacey sold 7 children's tickets and 14 adult tickets.

Step 6: **Reread and check**.

$$6(2x) + 4.5x = 115.50$$
$$6(2 \cdot 7) + 4.5(7) \overset{?}{=} 115.50$$
$$84 + 31.50 \overset{?}{=} 115.50$$
$$115.50 = 115.50$$

21. Step 1: **Read and list**. Known items: Jeff averages 55 mph and leaves at 11 am. Carla averages 65 mph and leaves at 1 pm. The distance to Lake Tahoe is 425 miles. Unknown items: How long it takes Jeff and Carla to drive to Lake Tahoe and who arrived first.

Step 2: **Assign a variable and translate information**.

Let x = time it takes Jeff and y = time it takes Carla

Step 3: **Reread and write an equation**.

Jeff $425 = 55x$ Carla $425 = 65y$

Step 4: **Solve the equation**.

$$\text{Jeff:} \qquad 425 = 55x$$
$$7\tfrac{8}{11}\ \text{hr} = x$$
$$11\ \text{am} + 7\tfrac{8}{11}\ \text{hours} = 6\tfrac{8}{11}\ \text{pm}$$

$$\text{Carla:}\ \ 425 = 65y$$
$$6\tfrac{7}{13} = y$$
$$1\ \text{pm} + 6\tfrac{7}{13}\ \text{hours} = 7\tfrac{7}{13}\ \text{pm}$$

Step 5: **Write the answer**. Jeff arrived at Lake Tahoe first.

Step 6: **Reread and check**.

Jeff arrived at $6\tfrac{8}{11}$ pm and Carla arrived at $7\tfrac{7}{13}$ pm. Jeff arrived first.

25. $12.4 - .2 = 12.2$ additional miles

$\frac{12.2}{.2} = 61$ how many additional $\tfrac{1}{5}$ of a mile

$1.25 + .25(61) = 1.25 + 15.25 = 16.50$

Yes, the meter worked correctly.

29. Step 1: **Read and list**. Known items: Dance lessons are $3.00 for members and $5.00 for non-members. Ike and Nancy receive half of the money collected. Unknown items: If the $80 covered half of the receipts.

Step 2: **Assign a variable and translate information**. Let x = members and $36 - x$ the non-members. The total of the members and non-members need to make $80.

Step 3: **Reread and write an equation**.

$$80 = \tfrac{1}{2}\ \text{(of money) for members and non-members}$$
$$80 = \tfrac{1}{2}[3x + 5(36 - x)]$$

Step 4: **Solve the equation**.

$$80 = \tfrac{1}{2}[3x + 5(36 - x)]$$
$$80 = \tfrac{1}{2}[3x + 180 - 5x]$$
$$80 = \tfrac{1}{2}[180 - 2x]$$
$$80 = 90 - x$$
$$x = 10$$

Step 5: **Write the answer**. This is correct if there were 10 members and 26 non-members.

Step 6: **Reread and check**.

$$80 = \tfrac{1}{2}[3 \cdot 10 + 5(36 - 10)]$$
$$80 = \tfrac{1}{2}[30 + 5(26)]$$
$$80 = \tfrac{1}{2}[30 + 130]$$
$$80 = \tfrac{1}{2}[160]$$
$$80 = 80 \qquad \text{Checks}$$

33. 12, −6, 3, $-\tfrac{3}{2}$ the next number in the sequence is $\tfrac{3}{4}$ because you multiply the previous number by $-\tfrac{1}{2}$.

37. 2, $\tfrac{3}{2}$, 1, $\tfrac{1}{2}$ the next number in the sequence is 0 because you subtract $\tfrac{1}{2}$ from the previous number.

SECTION 2.8

SEE ALL GRAPHS FOR SECTION 2.8 IN THE BACK OF THE TEXTBOOK.

1. $x - 5 < 7$

$x - 5 + 5 < 7 + 5$ Add 5 to both sides

$\qquad x < 12$

5. $x - 4.3 > 8.7$

$x - 4.3 + 4.3 > 8.7 + 4.3$ Add 4.3 to both sides

$\qquad x > 13.0$

9. $2 < x - 7$

$2 + 7 < x - 7 + 7$ Add 7 to both sides

$\qquad 9 < x$ (Remember, this inequality may also be read $x > 9$)

13. $5a \le 25$

$\tfrac{1}{5}(5a) \le \tfrac{1}{5}(25)$ Multiply each side by $\tfrac{1}{5}$

$\qquad a \le 5$

17. $-2x > 6$

$\downarrow$

$-\tfrac{1}{2}(2x) < -\tfrac{1}{2}(6)$ Multiply each side by $-\tfrac{1}{2}$ and reverse the direction of the inequality symbol

$\qquad x < -3$

21. $-\tfrac{x}{5} \le 10$

$\downarrow$

$-5\left(-\tfrac{x}{5}\right) \ge -5(10)$ Multiply each side by −5 and reverse the direction of the inequality symbol

$\qquad x \ge -50$

25. $2x - 3 < 9$

  $2x - 3 + 3 < 9 + 3$ Add **3** to both sides

    $2x < 12$

   $\frac{1}{2}(2x) < \frac{1}{2}(12)$ Multiply each side by $\frac{1}{2}$

     $x < 6$

29. $-4x + 1 > -11$

  $-4x + 1 - 1 > -11 - 1$ Add -1 to both sides

    $-4x > -12$

     $\downarrow$

  $-\frac{1}{4}(-4x) < -\frac{1}{4}(-12)$ Multiply each side by $-\frac{1}{4}$ and reverse the direction of the inequality symbol

     $x < 3$

33. $-\frac{2}{5}a - 3 > 5$

  $-\frac{2}{5}a - 3 + 3 > 5 + 3$ Add **3** to both sides

    $-\frac{2}{5}a > 8$

     $\downarrow$

  $-\frac{5}{2}\left(-\frac{2}{5}a\right) < -\frac{5}{2}(8)$ Multiply each side by $-\frac{5}{2}$ and reverse the direction of the inequality symbol

     $a < -20$

37. $0.3(a + 1) \le 1.2$

   $0.3a + 0.3 \le 1.2$ Distributive property

 $0.3a + 0.3 - 0.3 \le 1.2 - 0.3$ Add -0.3 to both sides

  $\frac{1}{0.3}(0.3a) \le \frac{1}{0.3}(0.9)$ Multiply each side by $\frac{1}{.0.3}$

     $a \le 3$

41. $3x - 5 > 8x$

  $3x - 3x - 5 > 8x - 3x$ Add $-3x$ to both sides

    $-5 > 5x$

   $\frac{1}{5}(-5) > \frac{1}{5}(5x)$ Multiply each side by $\frac{1}{5}$

    $-1 > x$ Remember this can also be read $x < -1$

45. Method 1
$$-0.4x + 1.2 < -2x - 0.4$$
$$-0.4x + 2x + 1.2 < -2x + 2x - 0.4 \qquad \text{Add } 2x \text{ to each side}$$
$$1.6x + 1.2 < -0.4$$
$$1.6x + 1.2 + (-1.2) < -0.4 + (-1.2) \qquad \text{Add } -1.2 \text{ to each side}$$
$$1.6x < -1.6$$
$$\frac{1.6x}{1.6} < \frac{-1.6}{1.6} \qquad \text{Divide by 1.6}$$
$$x < -1$$

Method 2
$$10(-0.4x + 1.2) < 10(-2x - 0.4) \qquad \text{Multiply each side by } 10$$
$$10(-0.4x) + 10(1.2) < 10(-2x) + 10(-0.4) \qquad \text{Distributive property}$$
$$-4x + 12 < -20x - 4$$
$$-4x + 20x + 12 < -20x + 20x - 4 \qquad \text{Add 20x to each side}$$
$$16x + 12 < -4$$
$$16x + 12 + (-12) < -4 + (-12) \qquad \text{Add } -12 \text{ to each side}$$
$$16x < -16$$
$$\frac{16x}{16} < \frac{-16}{16} \qquad \text{Divide each side by } 16$$
$$x < -1$$

49.
$$3 - 4(x - 2) \le -5x + 6$$
$$3 - 4x + 8 \le -5x + 6 \qquad \text{Distributive property}$$
$$-4x + 11 \le -5x + 6 \qquad \text{Simplify the left side}$$
$$-4x + 11 - 11 \le -5x + 6 - 11 \qquad \text{Add } -11 \text{ to both sides}$$
$$-4x \le -5x - 5$$
$$-4x + 5x \le -5x + 5x - 5 \qquad \text{Add } 5x \text{ to both sides}$$
$$x \le -5$$

53.
$$2x - 5y > 10$$
$$2x - 2x - 5y > -2x + 10 \qquad \text{Add } -2x \text{ to both sides}$$
$$-5y > -2x + 10$$
$$\downarrow$$
$$-\tfrac{1}{5}(-5y) < -\tfrac{1}{5}(-2x + 10) \qquad \text{Multiply each side by } -\tfrac{1}{5} \text{ and reverse the direction of the inequality symbol}$$
$$y < \tfrac{2}{5}x - 2 \qquad \text{Distributive property}$$

57.
$$2x - 4y \ge -4$$
$$2x - 2x - 4y \ge -2x - 4 \qquad \text{Add } -2x \text{ to both sides}$$
$$-4y \ge -2x - 4$$
$$\downarrow$$
$$-\tfrac{1}{4}(4y) \le -\tfrac{1}{4}(-2x - 4) \qquad \text{Multiply each side by } -\tfrac{1}{4} \text{ and reverse the direction of the inequality symbol}$$
$$y \le \tfrac{1}{2}x + 1 \qquad \text{Distributive property}$$

61. **Step 1:** **Read and list**: Known items: One number, four, eight, greater than, product and minus.
Unknown items: the number.

Step 2: **Assign a variable and translate information**. Let $x =$ the number, we will multiply the number and four and subtract the number and eight.

Step 3: **Reread and write an inequality**.

$$4x > x - 8$$

Step 4: **Solve the inequality**.

$$4x > x - 8$$
$$4x - \mathbf{x} > x - \mathbf{x} - 8 \qquad \text{Add } -\mathbf{x} \text{ to each side}$$
$$3x > -8$$
$$\tfrac{1}{3}(3x) > \tfrac{1}{3}(-8) \qquad \text{Multiply each side by } \tfrac{1}{3}$$
$$x > -\tfrac{8}{3}$$

Step 5: **Write the answer**. The number is greater than $-\tfrac{8}{3}$.

Step 6: **Reread and check**. If you substitute any number greater than $-\tfrac{8}{3}$, the inequality is true.

65. **Step 1:** **Read and list**. Known items: Difference, less than, five, seven, sum and three times.
Unknown items: the number.

Step 2: **Assign a variable and translate information**. Let $x =$ the number, we will subtract three times the number and five and add the number and seven.

Step 3: **Reread and write an inequality**.

$$3x - 5 < x + 7$$

Step 4: **Solve the inequality**.

$$3x - 5 < x + 7$$
$$3x - \mathbf{x} - 5 < x - \mathbf{x} + 7 \qquad \text{Add } -\mathbf{x} \text{ to each side}$$
$$2x - 5 < 7$$
$$2x - 5 + \mathbf{5} < 7 + \mathbf{5} \qquad \text{Add } \mathbf{5} \text{ to each side}$$
$$2x < 12$$
$$\tfrac{1}{2}(2x) < \tfrac{1}{2}(12) \qquad \text{Multiply each side by } \tfrac{1}{2}$$
$$x < 6$$

Step 5: **Write the answer**. The number is less than 6.

Step 6: **Reread and check**. If you substitute any number less than 6, the inequality is true.

69. Step 1: **Read and list**. Known items: Three sides of a triangle are three consecutive even integers. The perimeter is greater than 24. Unknown items: The value of the three sides of the triangle.

   Step 2: **Assign a variable and translate information**. Let x, $x+2$ and $x+4$ be the three consecutive even integers. The three sides added are greater than 24.

   Step 3: **Reread and write an inequality**.

   $$x+(x+2)+(x+4) > 24$$

   Step 4: **Solve the inequality**.

   $$x+(x+2)+(x+4) > 24$$

$3x+6 > 24$	Combine like terms
$3x+6-6 > 24-6$	Add -6 to each side
$3x > 18$	
$\frac{1}{3}(3x) > \frac{1}{3}(18)$	Multiply each side by $\frac{1}{3}$
$x > 6$	

   Step 5: **Write the answer**. The shortest side is even and greater than 6 inches.

   Step 6: **Reread and check**. If three consecutive even integers are added together, and the first is greater than 6 the sum is greater than 24.

73. a   (Distributive property)
77. $\{0, 2\}$ - Remember whole numbers are $\{0, 1, 2, 3, ...\}$
81. $\frac{130}{858} = \frac{2 \cdot 5 \cdot 13}{2 \cdot 3 \cdot 11 \cdot 13} = \frac{5}{3 \cdot 11} = \frac{5}{33}$

CHAPTER 2 REVIEW

1. $5x - 8x = (5-8)x = -3x$
2. $4x - 7x = (4-7)x = -3x$
3. $6x - 3 - 8x = 6x - 8x - 3 = -2x - 3$
4. $5x - 4 - 9x = 5x - 9x - 4 = -4x - 4$
5. $-a + 2 + 5a - 9 = -a + 5a + 2 - 9 = (-1+5)a + (-7) = 4a - 7$
6. $-a + 3 + 6a - 8 = -a + 6a + 3 - 8 = 5a - 5$
7. $5(2a-1) - 4(3a-2) = 10a - 5 - 12a + 8 = 10a - 12a - 5 + 8 = -2a + 3$
8. $6(4a+2) - 3(5a-1) = 24a + 12 - 15a + 3 = 9a + 15$
9. $6 - 2(3y+1) - 4 = 6 - 6y - 2 - 4 = -6y$
10. $7 - 5(2a-3) + 7 = 7 - 10a + 15 + 7 = -10a + 29$
11. $4 - 2(3x-1) - 5 = 4 - 6x + 2 - 5 = -6x + 1$
12. $6 - 2(4a+2) - 5 = 6 - 8a - 4 - 5 = -8a - 3$
13. $7x - 2$,   Letting $x = 3$ $7(3) - 2 = 19$
14. $8x - 3$,   Letting $x = 3$ $8(3) - 3 = 24 - 3 = 21$
15. $-4x - 5 + 2x$,   Letting $x = 3$ $-4(3) - 5 + 2(3) = -11$
16. $-3x + 7 + 5x = 2x + 7$,   Letting $x = 3$ $2(3) + 7 = 13$
17. $-x - 2x - 3x = -6x$,   Letting $x = 3$ $-6(3) = -18$

18. $-x - 4x - 2x = -7x$,   Letting $x = 3$ $-7(3) = -21$
19. $5x - 3$,   Letting $x = -2$ $5(-2) - 3 = -13$
20. $2x + 9$,   Letting $x = -2$ $2(-2) + 9 = 5$
21. $-3x + 2$,   Letting $x = -2$ $-3(-2) + 2 = 6 + 2 = 8$
22. $-4x - 5$,   Letting $x = -2$ $-4(-2) - 5 = 3$
23. $7 - x - 3$,   Letting $x = -2$ $7 - (-2) - 3 = 6$
24. $8 - x - 4 = -x + 4$,   Letting $x = -2$ $-(-2) + 4 = 6$

25. $\quad\quad x + 2 = -6$

$\quad x + 2 + (\mathbf{-2}) = -6 + (\mathbf{-2})$ Add $\mathbf{-2}$ to each side

$\quad\quad\quad\quad x = -8$

26. $\quad\quad x + 3 = -4$

$\quad x + 3 + (\mathbf{-3}) = -4 + (\mathbf{-3})$ Add $\mathbf{-3}$ to each side

$\quad\quad\quad\quad x = -7$

27. $\quad\quad x - \frac{1}{2} = \frac{4}{7}$

$\quad x - \frac{1}{2} + \frac{1}{2} = \frac{4}{7} + \frac{1}{2}$ Add $\frac{1}{2}$ to each side

$\quad\quad\quad x = \frac{8}{14} + \frac{7}{14}$ $\frac{1}{2} = \frac{7}{14}, \quad \frac{4}{7} = \frac{8}{14}$

$\quad\quad\quad x = \frac{15}{14}$

28. $\quad\quad x - \frac{3}{4} = \frac{5}{6}$

$\quad x - \frac{3}{4} + \frac{3}{4} = \frac{5}{6} + \frac{3}{4}$ Add $\frac{3}{4}$ to each side

$\quad\quad\quad x = \frac{10}{12} + \frac{9}{12}$ $\frac{5}{6} = \frac{10}{12}, \quad \frac{3}{4} = \frac{9}{12}$

$\quad\quad\quad x = \frac{19}{12}$

29. $\quad 10 - 3y + 4y = 12$

$\quad\quad\quad 10 + y = 12$ Combine like terms

$\quad 10 + (\mathbf{-10}) + y = 12 + (\mathbf{-10})$ Add $\mathbf{-10}$ to each side

$\quad\quad\quad\quad y = 2$

30. $8 - 2y + 3y = 12$

$\quad\quad\quad 8 + y = 12$ Combine like terms

$\quad 8 + (\mathbf{-8}) + y = 12 + (\mathbf{-8})$ Add $\mathbf{-8}$ to each side

$\quad\quad\quad\quad y = 4$

31. $-3-4=-y-2+2y$

$-7=y-2$ Combine like terms

$-7+2=y-2+2$ Add **2** to each side

$-5=y$

32. $-5-6=-y-3+2y$

$-11=y-3$ Combine like terms

$-11+3=y-3+3$ Add **3** to each side

$-8=y$

33. $2x=-10$

$\frac{1}{2}(2x)=\frac{1}{2}(-10)$ Multiply each side by $\frac{1}{2}$

$x=-5$

34. $3x=-18$

$\frac{1}{3}(3x)=\frac{1}{3}(-18)$ Multiply each side by $\frac{1}{3}$

$x=-6$

35. $3x=0$

$\frac{1}{3}(3x)=\frac{1}{3}(0)$ Multiply each side by $\frac{1}{3}$

$x=0$

36. $-2x=0$

$-\frac{1}{2}(-2x)=-\frac{1}{2}(0)$ Multiply each side by $-\frac{1}{2}$

$x=0$

37. $\frac{x}{3}=4$

$3\left(\frac{x}{3}\right)=3(4)$ Multiply each side by **3**

$x=12$

38. $\frac{x}{2}=5$

$2\left(\frac{x}{2}\right)=2(5)$ Multiply each side by **2**

$x=10$

39. $$-\frac{x}{4} = 2$$

$$-4\left(-\frac{x}{4}\right) = -4(2) \qquad \text{Multiply each side by } -4$$

$$x = -8$$

40. $$-\frac{x}{3} = 7$$

$$-3\left(-\frac{x}{3}\right) = -3(7) \qquad \text{Multiply each side by } -3$$

$$x = -21$$

41. $$3a - 2 = 5a$$

$$3a + (-3a) - 2 = 5a + (-3a) \qquad \text{Add } -3a \text{ to each side}$$

$$-2 = 2a$$

$$\tfrac{1}{2}(-2) = \tfrac{1}{2}(2a) \qquad \text{Multiply each side by } \tfrac{1}{2}$$

$$-1 = a$$

42. $$6a - 5 = 4a$$

$$6a + (-6a) - 5 = 4a + (-6a) \qquad \text{Add } -6a \text{ to each side}$$

$$-5 = -2a$$

$$-\tfrac{1}{2}(-5) = -\tfrac{1}{2}(-2a) \qquad \text{Multiply each side by } -\tfrac{1}{2}$$

$$\tfrac{5}{2} = a$$

43. $$\frac{7}{10}a = \frac{1}{5}a + \frac{1}{2}$$

$$\tfrac{7}{10}a + (-\tfrac{1}{5}a) = \tfrac{1}{5}a + (-\tfrac{1}{5}a) + \tfrac{1}{2} \quad \text{Add } -\tfrac{1}{5}a \text{ to each side}$$

$$\tfrac{7}{10}a - \tfrac{2}{10}a = \tfrac{1}{2} \qquad\qquad -\tfrac{1}{5}a = -\tfrac{2}{10}a$$

$$\tfrac{1}{2}a = \tfrac{1}{2}$$

$$2\left(\tfrac{1}{2}a\right) = 2\left(\tfrac{1}{2}\right) \qquad \text{Multiply each side by } 2$$

$$a = 1$$

44. $$\frac{2}{3}a = \frac{1}{6}a + 1$$

$$\tfrac{2}{3}a + (-\tfrac{1}{6}a) = \tfrac{1}{6}a + (-\tfrac{1}{6}a) + 1 \quad \text{Add } -\tfrac{1}{6}a \text{ to each side}$$

$$\tfrac{4}{6}a - \tfrac{1}{6}a = 1 \qquad\qquad \tfrac{2}{3}a = \tfrac{4}{6}a$$

$$\tfrac{1}{2}a = 1 \qquad\qquad\quad \tfrac{3}{6}a = \tfrac{1}{2}a$$

$$2\left(\tfrac{1}{2}a\right) = 2(1) \qquad \text{Multiply each side by } 2$$

$$a = 2$$

45. $\qquad 3x + 2 = 5x - 8$

$\qquad 3x + (-3\mathbf{x}) + 2 = 5x + (-3\mathbf{x}) - 8$ Add $-3\mathbf{x}$ to each side

$\qquad\qquad\qquad 2 = 2x - 8$

$\qquad\qquad 2 + 8 = 2x - 8 + 8$ Add $\mathbf{8}$ to each side

$\qquad\qquad\quad 10 = 2x$

$\qquad\quad \frac{1}{2}(10) = \frac{1}{2}(2x)$ Multiply each side by $\frac{1}{2}$

$\qquad\qquad\quad 5 = x$

46. $\qquad 4x - 3 = 8x + 5$

$\qquad 4x + (-4\mathbf{x}) - 3 = 8x + (-4\mathbf{x}) + 5$ Add $-4\mathbf{x}$ to each side

$\qquad\qquad\qquad -3 = 4x + 5$

$\qquad\quad -3 + (-5) = 4x + 5 + (-5)$ Add -5 to each side

$\qquad\qquad\qquad -8 = 4x$

$\qquad\qquad \frac{1}{4}(-8) = \frac{1}{4}(4x)$ Multiply each side by $\frac{1}{4}$

$\qquad\qquad\qquad -2 = x$

47. $\qquad 6x - 3 = 2x + 7$

$\qquad 6x + (-2\mathbf{x}) - 3 = 2x + (-2\mathbf{x}) + 7$ Add $-2\mathbf{x}$ to each side

$\qquad\qquad\quad 4x - 3 = 7$

$\qquad\quad 4x - 3 + 3 = 7 + 3$ Add $\mathbf{3}$ to each side

$\qquad\qquad\qquad 4x = 10$

$\qquad\quad \frac{1}{4}(4x) = \frac{1}{4}(10)$ Multiply each side by $\frac{1}{4}$

$\qquad\qquad\qquad x = \frac{5}{2}$ $\frac{10}{4} = \frac{5}{2}$

48. $\qquad 5x - 1 = 9x + 9$

$\qquad 5x + (-5\mathbf{x}) - 1 = 9x + (-5\mathbf{x}) + 9$ Add $-5\mathbf{x}$ to each side

$\qquad\qquad\qquad -1 = 4x + 9$

$\qquad\quad -1 + (-9) = 4x + 9 + (-9)$ Add -9 to each side

$\qquad\qquad\qquad -10 = 4x$

$\qquad\qquad \frac{1}{4}(-10) = \frac{1}{4}(4x)$ Multiply each side by $\dfrac{1}{4}$

$\qquad\qquad 4 - \frac{5}{2} = x$ $-\dfrac{10}{4} = -\dfrac{5}{2}$

49. $\qquad 0.7x - 0.1 = 0.5x - 0.1$

$\qquad 0.7x + (-0.5\mathbf{x}) - 0.1 = 0.5x + (-0.5\mathbf{x}) - 0.1$ Add $-0.5\mathbf{x}$ to each side

$\qquad 0.2x - 0.1 + 0.1 = -0.1 + 0.1$ Add 0.1 to each side

$\qquad\qquad\qquad 0.2x = 0$

$\qquad\quad \frac{1}{0.2}(0.2x) = \frac{1}{0.2}(0)$ Multiply each side by $\frac{1}{0.2}$

$\qquad\qquad\qquad x = 0$

50.
$$0.2x - 0.3 - 0.8x - 0.3$$

$0.2x + (-\mathbf{0.2x}) - 0.3 = 0.8x + (-\mathbf{0.2x}) - 0.3$ Add $-\mathbf{0.2x}$ to each side

$-0.3 = 0.6x - 0.3$

$-0.3 + \mathbf{0.3} = 0.6x - 0.3 + \mathbf{0.3}$ Add $\mathbf{0.3}$ to each side

$0 = 0.6x$

$\frac{1}{0.6}(0) = \frac{1}{0.6}(0.6x)$ Multiply each side by $\frac{1}{0.6}$

$0 = x$

51. $2(x-5) = 10$

$2x - 10 = 10$ Distributive property

$2x - 10 + \mathbf{10} = 10 + \mathbf{10}$ Add $\mathbf{10}$ to each side

$2x = 20$

$\frac{1}{2}(2x) = \frac{1}{2}(20)$ Multiply each side by $\frac{1}{2}$

$x = 10$

52. $3(x-4) = 9$

$3x - 12 = 9$ Distributive property

$3x - 12 + \mathbf{12} = 9 + \mathbf{12}$ Add $\mathbf{12}$ to each side

$3x = 21$

$\frac{1}{3}(3x) = \frac{1}{3}(21)$ Multiply each side by $\frac{1}{3}$

$x = 7$

53. $12 = 2(5x - 4)$

$12 = 10x - 8$ Distributive property

$12 + \mathbf{8} = 10x - 8 + \mathbf{8}$ Add $\mathbf{8}$ to each side

$20 = 10x$

$\frac{1}{10}(20) = \frac{1}{10}(10x)$ Multiply each side by $\frac{1}{10}$

$2 = x$

54. $18 = 3(2x - 2)$

$18 = 6x - 6$ Distributive property

$18 + \mathbf{6} = 6x - 6 + \mathbf{6}$ Add $\mathbf{6}$ to each side

$24 = 6x$

$\frac{1}{6}(24) = \frac{1}{6}(6x)$ Multiply each side by $\frac{1}{6}$

$4 = x$

55. $\frac{1}{2}(3t - 2) + \frac{1}{2} = \frac{5}{2}$

$\frac{1}{2}(3t) - \frac{1}{2}(2) + \frac{1}{2} = \frac{5}{2}$ Distributive property

$\frac{3}{2}t - 1 + \frac{1}{2} = \frac{5}{2}$ Simplify

$\frac{3}{2}t - \frac{1}{2} = \frac{5}{2}$ Combine like terms

$\frac{3}{2}t - \frac{1}{2} + \left(\frac{1}{2}\right) = \frac{5}{2} + \left(\frac{1}{2}\right)$ Add $\frac{1}{2}$ to each side

$\frac{3}{2}t = \frac{6}{2}$

$\frac{2}{3}\left(\frac{3}{2}t\right) = \frac{2}{3}\left(\frac{6}{2}\right)$ Multiply each side by $\frac{2}{3}$

$t = 2$ $\frac{2}{3}\left(\frac{6}{2}\right) = \frac{12}{6} = 2$

56. $\frac{1}{2}(4t - 1) + \frac{1}{3} = -\frac{25}{6}$

$\frac{1}{2}(4t) - \frac{1}{2}(1) + \frac{1}{3} = -\frac{25}{6}$ Distributive property

$2t - \frac{1}{2} + \frac{1}{3} = -\frac{25}{6}$ Simplify

$2t - \frac{3}{6} + \frac{2}{6} = -\frac{25}{6}$ $-\frac{1}{2} = -\frac{3}{6}, \quad \frac{1}{3} = \frac{2}{6}$

$2t - \frac{1}{6} = -\frac{25}{6}$ Combine like terms

$2t - \frac{1}{6} + \left(\frac{1}{6}\right) = -\frac{25}{6} + \left(\frac{1}{6}\right)$ Add $\frac{1}{6}$ to each side

$2t = -\frac{24}{6}$ $-\frac{24}{6} = -4$

$\frac{1}{2}(2t) = \frac{1}{2}(-4)$ Multiply each side by $\frac{1}{2}$

$t = -2$ $\frac{1}{2}(-4) = -\frac{4}{2} = -2$

57. $\frac{3}{5}(5x - 10) = \frac{2}{3}(9x + 3)$

$\frac{3}{5}(5x) - \frac{3}{5}(10) = \frac{2}{3}(9x) + \frac{2}{3}(3)$ Distributive property

$3x - 6 = 6x + 2$ Simplify

$3x + (-3x) - 6 = 6x + (-3x) + 2$ Add $-3x$ to each side

$-6 = 3x + 2$

$-6 + (-2) = 3x + 2 + (-2)$ Add -2 to each side

$-8 = 3x$

$\frac{1}{3}(-8) = (\frac{1}{3})3x$ Multiply each side by $\frac{1}{3}$

$-\frac{8}{3} = x$ $\left(\frac{1}{3}\right)(-8) = \left(\frac{1}{3}\right)\left(-\frac{8}{1}\right) = -\frac{8}{3}$

58. $\frac{3}{4}(8x-12)=\frac{1}{2}(4x+4)$

$\quad \frac{3}{4}(8x)-\frac{3}{4}(12)=\frac{1}{2}(4x)+\frac{1}{2}(4)$ Distributive property

$\quad\quad\quad 6x-9=2x+2$ Simplify

$\quad 6x+(-2\mathbf{x})-9=2x+(-2\mathbf{x})+2$ Add $-2\mathbf{x}$ to each side

$\quad\quad\quad 4x-9=2$

$\quad\quad 4x-9+9=2+9$ Add 9 to each side

$\quad\quad\quad\quad 4x=11$

$\quad\quad \frac{1}{4}(4x)=\frac{1}{4}(11)$ Multiply each side by $\frac{1}{4}$

$\quad\quad\quad\quad x=\frac{11}{4}$

59. $\quad 2(3x+7)=4(5x-1)+18$

$\quad\quad 6x+14=20x-4+18$ Distributive property

$\quad\quad 6x+14=20x+14$ Combine like terms

$\quad 6x+(-6\mathbf{x})+14=20x+(-6\mathbf{x})+14$ Add $-6\mathbf{x}$ to each side

$\quad\quad\quad 14=14x+14$

$\quad 14+(-14)=14x+14+(-14)$ Add -14 to each side

$\quad\quad\quad\quad 0=14x$

$\quad\quad \frac{1}{14}(0)=\frac{1}{14}(14x)$ Multiply each side by $\frac{1}{14}$

$\quad\quad\quad\quad 0=x$

60. $\quad 3(5x-1)=6(2x+3)-21$

$\quad\quad 15x-3=12x+18-21$ Distributive property

$\quad\quad 15x-3=12x-3$ Combine like terms

$\quad 15x+(-12\mathbf{x})-3=12x+(-12\mathbf{x})-3$ Add $-12\mathbf{x}$ to each side

$\quad\quad\quad 3x-3=-3$

$\quad\quad 3x-3+3=-3+3$ Add 3 to each side

$\quad\quad\quad\quad 3x=0$

$\quad\quad \frac{1}{3}(3x)=\frac{1}{3}(0)$ Multiply each side by $\frac{1}{3}$

$\quad\quad\quad\quad x=0$

61. $7-3(y+4)=10$

$\quad\quad 7-3y-12=10$ Distributive property

$\quad\quad -3y-5=10$ Simplify

$\quad -3y-5+5=10+5$ Add 5 to each side

$\quad\quad\quad -3y=15$

$\quad -\frac{1}{3}(-3y)=-\frac{1}{3}(15)$ Multiply each side by $-\frac{1}{3}$

$\quad\quad\quad y=-5$

$$8 - 2y - 8 = 12 \qquad \text{Distributive property}$$
$$-2y = 12 \qquad \text{Simplify}$$
$$-\tfrac{1}{2}(-2y) = -\tfrac{1}{2}(12) \qquad \text{Multiply each side by } -\tfrac{1}{2}$$
$$y = -6$$

63. $\quad 10 - 2(2x + 3) = -5x$
$$10 - 4x - 6 = -5x \qquad \text{Distributive property}$$
$$-4x + 4 = -5x \qquad \text{Simplify}$$
$$-4x + (\mathbf{4x}) + 4 = -5x + (\mathbf{4x}) \qquad \text{Add } \mathbf{4x} \text{ to each side}$$
$$4 = -1x$$
$$-1(4) = -1(-1x) \qquad \text{Multiply each side by } -1$$
$$-4 = x$$

64. $\quad 7 - 4(3x + 4) = -9x$
$$7 - 12x - 16 = -9x \qquad \text{Distributive property}$$
$$-12x - 9 = -9x$$
$$-12x + \mathbf{12x} - 9 = -9x + \mathbf{12x} \qquad \text{Add } \mathbf{12x} \text{ to each side}$$
$$-9 = 3x$$
$$\tfrac{1}{3}(-9) = \tfrac{1}{3}(3x) \qquad \text{Multiply each side by } \tfrac{1}{3}$$
$$-3 = x$$

65. When $x = 5$, the formula $4x - 5y = 20$ becomes
$$4(5) - 5y = 20$$
$$20 - 5y = 20$$
$$20 + (-\mathbf{20}) - 5y = 20 + (-\mathbf{20}) \qquad \text{Add } -\mathbf{20} \text{ to each side}$$
$$-5y = 0$$
$$-\tfrac{1}{5}(-5y) = -\tfrac{1}{5}(0) \qquad \text{Multiply each side by } -\tfrac{1}{5}$$
$$y = 0$$

66. When $x = 0$, the formula $4x - 5y = 20$ becomes
$$4(0) - 5y = 20$$
$$-5y = 20$$
$$-\tfrac{1}{5}(-5y) = -\tfrac{1}{5}(20) \qquad \text{Multiply each side by } -\tfrac{1}{5}$$
$$y = -4$$

67. When $x = -5$, the formula $4x - 5y = 20$ becomes

$$4(-5) - 5y = 20$$
$$-20 - 5y = 20$$
$$-20 + 20 - 5y = 20 + 20 \qquad \text{Add } 20 \text{ to each side}$$
$$-5y = 40$$
$$-\tfrac{1}{5}(-5y) = -\tfrac{1}{5}(40) \qquad \text{Multiply each side by } -\tfrac{1}{5}$$
$$y = -8$$

68. When $x = 10$, the formula $4x - 5y = 20$ becomes

$$4(10) - 5y = 20$$
$$40 - 5y = 20$$
$$40 + (-40) - 5y = 20 + (-40) \qquad \text{Add } -40 \text{ to each side}$$
$$-5y = -20$$
$$-\tfrac{1}{5}(-5y) = -\tfrac{1}{5}(-20) \qquad \text{Multiply each side by } -\tfrac{1}{5}$$
$$y = 4$$

69.
$$2x - 5y = 10$$
$$2x + (-2x) - 5y = (-2x) + 10 \qquad \text{Add } -2x \text{ to each side}$$
$$-5y = -2x + 10$$
$$-\tfrac{1}{5}(-5y) = -\tfrac{1}{5}(-2x + 10) \qquad \text{Multiply each side by } -\tfrac{1}{5}$$
$$y = -\tfrac{1}{5}(-2x) + -\tfrac{1}{5}(10) \qquad \text{Distributive property}$$
$$y = \tfrac{2x}{5} - 2 \qquad\qquad\quad -\tfrac{1}{5}(-2x) = \tfrac{-2x}{-5} = \tfrac{2x}{5}, \quad -\tfrac{1}{5}(10) = -\tfrac{10}{5} = -2$$
$$y = \tfrac{2}{5}x - 2$$

70.
$$5x - 2y = 10$$
$$5x + (-5x) - 2y = -5x + 10 \qquad \text{Add } -5x \text{ to each side}$$
$$-2y = -5x + 10$$
$$-\tfrac{1}{2}(-2y) = -\tfrac{1}{2}(-5x + 10) \qquad \text{Multiply each side by } -\tfrac{1}{2}$$
$$y = -\tfrac{1}{2}(-5x) + \left(-\tfrac{1}{2}\right)10$$
$$y = \tfrac{5}{2}x - 5 \qquad\qquad -\tfrac{1}{2}\left(\tfrac{-5x}{1}\right) = \tfrac{5}{2}x, \quad \left(-\tfrac{1}{2}\right)10 = -\tfrac{10}{2} = -5$$

71. $V = \pi r^2 h$
$$\frac{V}{\pi r^2} = \frac{\pi r^2 h}{\pi r^2} \qquad\qquad \text{Divide each side by } \pi r^2$$
$$\frac{V}{\pi r^2} = h$$

72.
$$P = 2l + 2w$$
$$P + (-2l) = 2l + (-2l) + 2w \qquad \text{Add } -2l \text{ to each side}$$
$$P - 2l = 2w$$
$$\frac{1}{2}(P - 2l) = \frac{1}{2}(2w) \qquad \text{Multiply each side by } \frac{1}{2}$$
$$\frac{P - 2l}{2} = w \qquad\qquad \frac{1}{2}(P - 2l) = \frac{1}{2}\left(\frac{P - 2l}{1}\right) = \frac{P - 2l}{2}$$

73. <u>What number</u> is 86% of 240?

$$
\begin{aligned}
x &= 0.86 \cdot 240 \\
\text{or} \quad x &= (0.86)(240) \\
x &= 206.4
\end{aligned}
$$

74. <u>What percent</u> of 2000 is 180?

$$
\begin{aligned}
P \quad \cdot \ 2000 &= 180 \\
\text{or} \qquad 2000P &= 180 \\
\frac{1}{2000}(2000P) &= \frac{1}{2000}(180) \qquad \text{Multiply each side by } \frac{1}{2000} \\
P &= \frac{180}{2000} \\
P &= 0.09 = 9\%
\end{aligned}
$$

75. Step 1: **Read and list**. Known items: Sum means add, twice means to multiply by 2. Unknown item: the number.

Step 2: **Assign a variable and translate information**.

Let $x =$ the number, after you add twice a number and 6 the answer is 28.

Step 3: **Reread and write an equation**.
$$2x + 6 = 28$$

Step 4: **Solve the equation**.

$$
\begin{aligned}
2x + 6 &= 28 \\
2x + 6 - 6 &= 28 - 6 \qquad\qquad \text{Add } -6 \text{ to each side} \\
2x &= 22 \\
\tfrac{1}{2}(2x) &= \tfrac{1}{2}(22) \qquad\qquad \text{Multiply each side by } \tfrac{1}{2} \\
x &= 11 \qquad\qquad\qquad \tfrac{1}{2}(22) = \tfrac{22}{2} = 11
\end{aligned}
$$

Step 5: **Write the answer**. The number is 11.

Step 6: **Reread and check**.

Replacing x with 11, we have

$$2(11) + 6 \overset{?}{=} 28$$
$$22 + 6 \overset{?}{=} 28$$
$$28 = 28$$

76. Step 1: **Read and list**. Known items: Sum means add, times means multiply and the total is 23.

   Unknown item: the number.

   Step 2: **Assign the variable and translate information**.

   Let x = the number, after you multiply the number by 3 and add 5, the answer is 23.

   Step 3: **Reread and write the equation**.

$$3x + 5 = 23$$

   Step 4: **Solve the equation**.

$$3x + 5 = 23$$
$$3x + 5 - 5 = 23 - 5 \qquad \text{Add } -5 \text{ to each side}$$
$$3x = 18$$
$$\tfrac{1}{3}(3x) = \tfrac{1}{3}(18) \qquad \text{Multiply each side by } \tfrac{1}{3}$$
$$x = 6$$

   Step 5: **Write the answer**. The number is 6.

   Step 6: **Reread and check**.

   Replacing x with 6, we have

$$3(6) + 5 \stackrel{?}{=} 23$$
$$18 + 5 \stackrel{?}{=} 23$$
$$23 = 23$$

77. Step 1: **Read and list**. Known items: Bob is 4 years older than Tom. In 3 years the sum of their ages is 40.

   Unknown items: Bob and Tom's ages.

   Step 2: **Assign a variable and translate information**.

	Now	In 3 years
Bob	$x+4$	$x+4+3 = x+7$
Tom	x	$x+3$

   Step 3: **Reread and write an equation**.

$$(x+7) + (x+3) = 40$$

   Step 4: **Solve the equation**.

$$(x+7) + (x+3) = 40$$
$$2x + 10 = 40 \qquad \text{Combine like terms}$$
$$2x = 30 \qquad \text{Add } -10 \text{ to each side}$$
$$x = 15 \qquad \text{Multiply each side by } \frac{1}{2}$$
$$x + 4 = 19$$

   Step 5: **Write the answer**. Tom is 15 and Bob is 19 now.

   Step 6: **Reread and check**.

   If Tom is 15, in three years he will be 18. If Bob is 19, in three years he will be 22. $22 + 18 = 40$

   The problem checks.

78. **Step 1:** **Read and list.** Known items: Debra is 6 years older than Tracey. In 5 years the sum of their ages will be 36. Unknown items: Debra and Tracey's ages.

Step 2: **Assign a variable and translate information.**

	Now	In 5 years
Debra	$x+6$	$x+6+5 = x+11$
Tracey	x	$x+5$

Step 3: **Reread and write an equation.**

$$(x+11)+(x+5) = 36$$

Step 4: **Solve the equation.**

$$(x+11)+(x+5) = 36$$
$$2x+16 = 36 \qquad \text{Combine like terms}$$
$$2x = 20 \qquad \text{Add } -16 \text{ to each side}$$
$$x = 10 \qquad \text{Multiply each side by } \frac{1}{2}$$
$$x+6 = 16$$

Step 5: **Write the answer.** Tracey is 10 and Debra is 16 now.

Step 6: **Reread and check.**

If Tracey is 10, then in 5 years she will be 15. If Debra is 16, then in 5 years she will be 21. $15+21 = 36$ The problem checks.

79. **Step 1:** **Read and list.** Known items: Perimeter of the rectangle is 60 meters. The length is 5 times the width. $P = 2l + 2w$ Unknown items: length and width.

Step 2: **Assign a variable and translate information.**

Let $x =$ width, Let $5x =$ length

Step 3: **Reread and write an equation.**

$$P = 2l + 2w$$
$$60 = 2(5x) + 2x$$

Step 4: **Solve the equation.**

$$P = 2l + 2w$$
$$60 = 2(5x) + 2x$$
$$60 = 12x \qquad \text{Combine like terms}$$
$$x = 5 \qquad \text{Multiply both sides by } \frac{1}{12}$$
$$5x = 25$$

Step 5: **Write the answer.** Width is 5 meters and length is 25 meters.

Step 6: **Reread and check.**

$$60 \stackrel{?}{=} 2(25) + 2(5)$$
$$60 \stackrel{?}{=} 50 + 10$$
$$60 = 60$$

80. Step 1: **Read and list**. Known items: Perimeter of the rectangle is 60 meters. The length is 4 times the width. $P = 2l + 2w$ Unknown items: Length and width.

Step 2: **Assign a variable and translate information**.

Let x = width, Let $4x$ = length

Step 3: **Reread and write an equation**.

$$P = 2l + 2w$$
$$60 = 2(4x) + 2x$$

Step 4: **Solve the equation**.

$$P = 2l + 2w$$
$$60 = 2(4x) + 2x$$
$$60 = 10x \qquad \text{Combine like terms}$$
$$x = 6 \qquad \text{Multiply both sides by } \tfrac{1}{10}$$
$$4x = 24$$

Step 5: **Write the answer**. Width is 6 meters and length is 24 meters.

Step 6: **Reread and check**.

$$60 \stackrel{?}{=} 2(24) + 2(6)$$
$$60 \stackrel{?}{=} 48 + 12$$
$$60 = 60$$

81. Step 1: **Read and list**. Known items: 15 coins worth $1.00. Unknown items: Number of dimes and nickels.

Step 2: **Assign the variables and translate information**.

	Dimes	Nickels
Number	x	$15 - x$
Value	$10x$	$5(15 - x)$

Step 3: **Reread and write an equation**.

Dimes	+	Nickels	=	Total money
$10x$		$5(15-x)$		100

Step 4: **Solve the equation**.

$$10x + 5(15 - x) = 100$$
$$10x + 75 - 5x = 100 \qquad \text{Distributive property}$$
$$5x + 75 = 100 \qquad \text{Combine like terms}$$
$$5x + 75 + (-75) = 100 + (-75) \qquad \text{Add } -75 \text{ to each side}$$
$$5x = 25$$
$$\tfrac{1}{5}(5x) = \tfrac{1}{5}(25) \qquad \text{Multiply each side by } \tfrac{1}{5}$$
$$x = 5$$
$$15 - x = 10$$

Step 5: **Write the answer**. There are 5 dimes and 10 nickels.

Step 6: **Reread and check**.

$$10(5) + 5(10) \overset{?}{=} 100$$
$$50 + 50 \overset{?}{=} 100$$
$$100 = 100$$

82. Step 1: **Read and list**. Known items: 15 coins worth $2.55. Unknown items: Number of dimes and quarters.

Step 2: **Assign the variables and translate information**.

	Dimes	Quarters
Number	x	$15 - x$
Value	$10x$	$25(15 - x)$

Step 3: **Reread and write an equation**.

Dimes	+	Quarters	=	Total money
$10x$		$25(15 - x)$		255

Step 4: **Solve the equation**.

$$10x + 25(15 - x) = 255$$
$$10x + 375 - 25x = 255 \qquad \text{Combine like terms}$$
$$-15x + 375 + (-375) = 255 + (-375) \qquad \text{Add } -375 \text{ to each side}$$
$$-15x = -120$$
$$-\tfrac{1}{15}(-15x) = -\tfrac{1}{15}(-120) \qquad \text{Multiply both sides by } -\tfrac{1}{15}$$
$$x = 8$$

Step 5: **Write the answer**. You have 8 dimes and 7 quarters.

Step 6: **Reread and check**.

$$10(8) + 25(7) \overset{?}{=} 255$$
$$80 + 175 \overset{?}{=} 255$$
$$255 = 255$$

83. Step 1: **Read and list**. Known items: A man invests money at 9% and 10%. He invets $300 more at 10% than he does at 9%. The interest in one year is $125.00. Unknown items: The amount of money invested at each rate.

Step 2: **Assign the variables and translate information**.

	Dollars invested at 9%	Dollars invested at 10%
Number of	x	$x + 300$
Interest on	$0.09x$	$0.10(x + 300)$

Step 3: **Reread and write an equation**.

Interest earned		Interest earned	=	Total interest
9%	+	10%	=	earned
$0.09x$	+	$0.10(x + 300)$	=	125

Step 4: **Solve the equation**.

$$0.09x + 0.10(x + 300) = 125$$
$$0.09x + 0.10x + 30 = 125$$
$$0.19x + 30 = 125$$
$$\frac{0.19x}{0.19} = \frac{95}{0.19}$$
$$x = \$500$$
$$x + 300 = \$800$$

Step 5: **Write the answer**. The man invested $500 at 9% and $800 at 10%.

Step 6: **Reread and check**.

$$(0.09)500 + (0.10)800 \overset{?}{=} 125$$
$$45 + 80 \overset{?}{=} 125$$
$$125 = 125$$

84. Step 1: **Read and list**. Known items: Mary invested money at 8% and 9%. She invested $400 more at 9% than she did at 8%. The interest in one year was $155.00. Unknown items: The amount of money invested at each rate.

Step 2: **Assign the variables and translate information**.

	Dollars invested at 8%	Dollars invested at 9%
Number of	x	$x + 400$
Interest on	$0.08x$	$0.09(x + 400)$

Step 3: **Reread and write an equation**.

Interest earned		Interest earned		Total interest
8%	+	9%	=	earned
$0.08x$	+	$0.09(x + 400)$	=	155

Step 4: **Solve the equation**.

$$0.08x + 0.09(x + 400) = 155$$
$$0.08x + 0.09x + 36 = 155 \qquad \text{Distributive property}$$
$$0.17x + 36 = 155$$
$$0.17x = 119 \qquad \text{Add } -36 \text{ to each side}$$
$$x = \$700$$
$$x + 400 = \$1,100$$

Step 5: **Write the answer**. Mary invested $700 at 8% and $1,100 at 9%.

Step 6: **Reread and check**.

$$0.08(700) + 0.09(1,100) \overset{?}{=} 155$$
$$56 + 99 \overset{?}{=} 155$$
$$155 = 155$$

85. $-2x < 4$

$\downarrow$

$-\dfrac{1}{2}(-2x) > -\dfrac{1}{2}(4)$ Multiply each side by $-\dfrac{1}{2}$ and reverse the direction of the inequality symbol

$x > -2$ $-\dfrac{1}{2}(4) = -\dfrac{4}{2} = -2$

86. $-3x < 9$

$\downarrow$

$-\dfrac{1}{3}(-3x) > -\dfrac{1}{3}(9)$ Multiply each side by $-\dfrac{1}{3}$ and reverse the direction of the inequality symbol

$x > -3$ $-\dfrac{1}{3}(9) = -\dfrac{9}{3} = -3$

87. $-5x > -10$

$\downarrow$

$-\dfrac{1}{5}(-5x) < -\dfrac{1}{5}(-10)$ Multiply each side by $-\dfrac{1}{5}$ and reverse the direction of the inequality symbol

$x < 2$ $-\dfrac{1}{5}(-10) = \dfrac{10}{5} = 2$

88. $-2x > -8$

$\downarrow$

$-\dfrac{1}{2}(-2x) < -\dfrac{1}{2}(-8)$ Multiply each side by $-\dfrac{1}{2}$ and reverse the direction of the inequality symbol

$x < 4$ $-\dfrac{1}{2}(-8) = \dfrac{8}{2} = 4$

89. $-\dfrac{a}{2} \le -3$

$\downarrow$

$-2\left(-\dfrac{a}{2}\right) \ge -2(-3)$ Multiply each side by -2 and reverse the direction of the inequality symbol

$a \ge 6$

90. $-\dfrac{a}{3} \le -2$

$\downarrow$

$-3\left(-\dfrac{a}{3}\right) \ge -3(-2)$ Multiply each side by -3 and reverse the direction of the inequality symbol

$a \ge 6$

91. $\qquad -\dfrac{a}{3} > 5$

$$\downarrow$$

$-3\left(-\dfrac{a}{3}\right) < -3(5)$ Multiply each side by -3 and reverse the direction of the inequality symbol

$\qquad a < -15$

92. $\qquad -\dfrac{a}{6} > 4$

$$\downarrow$$

$-6\left(-\dfrac{a}{6}\right) < -6(4)$ Multiply each side by -6 and reverse the direction of the inequality symbol

$\qquad a < -24$

93. $\qquad -4x + 5 > 37$

$-4x + 5 + (-5) > 37 + (-5)$ Add -5 to each side

$\qquad -4x > 32$

$$\downarrow$$

$-\dfrac{1}{4}(-4x) < -\dfrac{1}{4}(32)$ Multiply each side by $-\dfrac{1}{4}$ and reverse the direction of the inequality symbol

$\qquad x < -8 \qquad -\dfrac{1}{4}(32) = -\dfrac{32}{4} = -8$

See the graph in the back of the textbook.

94. $\qquad -0.3x + 0.7 \leq -2$

$-0.3x + 0.7 + (-0.7) \leq -2 + (-0.7)$ Add -0.7 to each side

$\qquad -0.3x \leq -2.7$

$$\downarrow$$

$-\dfrac{1}{0.3}(-0.3x) \geq -\dfrac{1}{0.3}(-2.7)$ Multiply each side by $-\dfrac{1}{0.3}$ and reverse the direction of the inequality symbol

$\qquad x \geq 9 \qquad -\dfrac{1}{0.3}(-2.7) = \dfrac{2.7}{0.3} = 9$

See the graph in the back of the textbook.

95.
$$\frac{1}{3} - \frac{1}{4}x < \frac{19}{12}$$

$\frac{1}{3} + \left(-\frac{1}{3}\right) - \frac{1}{4}x < \frac{19}{12} + \left(-\frac{1}{3}\right)$ Add $-\frac{1}{3}$ to each side

$-\frac{1}{4}x < \frac{19}{12} + \left(-\frac{4}{12}\right)$ $-\frac{1}{3} = -\frac{4}{12}$

$-\frac{1}{4}x < \frac{15}{12}$

$\downarrow$

$-4\left(-\frac{1}{4}x\right) > -4\left(\frac{15}{12}\right)$ Multiply each side by -4 and reverse the direction of the inequality symbol

$x > -5$ $-4\left(\frac{15}{12}\right) = -\frac{60}{12} = -5$

See the graph in the back of the textbook.

96.
$$8 - 2x \geq 0$$

$8 + (-8) - 2x \geq 0 + (-8)$ Add -8 to each side

$-2x \geq -8$

$\downarrow$

$-\frac{1}{2}(-2x) \leq -\frac{1}{2}(-8)$ Multiply each side by $-\frac{1}{2}$ and reverse the direction of the inequality symbol

$x \leq 4$ $-\frac{1}{2}(-8) = \frac{8}{2} = 4$

See the graph in the back of the textbook.

97.
$$2x + 10 < 5x - 11$$

$2x + (-2x) + 10 < 5x + (-2x) - 11$ Add $-2x$ to each side

$10 < 3x - 11$

$10 + 11 < 3x - 11 + 11$ Add 11 to each side

$21 < 3x$

$\frac{1}{3}(21) < \frac{1}{3}(3x)$ Multiply each side by $\frac{1}{3}$

$7 < x$ or $x > 7$ $\frac{1}{3}(21) = \frac{21}{3} = 7$

See the graph in the back of the textbook.

98. $5x + 10 \le 7x - 14$

    $5x + (-5x) + 10 \le 7x + (-5x) - 14$ Add $-5x$ to each side

           $10 \le 2x - 14$

      $10 + 14 \le 2x - 14 + 14$ Add 14 to each side

          $24 \le 2x$

      $\dfrac{1}{2}(24) \le \dfrac{1}{2}(2x)$ Multiply each side by $\dfrac{1}{2}$

        $12 \le x$ or $x \ge 12$ $\dfrac{1}{2}(24) = \dfrac{24}{2} = 12$

See the graph in the back of the textbook.

99. $2(3t + 1) + 6 \ge 5(2t + 4)$

        $6t + 2 + 6 \ge 10t + 20$ Distributive property

    $6t + (-6t) + 8 \ge 10t + (-6t) + 20$ Combine like terms and add $-6t$ to each side

            $8 \ge 4t + 20$

    $8 + (-20) \ge 4t + 20 + (-20)$ Add -20 to each side

         $-12 \ge 4t$

      $\dfrac{1}{4}(-12) \ge \dfrac{1}{4}(4t)$ Multiply each side by $\dfrac{1}{4}$

        $-3 \ge t$ or $t \le -3$ $\dfrac{1}{4}(-12) = \dfrac{-12}{4} = -3$

See the graph in the back of the textbook.

100. $3(2t - 5) - 7 \le 5(3t + 1)$

        $6t - 15 - 7 \le 15t + 5$ Distributive property

          $6t - 22 \le 15t + 5$ Combine like terms

   $6t + (-6t) - 22 \le 15t + (-6t) + 5$ Add $-6t$ to each side

           $-22 \le 9t + 5$

    $-22 + (-5) \le 9t + 5 + (-5)$ Add -5 to each side

          $-27 \le 9t$

      $\dfrac{1}{9}(-27) \le \dfrac{1}{9}(9t)$ Multiply each side by $\dfrac{1}{9}$

        $-3 \le t$ or $t \ge -3$ $\dfrac{1}{9}(-27) = \dfrac{-27}{9} = -3$

See the graph in the back of the textbook.

CHAPTER 2 TEST

1. $3x + 2 - 7x + 3 = 3x - 7x + 2 + 3$ Commutative property

$$= (3x - 7x) + (2 + 3)$$ Associative property

$$= -4x + 5$$ Add

2. $4a - 5 - a + 1 = 4a - a - 5 + 1$ Commutative property

$$= (4a - a) + (-5 + 1)$$ Associative property

$$= 3a - 4$$

3. $7 - 3(y + 5) - 4 = 7 - 3y - 15 - 4$ Distributive property

$$= -3y + 7 - 15 - 4$$ Commutative property

$$= -3y - 12$$ Add

4. $8(2x + 1) - 5(x - 4)$

$$= 16x + 8 - 5x + 20$$ Distributive property

$$= 16x - 5x + 8 + 20$$ Commutative property

$$= (16x - 5x) + (8 + 20)$$ Associative property

$$= 11x + 28$$ Add

5. $2x - 3 - 7x$ when $x = -5$

$$2x - 3 - 7x = 2(-5) - 3 - 7(-5)$$

$$= -10 - 3 + 35$$

$$= 22$$

6. $x^2 + 2xy + y^2$ when $x = 2$ and $y = 3$

$$x^2 + 2xy + y^2 = 2^2 + 2 \cdot 2 \cdot 3 + 3^2$$

$$= 4 + 2 \cdot 2 \cdot 3 + 9$$ Definition of exponents

$$= 4 + 12 + 9$$ Multiply

$$= 25$$ Add

7. $2x - 5 = 7$

$$2x - 5 + 5 = 7 + 5$$ Add **5** to both sides

$$2x = 12$$

$$\frac{1}{2}(2x) = \frac{1}{2}(12)$$ Multiply each side by $\frac{1}{2}$

$$x = 6$$

8. $2y + 4 = 5y$

$2y - \mathbf{2y} + 4 = 5y - \mathbf{2y}$ Add $-\mathbf{2y}$ to both sides

$4 = 3y$

$\dfrac{\mathbf{1}}{\mathbf{3}}(4) = \dfrac{\mathbf{1}}{\mathbf{3}}(3y)$ Multiply each side by $\dfrac{\mathbf{1}}{\mathbf{3}}$

$\dfrac{4}{3} = y$

9. $\dfrac{1}{2}x - \dfrac{1}{10} = \dfrac{1}{5}x + \dfrac{1}{2}$

First multiply both sides by LCD which is 10.

$$10\left(\dfrac{1}{2}x - \dfrac{1}{10}\right) = 10\left(\dfrac{1}{5}x + \dfrac{1}{2}\right)$$

$$10\left(\dfrac{1}{2}x\right) - 10\left(\dfrac{1}{10}\right) = 10\left(\dfrac{1}{5}x\right) + 10\left(\dfrac{1}{2}\right)$$

$5x - 1 = 2x + 5$

$5x + (-\mathbf{2x}) - 1 = 2x + (-\mathbf{2x}) + 5$ Add $-\mathbf{2x}$ to each side

$3x - 1 = 5$

$3x - 1 + \mathbf{1} = 5 + \mathbf{1}$ Add $\mathbf{1}$ to each side

$3x = 6$

$\dfrac{\mathbf{1}}{\mathbf{3}}(3x) = \dfrac{\mathbf{1}}{\mathbf{3}}(6)$ Multiply each side by $\dfrac{\mathbf{1}}{\mathbf{3}}$

$x = 2$

10. $\dfrac{2}{5}(5x - 10) = -5$

$\dfrac{2}{5}(5x) - \dfrac{2}{5}(10) = -5$ Distributive property

$2x - 4 = -5$ $\dfrac{2}{5}(5) = \dfrac{2}{5}\left(\dfrac{5}{1}\right) = \dfrac{10}{5} = 2,\ -\dfrac{2}{5}(10) = -\dfrac{2}{5}\left(\dfrac{10}{1}\right) = -\dfrac{20}{5} = -4$

$2x - 4 + \mathbf{4} = -5 + \mathbf{4}$ Add $\mathbf{4}$ to each side

$2x = -1$

$\dfrac{\mathbf{1}}{\mathbf{2}}(2x) = \dfrac{\mathbf{1}}{\mathbf{2}}(-1)$ Multiply each side by $\dfrac{\mathbf{1}}{\mathbf{2}}$

$x = -\dfrac{1}{2}$

11. $-5(2x + 1) - 6 = 19$

$-10x - 5 - 6 = 19$ Distributive property

$-10x - 11 = 19$ Simplify the left side

$-10x - 11 + \mathbf{11} = 19 + \mathbf{11}$ Add $\mathbf{11}$ to both sides

$-10x = 30$

$-\dfrac{\mathbf{1}}{\mathbf{10}}(-10x) = -\dfrac{\mathbf{1}}{\mathbf{10}}(30)$ Multiply each side by $-\dfrac{\mathbf{1}}{\mathbf{10}}$

$x = -3$

12. $0.04x + 0.06(100 - x) = 4.6$

$0.04x + 0.06(100) - 0.06(x) = 4.6$ Distributive property

$0.04x + 6 - 0.06x = 4.6$ Simplify the left side

$-0.02x + 6 + (-6) = 4.6 + (-6)$ Add -6 to each side

$-0.02x = -1.4$

$-\dfrac{1}{0.02}(-0.02x) = -\dfrac{1}{0.02}(-1.4)$ Multiply each side by $-\dfrac{1}{0.02}$

$x = 70$ $-\dfrac{1}{0.02}(-1.4) = \dfrac{1.4}{0.02} = 70$

13. $2(t - 4) + 3(t + 5) = 2t - 2$

$2t - 8 + 3t + 15 = 2t - 2$ Distributive property

$5t + 7 = 2t - 2$ Simplify the left side

$5t + 7 - 7 = 2t - 2 - 7$ Add -7 to both sides

$5t = 2t - 9$

$5t - 2t = 2t - 2t - 9$ Add $-2t$ to both sides

$3t = -9$

$\dfrac{1}{3}(3t) = \dfrac{1}{3}(-9)$ Multiply each side by $\dfrac{1}{3}$

$t = -3$

14. $2x - 4(5x + 1) = 3x + 17$

$2x - 20x - 4 = 3x + 17$ Distributive property

$-18x - 4 = 3x + 17$ Simplify the left side

$-18x - 4 - 17 = 3x + 17 - 17$ Add -17 to both sides

$-18x - 21 = 3x$

$-18x + 18x - 21 = 3x + 18x$ Add $18x$ to both sides

$-21 = 21x$

$\dfrac{1}{21}(-21) = \dfrac{1}{21}(21x)$ Multiply each side by $\dfrac{1}{21}$

$-1 = x$

15. <u>What number</u> is 15% of 38?

$x \qquad = 0.15 \ \cdot \ 38$

$x = 0.15 \cdot 38$

$x = 5.7$

16. 240 is 12% of what number?

$$240 = 0.12 \cdot x$$

$$240 = 0.12x$$

$$\frac{1}{0.12}(240) = \frac{1}{0.12}(0.12x) \quad \text{Multiply each side by } \frac{1}{0.12}$$

$$2000 = x \qquad \frac{1}{0.12}(240) = \frac{240}{0.12} = 2000$$

17. $\qquad 2x - 3y = 12 \qquad y = -2$

$$2x - 3(-2) = 12$$

$$2x + 6 = 12$$

$$2x + 6 - 6 = 12 - 6 \qquad \text{Add } -6 \text{ to both sides}$$

$$2x = 6$$

$$\frac{1}{2}(2x) = \frac{1}{2}(6) \qquad \text{Multiply each side by } \frac{1}{2}$$

$$x = 3$$

18. $V = \frac{1}{3}\pi r^2 h \quad$ Find h if $V = 88$ cubic inches, $\pi = \frac{22}{7}$, $r = 3$ inches

$$88 = \frac{1}{3} \cdot \frac{22}{7} \cdot 3^2 h$$

$$88 = \frac{1}{3} \cdot \frac{22}{7} \cdot 9h \qquad \text{Definition of exponents}$$

$$88 = \frac{1}{3} \cdot 9 \cdot \frac{22}{7} h \qquad \text{Commutative property}$$

$$88 = 3 \cdot \frac{22}{7} h \qquad \text{Simplify right side}$$

$$88 = \frac{66}{7} h$$

$$\frac{7}{66}(88) = \frac{7}{66}\left(\frac{66}{7}\right)h \qquad \text{Multiply each side by } \frac{7}{66}$$

$$\frac{7}{3}(4) = h$$

$$\frac{28}{3} \text{ inches} = h$$

19. $\qquad 2x + 5y = 20$ for y

$\qquad 2x - 2x + 5y = -2x + 20 \qquad$ Add $-2x$ to both sides

$\qquad\qquad\qquad 5y = -2x + 20$

$\qquad \dfrac{1}{5}(5y) = \dfrac{1}{5}(-2x + 20) \qquad$ Multiply each side by $\dfrac{1}{5}$

$\qquad\qquad y = -\dfrac{2}{5}x + 4 \qquad$ Distributive property

20. $\qquad\qquad h = x + vt + 16t^2$ for v

$\qquad h - x - 16t^2 = x - x + vt + 16t^2 - 16t^2 \qquad$ Add $-x$ and $-16t^2$ to both sides

$\qquad h - x - 16t^2 = vt$

$\qquad \dfrac{1}{t}\left(h - x - 16t^2\right) = \dfrac{1}{t}(vt) \qquad\qquad$ Multiply both sides by $\dfrac{1}{t}$

$\qquad\qquad \dfrac{h - x - 16t^2}{t} = v$

21. **Step 1:** **Read and list**. Known items: Dave is twice as old as Rick. Ten years ago the sum of their ages was 40. Unknown items: Dave and Rick's ages.

Step 2: **Assign the variables and translate information**.

	Age now	Ten years ago
Dave	$2x$	$2x$ - 10
Rick	x	x - 10

Step 3: **Write an equation**.

$\qquad\qquad$ Dave's age $\qquad + \qquad$ Rick's age $\qquad = \quad 40$

$\qquad\qquad (2x - 10) \qquad + \qquad (x - 10) \qquad = \quad 40$

Step 4: **Solve the equation**.

$\qquad\qquad (2x - 10) + (x - 10) = 40$

$\qquad\qquad (2x + x) + (-10 - 10) = 40 \qquad$ Commutative and Associative properties

$\qquad\qquad\qquad\qquad 3x - 20 = 40 \qquad$ Simplify the left side

$\qquad\qquad\qquad 3x - 20 + 20 = 40 + 20 \qquad$ Add **20** to both sides

$\qquad\qquad\qquad\qquad\qquad 3x = 60$

$\qquad\qquad\qquad \dfrac{1}{3}(3x) = \dfrac{1}{3}(60) \qquad$ Multiply each side by $\dfrac{1}{3}$

$\qquad\qquad\qquad\qquad\qquad x = 20$

$\qquad\qquad\qquad\qquad 2x = 40$

Step 5: **Write the answer**. Rick is 20 years old and Dave is 40 years old.

Step 6: **Reread and check**. Ten years ago, Rick was 10 years old and Dave was 30 years old.

$\qquad\qquad\qquad\qquad 10 + 30 = 40$

22. Step 1: **Read and list**. Known items: Perimeter of the rectangle is 60 inches. The length is twice the width.
$P = 2l + 2w$ Unknown items: length and width.

Step 2: **Assign a variable and translate information**.

Let x = width and $2x$ = length

Step 3: **Write an equation**.

2 lengths	+	2 widths	=	Perimeter
$2(2x)$	+	$2x$	=	60

Step 4: **Solve the equation**.

$$2(2x) + 2x = 60$$
$$4x + 2x = 60 \qquad \text{Multiply}$$
$$6x = 60 \qquad \text{Simplify the left side}$$
$$\frac{1}{6}(6x) = \frac{1}{6}(60) \qquad \text{Multiply each side by } \frac{1}{6}$$
$$x = 10$$
$$2x = 20$$

Step 5: **Write the answer**. Width is 10 inches and length is 20 inches.

Step 6: **Reread and check**.

$$2(20) + 2(10) \overset{?}{=} 60$$
$$40 + 20 \overset{?}{=} 60$$
$$60 = 60$$

23. Step 1: **Read and list**. Known items: A man has dimes and quarters that equal $3.50. He has 7 more dimes than quarters. Unknown items: How many dimes and how many quarters.

Step 2: **Assign the variables and translate information**.

	Dimes	Quarters
Number	$x + 7$	x
Value	$0.10(x + 7)$	$0.25x$

Step 3: **Write the equation**.

Dimes	+	Quarters	=	3.50
$0.10(x + 7)$	+	$0.25x$	=	3.50

Step 4: **Solve the equation**.

$$0.10(x + 7) + 0.25x = 3.50$$

$0.10x + 0.70 + 0.25x = 3.50$	Distributive property
$0.35x + 0.70 = 3.50$	Commutative and Associative properties
$0.35x + 0.70 - \mathbf{0.70} = 3.50 - \mathbf{0.70}$	Add $-\mathbf{0.70}$ to both sides
$0.35x = 2.80$	

$$\frac{1}{\mathbf{0.35}}(-0.35x) = \frac{1}{\mathbf{0.35}}(2.80) \qquad \text{Multiply each side by } \frac{1}{\mathbf{0.35}}$$

$$x = 8$$

$$x + 7 = 15$$

Step 5: **Write the answer**. The man has 15 dimes and 8 quarters.

Step 6: **Reread and check**.

$$0.10(15) + 0.25(8) \overset{?}{=} 3.50$$

$$1.50 + 2.00 \overset{?}{=} 3.50$$

$$3.50 = 3.50$$

24. Step 1: **Read and list**. Known items: A woman invests money at 7% and 9%. She has $600 more invested at 9% than she does at 7%. The interest for 1 year is $182.00. Unknown items: The amount of money invested at each rate.

Step 2: **Assign the variables and translate information**.

	Dollars invested at 7%	Dollars invested at 9%
Number of	x	$x + 600$
Interest on	$0.07x$	$0.09(x + 600)$

Step 3: **Reread and write an equation**.

Money at 7%	+	Money at 9%	=	182
$0.07x$	+	$0.09(x + 600)$	=	182

Step 4: **Solve the equation**.

$$0.07x + 0.09(x + 600) = 182$$

$0.07x + 0.09x + 54 = 182$	Distributive property
$0.16x + 54 = 182$	Simplify the left side
$0.16x + 54 - \mathbf{54} = 182 - \mathbf{54}$	Add $-\mathbf{54}$ to both sides
$0.16x = 128$	

$$\frac{1}{\mathbf{0.16}}(0.16x) = \frac{1}{\mathbf{0.16}}(128) \qquad \text{Multiply each side by } \frac{1}{\mathbf{0.16}}$$

$$x = 800$$

$$x + 600 = 1400$$

Step 5: **Write the answer**. The woman invests $800 at 7% and $1,400 at 9%.

Step 6. **Reread and check**.

$$0.07(800) + 0.09(1400) \overset{?}{=} 182$$

$$56 + 126 \overset{?}{=} 182$$

$$182 = 182$$

25. $2x + 3 < 5$

$2x + 3 - 3 < 5 - 3$ Add -3 to both sides

   $2x < 2$

$\dfrac{1}{2}(2x) < \dfrac{1}{2}(2)$ Multiply each side by $\dfrac{1}{2}$

   $x < 1$

See the graph in the back of the textbook.

26. $-5a > 20$

$-\dfrac{1}{5}(-5a) < -\dfrac{1}{5}(20)$ Multiply by $-\dfrac{1}{5}$, and reverse the direction of the inequality symbol

   $a < -4$

See the graph in the back of the textbook.

27.

   $0.4 - 0.2x \geq 1$

$0.4 + (-0.4) - 0.2x \geq 1 + (-0.4)$ Add -0.4 to each side

   $-0.2x \geq 0.6$

    $\downarrow$

$-\dfrac{1}{0.2}(-0.2x) \leq -\dfrac{1}{0.2}(0.6)$ Multiply each side by $-\dfrac{1}{0.2}$ and reverse the direction of the inequality symbol

   $x \leq -3$ $-\dfrac{1}{0.2}(0.6) = -\dfrac{0.6}{0.2} = -3$

See the graph in the back of the textbook.

28. $4 - 5(m + 1) \leq 9$

   $4 - 5m - 5 \leq 9$ Distributive property

   $-5m - 1 \leq 9$ Commutative and Associative properties

$-5m - 1 + 1 \leq 9 + 1$ Add 1 to both sides

   $-5m \leq 10$

    $\downarrow$

$-\dfrac{1}{5}(-5m) \geq -\dfrac{1}{5}(10)$ Multiply by $-\dfrac{1}{5}$ and reverse the direction of the inequality symbol

   $m \geq -2$

See the graph in the back of the textbook.

CHAPTER 3

SECTION 3.1

1. See the histogram in the back of the textbook.

5. See the scatter diagram in the back of the textbook.

9. See the scatter diagram in the back of the textbook.

13, 17, 21, 25, & 29.

   See the graph of these ordered pairs in the back of the textbook.

33. $(-4, 2)$

37. $(2, -2)$

41. Yes, see the graph in the back of the textbook.

45. Yes, see the graph in the back of the textbook.

49. Yes, see the graph in the back of the textbook.

53. No, see the graph in the back of the textbook.

57. Every point on this line has a y-coordinate of -3. See the graph in the back of the textbook.

61. 3, 10, 17, 24, 31   Add seven to the previous number to produce the new number.

65. 7, 4, 1, -2, -5 Add -3 to the previous number to produce the new number.

69. 5, 6, 8, 11, 15   Add the next consecutive counting number to the previous number to produce the new number.

SECTION 3.2

1. To complete $(0, \ )$, substitute $x = 0$ into $2x + y = 6$.

$$2(0) + y = 6$$
$$y = 6$$

   The ordered pair is $(0, \ 6)$.

   To complete $(3, \ )$, substitute $x = 3$ into $2x + y = 6$.

$$2(3) + y = 6$$
$$6 + y = 6$$
$$y = 0$$

   The ordered pair is $(3, \ 0)$.

   To complete $(\ , -6)$, substitute $y = -6$ into $2x + y = 6$

$$2x - 6 = 6$$
$$2x = 12$$
$$x = 6$$

   The ordered pair is $(6, -6)$.

5. To complete $(1, \ )$, substitute $x = 1$ into $y = 4x - 3$.

$$y = 4(1) - 3$$
$$y = 1$$

The ordered pair is $(1, 1)$.

To complete $(\ , 0)$, substitute $y = 0$ into $y = 4x - 3$.

$$0 = 4x - 3$$
$$3 = 4x$$
$$\tfrac{3}{4} = x$$

The ordered pair is $\left(\tfrac{3}{4}, 0 \right)$.

To complete $(5, \ )$, substitute $x = 5$ into $y = 4x - 3$.

$$y = 4(5) - 3$$
$$y = 20 - 3$$
$$y = 17$$

The ordered pair is $(5, \ 17)$.

9. Let $x = -5$ in each ordered pair.

13. When $x = 0$, we have

$$y = 4(0)$$
$$y = 0$$

When $x = -3$, we have

$$y = 4(-3)$$
$$y = -12$$

When $y = -2$, we have

$$-2 = 4x$$
$$-\tfrac{1}{2} = x$$

When $y = 12$, we have

$$12 = 4x$$
$$3 = x$$

See the table in the back of the textbook.

17. When $y = 0$, we have

$$2x - 0 = 4$$
$$2x = 4$$
$$x = 2$$

When $x = 1$, we have

$$2(1) - y = 4$$
$$2 - y = 4$$
$$-y = 2$$
$$y = -2$$

When $y = 2$, we have

$$2x - 2 = 4$$
$$2x = 6$$
$$x = 3$$

When $x = -3$, we have

$$2(-3) - y = 4$$
$$-6 - y = 4$$
$$-y = 10$$
$$y = -10$$

See the table in the back of the textbook.

21. Try $(2, 3)$ in $2x - 5y = 10$

$$2(2) - 5(3) = 10$$
$$4 - 15 = 10$$
$$-11 = 10 \quad \text{A false statement.}$$

Try $(0, -2)$ in $2x - 5y = 10$

$$2(0) - 5(-2) = 10$$
$$0 + 10 = 10$$
$$10 = 10 \quad \text{A true statement.}$$

Try $\left(\dfrac{5}{2}, 1\right)$ in $2x - 5y = 10$

$$2\left(\dfrac{5}{2}\right) - 5(1) = 10$$
$$5 - 5 = 10$$
$$0 = 10 \quad \text{A false statement.}$$

The ordered pair $(0, -2)$ is a solution to the equation $2x - 5y = 10$; $(2, 3)$ and $\left(\frac{5}{2}, 1\right)$ are not.

25. Try $(1, 6)$ in $y = 6x$

$$6 = 6(1)$$
$$6 = 6 \quad \text{A true statement.}$$

Try $(-2, -12)$ in $y = 6x$

$$-12 = 6(-2)$$
$$-12 = -12 \quad \text{A true statement.}$$

Try $(0, 0)$ in $y = 6x$

$$0 = 6(0)$$
$$0 = 0 \quad \text{A true statement.}$$

The ordered pairs $(1, 6)$, $(-2, -12)$, $(0, 0)$ are solutions to the equation $y = 6x$.

29. Try $(3, 0)$ in $x = 3$

$$3 = 3 \quad \text{A true statement.}$$

Try $(3, -3)$ in $x = 3$

$$3 = 3 \quad \text{A true statement.}$$

Try $(5, 3)$ in $x = 3$

$$5 = 3 \quad \text{A false statement.}$$

The ordered pairs $(3, 0)$, $(3, -3)$ are solutions to the equation $x = 3$; $(5, 3)$ is not.

33. When $x = 5$ the ordered pair $(x, 2x)$ becomes $(5, 2 \cdot 5) = (5, 10)$

37. $y = 13 + 1.5x$ See the table and histogram in the back of the textbook.

41. When $x = 4$, the equation $3x + 2y = 6$ becomes,

$$3(4) + 2y = 6$$
$$12 + 2y = 6$$
$$2y = -6$$
$$y = -3$$

45. When $x = 2$, the equation $y = \frac{3}{2}x - 3$ becomes,

$$y = \frac{3}{2}(2) - 3$$
$$y = 3 - 3$$
$$y = 0$$

49. $3x - 2y = 6$

$$-2y = -3x + 6$$
$$-\frac{1}{2}(-2y) = -\frac{1}{2}(-3x + 6)$$
$$y = \frac{3}{2}x - 3$$

SECTION 3.3

1. $x + y = 4$ $(0,)$ $x + y = 4$ $(2,)$ $x + y = 4$ $(, 0)$
   $0 + y = 4$ $2 + y = 4$ $x + 0 = 4$
   $\quad y = 4$ $(0, 4)$ $y = 2$ $(2, 2)$ $x = 4$ $(4, 0)$

See the graph in the back of the textbook.

5. $y = 2x$ $(0,)$ $y = 2x$ $(-2,)$ $y = 2x$ $(2,)$
   $y = 2(0)$ $y = 2(-2)$ $y = 2(2)$ $(2, 4)$
   $y = 0$ $(0, 0)$ $y = -4$ $(-2, -4)$

See the graph in the back of the textbook.

9. $y = 2x + 1$ $(0,)$ $y = 2x + 1$ $(-1,)$ $y = 2x + 1$ $(1,)$
   $y = 2(0) + 1$ $y = 2(-1) + 1$ $y = 2(1) + 1$
   $y = 0 + 1$ $y = -2 + 1$ $y = 2 + 1$
   $y = 1$ $(0, 1)$ $y = -1$ $(-1, -1)$ $y = 3$ $(1, 3)$

See the graph in the back of the textbook.

13. $y = \dfrac{1}{2}x + 3 \quad (-2, \ )$ $y = \dfrac{1}{2}x + 3 \quad (0, \ )$ $y = \dfrac{1}{2}x + 3 \quad (2, \ )$

$y = \dfrac{1}{2}(-2) + 3$ $y = \dfrac{1}{2}(0) + 3$ $y = \dfrac{1}{2}(2) + 3$

$y = -1 + 3$ $y = 0 + 3$ $y = 1 + 3$

$y = 2 \qquad (-2, 2)$ $y = 3 \qquad (0, 3)$ $y = 4 \qquad (2, 4)$

See the graph in the back of the textbook.

17. $2x + y = 3 \quad (-1, \ )$ $2x + y = 3 \quad (0, \ )$ $2x + y = 3 \quad (1, \ )$

$2(-1) + y = 3$ $2(0) + y = 3$ $2(1) + y = 3$

$-2 + y = 3$ $y = 3 \quad (0, 3)$ $2 + y = 3$

$y = 5 \quad (-1, 5)$ $y = 1 \quad (1, 1)$

See the graph in the back of the textbook.

21. $-x + 2y = 6 \quad (-2, \ )$ $-x + 2y = 6 \quad (0, \ )$ $-x + 2y = 6 \quad (2, \ )$

$-(-2) + 2y = 6$ $0 + 2y = 6$ $-2 + 2y = 6$

$2 + 2y = 6$ $2y = 6$ $2y = 8$

$2y = 4$ $y = 3 \quad (0, 3)$ $y = 4 \quad (2, 4)$

$y = 2 \quad (-2, 2)$

See the graph in the back of the textbook.

25. To find three solutions to $y = 3x - 1$ we can let $x = 1$, $x = 0$, and $x = -1$, and find the corresponding values of y.

When $x = 1$, $y = 3(1) - 1 = 2$. The ordered pair $(1, \ 2)$ is one solution.

When $x = 0$, $y = 3(0) - 1 = -1$. The ordered pair $(0, \ -1)$ is a second solution.

When $x = -1$, $y = 3(-1) - 1 = -4$. The ordered pair $(-1, \ -4)$ is a third solution.

See the graph in the back of the textbook.

29. To find three solutions to $3x + 4y = 8$ we can let $x = 0$, $x = 1$, and $y = 0$.

When $x = 0$, $3(0) + 4y = 8$

$4y = 8$

$y = 2$ so $(0, \ 2)$ is one solution.

When $x = 1$, $3(1) + 4y = 8$

$$3 + 4y = 8$$
$$4y = 5$$
$$y = \tfrac{5}{4} \qquad \text{so } \left(1, \tfrac{5}{4}\right) \text{ is a second solution.}$$

When $y = 0$, $3x + 4(0) = 8$

$$3x = 8$$
$$x = \tfrac{8}{3} \qquad \text{so } \left(\tfrac{8}{3}, 0\right) \text{ is a third solution.}$$

See the graph in the back of the textbook.

33. $y = 2$ x may be equal to any real number.

See the graph in the back of the textbook.

37. Some suggested points: (0, 0), (1, 1), (-1, -1), (2, 2)

See the graph in the back of the textbook.

41. The lines cross the y-axis at the number that follows x in the equations.

See the graph in the back of the textbook.

45. $2(3x - 1) + 4 = -10$

$$6x - 2 + 4 = -10 \qquad \text{Distributive property}$$
$$6x + 2 = -10$$
$$6x = -12 \qquad \text{Add } -2 \text{ to both sides}$$
$$x = -2 \qquad \text{Multiply each side by } \frac{1}{6}$$

49. $\dfrac{1}{2}x + 4 = \dfrac{2}{3}x + 5$

Method 1: Working with the fractions.

$$\frac{1}{2}x + \left(-\frac{1}{2}x\right) + 4 = \frac{2}{3}x + \left(-\frac{1}{2}x\right) + 5$$

$$4 = \frac{1}{6}x + 5 \qquad\qquad \frac{2}{3} - \frac{1}{2} = \frac{4}{6} - \frac{3}{6} = \frac{1}{6}$$

$$-1 = \frac{1}{6}x \qquad\qquad\quad \text{Add } -5 \text{ to both sides}$$

$$-6 = x \qquad\qquad\qquad \text{Multiply each side by } 6$$

Method 2: Eliminating the fractions in the beginning.

$$6\left(\frac{1}{2}x+4\right)=6\left(\frac{2}{3}x+5\right)$$ Multiply each side by the LCD 6

$$6\left(\frac{1}{2}x\right)+6(4)=6\left(\frac{2}{3}x\right)+6(5)$$ Distributive property

$$3x+24=4x+30$$ Multiply

$$24=x+30$$

$$-6=x$$

SECTION 3.4

1. **x – Intercept**

   When $y=0$, the equation $2x+y=4$ becomes

   $$2x+0=4$$

   $$2x=4$$

   $$x=2$$ Multiply each side by $\dfrac{1}{2}$

   The x – intercept is 2, so the point $(2,0)$ is on the graph.

   y – Intercept

   When $x=0$, the equation $2x+y=4$ becomes

   $$2(0)+y=4$$

   $$y=4$$

   The y – intercept is 4, so the point $(0,4)$ is on the graph.

   See the graph in the back of the textbook.

5. **x – Intercept**

   When $y=0$, the equation $-x+2y=2$ becomes

   $$-x+2(0)=2$$

   $$-x=2$$

   $$x=-2$$

   The x – intercept is -2, so the point $(-2,0)$ is on the graph.

   y – Intercept

   When $x=0$, the equation $-x+2y=2$ becomes

   $$-(0)+2y=2$$

   $$y=1$$

   The y – intercept is 1, so the point $(0,1)$ is on the graph.

   See the graph in the back of the textbook.

9. **x – Intercept**

When $y = 0$, the equation $4x - 2y = 8$ becomes

$$4x - 2(0) = 8$$
$$4x = 8$$
$$x = 2 \qquad \text{Multiply each side by } \frac{1}{4}$$

The x – intercept is 2, so the point $(2, 0)$ is on the graph.

y – Intercept

When $x = 0$, the equation $4x - 2y = 8$ becomes

$$4(0) - 2y = 8$$
$$-2y = 8$$
$$y = -4 \qquad \text{Multiply each side by } -\frac{1}{2}$$

The y – intercept is -4, so the point $(0, -4)$ is on the graph.

See the graph in the back of the textbook.

13. **x – Intercept**

When $y = 0$, the equation $y = 2x - 6$ becomes

$$0 = 2x - 6$$
$$6 = 2x$$
$$3 = x \qquad \text{Multiply each side by } \frac{1}{2}$$

The x – intercept is 3, so the point $(3, 0)$ is on the graph.

y – Intercept

When $x = 0$, the equation $y = 2x - 6$ becomes

$$y = 2(0) - 6$$
$$y = -6$$

The y – intercept is -6, so the point $(0, -6)$ is on the graph.

See the graph in the back of the textbook.

17. **x – Intercept**

When $y = 0$, the equation $y = 2x - 1$ becomes

$$0 = 2x - 1$$
$$1 = 2x$$
$$\frac{1}{2} = x \qquad \text{Multiply each side by } \frac{1}{2}$$

The x – intercept is $\frac{1}{2}$, so the point $\left(\frac{1}{2}, 0\right)$ is on the graph.

y – Intercept

When $x = 0$, the equation $y = 2x - 1$ becomes

$$y = 2(0) - 1$$
$$y = -1$$

The y – intercept is -1, so the point $(0, -1)$ is on the graph.

See the graph in the back of the textbook.

21. **x – Intercept**

When $y = 0$, the equation $y = -\frac{1}{3}x - 2$ becomes

$$0 = -\frac{1}{3}x - 2$$
$$2 = -\frac{1}{3}x$$
$$-6 = x \qquad \text{Multiply each side by } -3$$

The x – intercept is -6, so the point $(-6, 0)$ is on the graph.

y – Intercept

When $x = 0$, the equation $y = -\frac{1}{3}x - 2$ becomes

$$y = -\frac{1}{3}(0) - 2$$
$$y = -2$$

The y – intercept is -2, so the point $(0, -2)$ is on the graph.

See the graph in the back of the textbook.

25. See the graph in the back of the textbook.

29. See the graph in the back of the textbook.

33. The y-intercept is −4. See the graph in the back of the textbook.

37. The x-intercept is 3 and the y-intercept is 3. See the graph in the back of the textbook.

41. The y-intercept is 4. See the graph in the back of the textbook.

45. $5 - 2 \cdot 6 = 5 - 12$ Multiply

        $= -7$ Subtract

49. $2(3)^2 - 4(3)^2 = 2(9) - 4(9)$ Definition of exponents

              $= 18 - 36$ Multiply

              $= -18$ Subtract

SECTION 3.5

1. We can let $(x_1, y_1) = (2, 1)$ and $(x_2, y_2) = (4, 4)$

    then $m = \dfrac{y_2 - y_1}{x_2 - x_1} = \dfrac{4-1}{4-2} = \dfrac{3}{2}$

    The slope is $\dfrac{3}{2}$. For every vertical change of 3 units,
    there will be a corresponding horizontal change of 2 units.
    See the graph in the back of the textbook.

5. We can let $(x_1, y_1) = (1, -3)$ and $(x_2, y_2) = (4, 2)$

    then $m = \dfrac{y_2 - y_1}{x_2 - x_1} = \dfrac{2-(-3)}{4-1} = \dfrac{5}{3}$

    The slope is $\dfrac{5}{3}$. For every vertical change of 5 units,
    there will be a corresponding horizontal change of 3 units.

    See the graph in the back of the textbook.

9. We can let $(x_1, y_1) = (-3, 2)$ and $(x_2, y_2) = (3, -2)$

    then $m = \dfrac{y_2 - y_1}{x_2 - x_1} = \dfrac{-2-2}{3-(-3)} = -\dfrac{4}{6} = -\dfrac{2}{3}$

    The slope is $-\dfrac{2}{3}$. For every vertical change of -2 units,
    there will be a corresponding horizontal change of 3 units.

    See the graph in the back of the textbook.

13. Since the y-intercept is 1, we begin at $(0, 1)$. Since the slope is $\dfrac{2}{3}$, we

    will move from point $(0, 1)$ vertically 2 units, then horizontally 3 units.
    See the graph in the back of the textbook.

17. Since the y-intercept is 5, we begin at $(0, 5)$. Since the slope is $-\dfrac{4}{3}$, we

    will move from point $(0, 5)$ horizontally 3 units, then down vertically 4 units.
    See the graph in the back of the textbook.

21. Since the y-intercept is -1, we begin at $(0, -1)$. Since the slope is $3 = \dfrac{3}{1}$, we

    will move from point $(0, -1)$ horizontally 1 unit, then vertically 3 units.
    See the graph in the back of the textbook.

25. The x-intercept is the point $(4, 0)$ and the y-intercept is
    the point $(0, 2)$. If you move from $(0, 2)$ to $(4, 0)$ the rise to
    run is $-\dfrac{2}{4} = -\dfrac{1}{2}$.

    See the graph in the back of the textbook.

29.

x	y
0	1
2	2
-2	0

$y = \frac{1}{2}x + 1$

The slope is $\frac{1}{2}$ and the y-intercept is 1.

33. $V = IR$

$\dfrac{V}{R} = \dfrac{IR}{R}$

$\dfrac{V}{R} = I$ or $I = \dfrac{V}{R}$

37. $\qquad -2x + y = -4 \qquad\qquad$ for y

$\quad -2x + 2x + y = 2x - 4 \qquad$ Add $2x$ to both sides

$\qquad\qquad\quad y = 2x - 4$

SECTION 3.6

1. Substituting $m = \dfrac{2}{3}$ and $b = 1$ into the equation $y = mx + b$,

   we have: $y = \dfrac{2}{3}x + 1$

5. Substituting $m = -\dfrac{2}{5}$ and $b = 3$ into the equation $y = mx + b$,

   we have: $y = -\dfrac{2}{5}x + 3$

9. Substituting $m = -3$ and $b = 2$ into the equation $y = mx + b$,
   we have: $y = -3x + 2$

13. $3x + y = 3$ Original equation

$-3x + 3x + y = -3x + 3$ Add $-3x$ to both sides

$y = -3x + 3$

Slope $= -3$ $y-$ intercept $= 3$

See the graph in the back of the textbook.

17. $4x - 5y = 20$ Original equation

$4x - 4x - 5y = -4x + 20$ Add $-4x$ to both sides

$-5y = -4x + 20$

$-\dfrac{1}{5}(-5y) = -\dfrac{1}{5}(-4x + 20)$ Multiply each side by $-\dfrac{1}{5}$

$y = \dfrac{4}{5}x - 4$

Slope $= \dfrac{4}{5}$ $y-$ intercept $= -4$

See the graph in the back of the textbook.

21. Using $(x_1,\ y_1) = (-2,\ -5)$ and $m = 2$ in $y - y_1 = m(x - x_1)$

gives us $y - (-5) = 2(x + 2)$

$y + 5 = 2x + 4$ Multiply out right side

$y = 2x - 1$ Add -5 to each side

25. Using $(x_1,\ y_1) = (2,\ -3)$ and $m = \dfrac{3}{2}$ in $y - y_1 = m(x - x_1)$

gives us $y - (-3) = \dfrac{3}{2}(x - 2)$

$y + 3 = \dfrac{3}{2}x - 3$ Multiply out right side

$y = \dfrac{3}{2}x - 6$ Add -3 to each side

29. Using $(x_1,\ y_1) = (2,\ 4)$ and $m = 1$ in $y - y_1 = m(x - x_1)$

gives us $y - (4) = 1(x - 2)$

$y - 4 = x - 2$ Multiply out right side

$y = x + 2$ Add 4 to each side

33. We begin by finding the slope of the line.
$$m = \frac{-5-1}{-1-2} = \frac{-6}{-3} = 2$$
Using $(x_1, y_1) = (-1, -5)$ and $m = 2$ in $y - y_1 = m(x - x_1)$ yields

$y - (-5) = 2(x + 1)$ Note: $x - (-1) = x + 1$

$y + 5 = 2x + 2$ Multiply out right side

$y = 2x - 3$ Add -5 to each side

Note: We could have used $(2, 1)$ to get the same result.

37. We begin by finding the slope of the line.
$$m = \frac{-1-(-5)}{-3-3} = \frac{4}{-6} = -\frac{2}{3}$$
Using $(x_1, y_1) = (-3, -1)$ and $m = -\frac{2}{3}$ in $y - y_1 = m(x - x_1)$ yields

$y - (-1) = -\frac{2}{3}(x + 3)$ Note: $x - (-3) = x + 3$

$y + 1 = -\frac{2}{3}x - 2$ Multiply out right side

$y = -\frac{2}{3}x - 3$ Add -1 to each side

Note: We could have used $(3, -5)$ to get the same result.

41. x – intercept 3 is $(3, 0)$ and y – intercept 2 is $(0, 2)$.
To find the slope of the line:
$$m = \frac{0-2}{3-0} = -\frac{2}{3}$$
With $m = -\frac{2}{3}$ and $b = 2$, we have
$$y = -\frac{2}{3}x + 2$$

45. $x = 3$ because the first coordinate in each ordered pair is 3.

49. What number is 25% of 300?

$N = 0.25 \quad 300$

$N = 75$

53. 60 is 15% of what number?

$60 = 0.15 \quad \cdot \quad N$

$\dfrac{60}{0.15} = \dfrac{0.15N}{0.15} \qquad$ Divide each side by 0.15

$400 = N$

57. $y = 2x - 3 \qquad$ Slope $= 2, \qquad y-\text{intercept} = -3$

See the graph in the back of the textbook.

SECTION 3.7

SEE THE GRAPHS FOR PROBLEMS 1 - 29 IN THE BACK OF THE TEXTBOOK.

1. To graph $x + y = 3$ we find its intercepts.

x – Intercept	**y – Intercept**
When $\qquad y = 0$	When $\qquad x = 0$
the equation $x + y = 3$	the equation $x + y = 3$
becomes $\quad x + 0 = 3$	becomes $\quad 0 + y = 3$
$x = 3$	$y = 3$
$(3, 0)$	$(0, 3)$

To graph $x - y = 1$ we find its intercepts.

x – Intercept	**y – Intercept**
When $\qquad y = 0$	When $\qquad x = 0$
the equation $x - y = 1$	the equation $x - y = 1$
becomes $\quad x - 0 = 1$	becomes $\quad 0 - y = 1$
$x = 1$	$-y = 1$
$(1, 0)$	$y = -1$
	$(0, -1)$

The solution to the system is the point $(2, 1)$ since the point satisfies both equations.

5. To graph $x + y = 8$ we find its intercepts.

x – Intercept	**y – Intercept**
When $\qquad y = 0$	When $\qquad x = 0$
the equation $x + y = 8$	the equation $x + y = 8$
becomes $\quad x + 0 = 8$	becomes $\quad 0 + y = 8$
$x = 8$	$y = 8$
$(8, 0)$	$(0, 8)$

To graph $-x + y = 2$ we find its intercepts.

x – Intercept		**y – Intercept**	
When	$y = 0$	When	$x = 0$
the equation	$-x + y = 2$	the equation	$-x + y = 2$
becomes	$-x = 2$	becomes	$0 + y = 2$
	$x = -2$		$y = 2$
	$(-2, 0)$		$(0, 2)$

The solution to the system is the point $(3, 5)$ since the point satisfies both equations.

9. To graph $6x - 2y = 12$ we find its intercepts.

x – Intercept		**y – Intercept**	
When	$y = 0$	When	$x = 0$
the equation	$6x - 2y = 12$	the equation	$6x - 2y = 12$
becomes	$6x - 2(0) = 12$	becomes	$6(0) - 2y = 12$
	$6x = 12$		$-2y = 12$
	$x = 2$		$y = -6$
	$(2, 0)$		$(0, -6)$

To graph $3x + y = -6$ we find its intercepts.

x – Intercept		**y – Intercept**	
When	$y = 0$	When	$x = 0$
the equation	$3x + y = -6$	the equation	$3x + y = -6$
becomes	$3x + 0 = -6$	becomes	$3(0) + y = -6$
	$3x = -6$		$y = -6$
	$x = -2$		$(0, -6)$
	$(-2, 0)$		

The solution to the system is the point $(0, -6)$, since the point satisfies both equations.

13. To graph $x + 2y = 0$ we find the x-intercept and another point.

x – Intercept		**Another Point**	
When	$y = 0$	When	$y = 1$
the equation	$x + 2y = 0$	the equation	$x + 2y = 0$
becomes	$x + 2(0) = 0$	becomes	$x + 2(1) = 0$
	$x = 0$		$x + 2 = 0$
	$(0, 0)$		$x = -2$

so $(-2, 1)$ is a point on the graph.

To graph $2x - y = 0$ we find the x-intercept and another point.

x – Intercept   **Another Point**

When $y = 0$ When $x = 1$

the equation $2x - y = 0$ the equation $2x - y = 0$

becomes $2x = 0$ becomes $2(1) - y = 0$

                              $x = 0$ $-y = -2$

                              $(0, 0)$ $y = 2$

so $(1, 2)$ is a point on the graph.

The solution to the system is the point $(0, 0)$, since the point satisfies both equations.

17. To graph $y = 2x + 1$ we find the intercepts.

x – Intercept   **y – Intercept**

When $y = 0$ When $x = 0$

the equation $y = 2x + 1$ the equation $y = 2x + 1$

becomes $0 = 2x + 1$ becomes $y = 2(0) + 1$

                            $-1 = 2x$ $y = 1$

                            $-\dfrac{1}{2} = x$

                            $\left(-\dfrac{1}{2}, 0\right)$ $(0, 1)$

To graph $y = -2x - 3$ we find the intercepts.

x – Intercept   **y – Intercept**

When $y = 0$ When $x = 0$

the equation $y = -2x - 3$ the equation $y = -2x - 3$

becomes $0 = -2x - 3$ becomes $y = -2(0) - 3$

                              $3 = -2x$ $y = -3$

                            $-\dfrac{3}{2} = x$

                            $\left(-\dfrac{3}{2}, 0\right)$ $(0, -3)$

The solution to the system is the point $(-1, -1)$, since the point satisfies both equations.

21. For $x + y = 2$, the intercepts are:

x – Intercept   **y – Intercept**

When $y = 0$ When $x = 0$

the equation $x + y = 2$ the equation $x + y = 2$

becomes $x + 0 = 2$ becomes $0 + y = 2$

                              $x = 2$ $y = 2$

                              $(2, 0)$ $(0, 2)$

$x = -3$ is a vertical line. y is all real numbers.

The solution to the system is the point $(-3, 5)$.

25. For $x + y = 4$, the intercepts are:

x – Intercept

When $y = 0$

the equation $x + y = 4$

becomes $x + 0 = 4$

$x = 4$

$(4, 0)$

y – Intercept

When $x = 0$

the equation $x + y = 4$

becomes $0 + y = 4$

$y = 4$

$(0, 4)$

For $2x + 2y = -6$, the intercepts are:

x – Intercept

When $y = 0$

the equation $2x + 2y = -6$

becomes $2x + 2(0) = -6$

$x = -3$

$(-3, 0)$

y – Intercept

When $x = 0$

the equation $2x + 2y = -6$

becomes $2(0) + 2y = -6$

$y = -3$

$(0, -3)$

Lines are parallel. There is no solution to the system.

29. For $y = -2x + 2$, the intercepts are:

x – Intercept

When $y = 0$

the equation $y = -2x + 2$

becomes $0 = -2x + 2$

$-2 = -2x$

$1 = x$

$(1, 0)$

y – Intercept

When $x = 0$

the equation $y = -2x + 2$

becomes $y = -2(0) + 2$

$y = 0 + 2$

$y = 2$

$(0, 2)$

For $y = 4x - 1$, the intercepts are:

x – Intercept

When $y = 0$

the equation $y = 4x - 1$

becomes $0 = 4x - 1$

$1 = 4x$

$\dfrac{1}{4} = x$

$\left(\dfrac{1}{4}, 0 \right)$

y – Intercept

When $x = 0$

the equation $y = 4x - 1$

becomes $y = 4(0) - 1$

$y = 0 - 1$

$y = -1$

$(0, -1)$

The solution to the system is the point $\left(\dfrac{1}{2}, 1 \right)$, since the point satisfies both equations.

33. Substituting $x = -3$ into $2x - 9$ we have $2(-3) - 9 = -6 - 9 = -15$

37. Substituting $x = -3$ into $4(3x + 2) + 1$, we have

$$\begin{aligned} 4(3x + 2) + 1 &= 12x + 8 + 1 \\ &= 12x + 9 \\ &= 12(-3) + 9 \\ &= -36 + 9 \\ &= -27 \end{aligned}$$

SECTION 3.8

1. $x + y = 3$

$\underline{x - y = 1}$ Add together

   $2x = 4$

     $x = 2$ Divide by 2

Substituting $x = 2$ into $x + y = 3$ gives us

$2 + y = 3$

    $y = 1$

The solution to our system is the ordered pair $(2, 1)$.

5. $x - y = 7$

$\underline{-x - y = 3}$ Add together

   $-2y = 10$

     $y = -5$ Divide by -2

Substituting $y = -5$ into $x - y = 7$ gives us

$x - (-5) = 7$

   $x + 5 = 7$

     $x = 2$

The solution to our system is the ordered pair $(2, -5)$.

9. $3x + 2y = 1$

$\underline{-3x + -2y = -1}$ Add together

     $0 = 0$

The lines coincide and there is an infinite number of solutions to the system.

13. $5x - 3y = -2$ $\underline{\text{No Change}}$ $\rightarrow$ $5x - 3y = -2$

    $10x - y = 1$ $\underline{\text{Multiply by } -3}$ $\rightarrow$ $\dfrac{-30x + 3y = -3}{}$

                                      $-25x = -5$

                                          $x = \dfrac{1}{5}$

Substituting $x = \frac{1}{5}$ into $10x - y = 1$ gives us

$$10\left(\frac{1}{5}\right) - y = 1$$
$$2 - y = 1$$
$$-y = -1$$
$$y = 1$$

The solution to our system is the ordered pair $\left(\frac{1}{5},\ 1\right)$.

17. $3x - 5y = 7$ No change $\rightarrow$ $3x - 5y = 7$

     $-x + y = -1$ Multiply by 3 $\rightarrow$ $-3x + 3y = -3$

$$-2y = 4$$
$$y = -2$$

Substituting $y = -2$ into $-x + y = -1$ gives us

$$-x - 2 = -1$$
$$-x = 1$$
$$x = -1$$

The solution to our system is the ordered pair $(-1, -2)$.

21. $-3x - y = 7$ Multiply by 2 $\rightarrow$ $-6x - 2y = 14$

     $6x + 7y = 11$ No change $\rightarrow$ $6x + 7y = 11$

$$5y = 25$$
$$y = 5$$

Substituting $y = 5$ into $-3x - y = 7$ gives us

$$-3x - 5 = 7$$
$$-3x = 12$$
$$x = -4$$

The solution to our system is the ordered pair $(-4,\ 5)$.

25. $x + 3y = 9$ No change $\rightarrow$ $x + 3y = 9$

     $2x - y = 4$ Multiply by 3 $\rightarrow$ $6x - 3y = 12$

$$7x = 21$$
$$x = 3$$

Substituting $x = 3$ into $x + 3y = 9$ gives us

$$3 + 3y = 9$$
$$3y = 6$$
$$y = 2$$

The solution to our system is the ordered pair $(3,\ 2)$.

29. $2x + 9y = 2$ <u>No change</u> $\rightarrow$ $2x + 9y = 2$

    $5x + 3y = -8$ <u>Multiply by -3</u> $\rightarrow$ $-15x - 9y = 24$

$$-13x = 26$$
$$x = -2$$

Substituting $x = -2$ into $2x + 9y = 2$ gives us

$$2(-2) + 9y = 2$$
$$-4 + 9y = 2$$
$$9y = 6$$
$$y = \frac{6}{9}$$
$$y = \frac{2}{3}$$

The solution to our system is the ordered pair $\left(-2, \frac{2}{3}\right)$.

33. $3x + 2y = -1$ <u>Multiply by -2</u> $\rightarrow$ $-6x - 4y = 2$

    $6x + 4y = 0$ <u>No change</u> $\rightarrow$ $6x + 4y = 0$

$$0 = 2$$

The lines are parallel and there is no solution to the system.

37. $\frac{1}{2}x + \frac{1}{6}y = \frac{1}{3}$ <u>Multiply by 6</u> $\rightarrow$ $3x + y = 2$

    $-x - \frac{1}{3}y = -\frac{1}{6}$ <u>Multiply by 6</u> $\rightarrow$ $-6x - 2y = -1$

    $3x + y = 2$ <u>Multiply by 2</u> $\rightarrow$ $6x + 2y = 4$

    $-6x - 2y = -1$ <u>No change</u> $\rightarrow$ $-6x - 2y = -1$

$$0 = 3$$

The lines are parallel and thee is no solution to the system.

41. $x + \quad y = 22$ <u>No change</u> $\rightarrow$ $x + \quad y = 22$

    $0.05x + 0.10y = 1.70$ <u>Multiply by 100</u> $\rightarrow$ $5x + 10y = 170$

    $x + \quad y = 22$ <u>Multiply by -5</u> $\rightarrow$ $-5x - 5y = -110$

    $5x + 10y = 170$ <u>No change</u> $\rightarrow$ $5x + 10y = 170$

$$5y = 60$$
$$y = 12$$

Substituting $y = 12$ into $x + y = 22$ gives us

$$x + 12 = 22$$
$$x = 10$$

The solution to our system is the ordered pair $(10, \ 12)$.

45. $\quad -3x \geq 12$

$\quad -\dfrac{1}{3}(-3x) \leq -\dfrac{1}{3}(12)$ $\quad$ Multiply each side by $-\dfrac{1}{3}$ and reverse the direction of the inequality symbol.

$\quad\quad x \leq -4$

49. $\quad -4x + 1 < 17$

$\quad\quad -4x < 16$ $\quad\quad\quad$ Add -1 to both sides

$\quad -\dfrac{1}{4}(-4x) > -\dfrac{1}{4}(16)$ $\quad$ Multiply each side by $-\dfrac{1}{4}$ and reverse the direction of the inequality symbol.

$\quad\quad x > -4$

SECTION 3.9

1. $x + y = 11$

$\quad\quad y = 2x - 1$

Substituting the expression $2x - 1$ for y from the second equation into the first equation, we have

$x + (2x - 1) = 11$

$\quad\quad 3x - 1 = 11$

$\quad\quad\quad 3x = 12$

$\quad\quad\quad\quad x = 4$

When $x = 4$, $\quad y = 2x - 1$

becomes $\quad\quad y = 2(4) - 1 = 8 - 1 = 7$

The solution to our system is $(4, 7)$.

5. $-2x + y = -1$

$\quad\quad\quad y = -4x + 8$

Substituting the expression $-4x + 8$ for y from the second equation into the first equation, we have

$-2x + (-4x + 8) = -1$

$\quad\quad -6x + 8 = -1$

$\quad\quad\quad -6x = -9$

$\quad\quad\quad\quad x = \dfrac{-9}{-6}$

$\quad\quad\quad\quad x = \dfrac{3}{2}$

When $x = \dfrac{3}{2}$, $\quad\quad y = -4x + 8$

becomes $\quad -4\left(\dfrac{3}{2}\right) + 8 = -6 + 8 = 2$

The solution to our system is $\left(\dfrac{3}{2}, 2\right)$.

9. $5x - 4y = -16$

$y = 4$

Substituting the expression 4 for y from the second equation into the first equation, we have

$5x - 4(4) = -16$

$5x - 16 = -16$

$5x = 0$

$x = 0$

When $x = 0$, $y = 4$, and the solution to our system is $(0, 4)$.

13. $x + 3y = 4$

$x - 2y = -1$

Solving the second equation for x gives us $x = 2y - 1$.

Substituting the expression $2y - 1$ for x into the first equation, we have

$(2y - 1) + 3y = 4$

$5y - 1 = 4$

$5y = 5$

$y = 1$

When $x = 2y - 1$

becomes $x = 2(1) - 1 = 2 - 1 = 1$

The solution to our system is $(1, 1)$.

17. $3x + 5y = -3$

$x - 5y = -5$

Solving the second equation for x gives us $x = 5y - 5$.

Substituting the expression $5y - 5$ for x into the first equation, we have

$3(5y - 5) + 5y = -3$

$15y - 15 + 5y = -3$

$20y - 15 = -3$

$20y = 12$

$y = \dfrac{12}{20}$

$y = \dfrac{3}{5}$

When $y = \dfrac{3}{5}$, $x - 5y = -5$

becomes $x - 5\left(\dfrac{3}{5}\right) = -5$

$x - 3 = -5$

$x = -2$

The solution to our system is $\left(-2, \dfrac{3}{5}\right)$.

21. $-3x - 9y = 7$

$\quad x + 3y = 12$

Solving the second equation for x gives us $x = -3y + 12$.

Substituting the expression $-3y + 12$ for x into the first equation, we have

$-3(-3y + 12) - 9y = 7$

$\quad 9y - 36 - 9y = 7$

$\qquad\qquad -36 = 7 \quad$ False

The lines are parallel and there is no solution to the system.

25. $7x - 6y = -1$

$\quad x = 2y - 1$

Substituting the expression $2y - 1$ for x into the first equation, we have

$7(2y - 1) - 6y = -1$

$\quad 14y - 7 - 6y = -1$

$\qquad 8y - 7 = -1$

$\qquad\quad 8y = 6$

$\qquad\quad y = \dfrac{6}{8}$

$\qquad\quad y = \dfrac{3}{4}$

When $\quad y = \dfrac{3}{4}, \quad x = 2y - 1$

becomes $\quad x = 2\left(\dfrac{3}{4}\right) - 1 = \dfrac{3}{2} - 1 = \dfrac{1}{2}$

The solution to our system is $\left(\dfrac{1}{2}, \dfrac{3}{4}\right)$.

29. $5x - 6y = -4$

$\quad x = y$

Substituting the expression y for x from the second equation into the first equation, we have

$5y - 6y = -4$

$\quad -y = -4$

$\quad y = 4$

Since x and y are equal x must be 4 also.

The solution to our system is $(4, 4)$.

33. $7x - 11y = 16$

$\quad\quad\quad y = 10$

Substituting the expression 10 for y from the second equation, we have

$7x - 11(10) = 16$

$\quad 7x - 110 = 16$

$\quad\quad\quad 7x = 126$

$\quad\quad\quad\quad x = 18$

The solution to our system is $(18, 10)$.

37. $0.05x + 0.10y = 1.70$

$\quad\quad\quad\quad y = 22 - x$

Substituting $y = 22 - x$ into the first equation, we have

$\quad 0.05x + 0.10(22 - x) = 1.70$

$\quad 0.05x + 2.2 - 0.10x = 1.70$ Distributive property

$\quad\quad\quad 2.2 - 0.05x = 1.70$

$\quad\quad\quad\quad\quad -0.05x = -0.5$ Add -2.2 to both sides

$\quad\quad\quad\quad\quad\quad\quad x = 10$ Divide both sides by -0.05

By first eliminating the decimals, we have

$\quad 5x + 10y = 170$

Substituting $y = 22 - x$ into this equation, we have

$\quad 5x + 10(22 - x) = 170$

$\quad 5x + 220 - 10x = 170$ Distributive property

$\quad\quad\quad 220 - 5x = 170$

$\quad\quad\quad\quad\quad -5x = -50$ Add -220 to both sides

$\quad\quad\quad\quad\quad\quad x = 10$ Divide both sides by -5

By substituting $x = 10$ into $y = 22 - x$, we have

$\quad y = 22 - 10$

$\quad y = 12$

The solution is $(10, 12)$.

41. Let $x =$ width, then $3x =$ length. The perimeter is 24 meters.

$$P = 2l + 2w$$

$$24 = 2(3x) + 2x$$

$$24 = 6x + 2x$$

$$24 = 8x$$

$$x = 3 \text{ meters}$$

$$3x = 9 \text{ meters}$$

The width is 3 meters and the length is 9 meters.

45. What number is 8% of 6000?

$$N = 0.08 \cdot 6000$$
$$N = 480$$

SECTION 3.10

1. Step 1: We know that the two numbers have a sum of 25 and one of the numbers is five more than the other. We don't know what the numbers themselves are.

   Step 2: Let x represent one of the numbers and y represent the other. "One number is five more than the other" translates to

   $$y = x + 5$$

   "Their sum is 25" translates to

   $$x + y = 25$$

   Step 3: The system that describes the situation must be

   $$x + y = 25$$
   $$y = x + 5$$

   Step 4: We can solve this system by substituting the expression $x + 5$ in the second equation for y in the first equation

   $$x + x + 5 = 25$$
   $$2x + 5 = 25$$
   $$2x = 20$$
   $$x = 10$$

   Using $x = 10$ in either of the first two equations and then solving for y, we get $y = 15$.

   Step 5: So 10 and 15 are the numbers we are looking for.

   Step 6: The number 15 is 5 more than 10 and the sum of 10 and 15 is 25.

5. Step 1: We know the two numbers are positive, their difference is 5 and the larger number is one more than twice the smaller.

   Step 2: Let x represent one of the numbers and y represent the other. "A difference of 5" translates to

   $$x - y = 5$$

   "The larger one is one more than twice the smaller" translates to

   $$x = 2y + 1$$

   Step 3: The system that describes the situation must be

   $$x - y = 5$$
   $$x = 2y + 1$$

Step 4: We can solve this system by substituting the expression $2y+1$ in the second equation for x in the first equation.

$$2y+1-y=5$$
$$y+1=5$$
$$y=4$$

Using $y=4$ in either of the first two equations and then solving for x, we get $x=9$.

Step 5: So 9 and 4 are the numbers we are looking for.

Step 6: The difference of the positive numbers 9 and 4 is 5. 9 is one more than twice 4.

9. Step 1: We do not know the specific amounts invested in the two accounts. We do know that their sum is $20,000 and the interest rates on the two accounts are 6% and 8%. We also know the total interest earned is $1,380. (Remember, Step 1 is done mentally.)

Step 2: Let $x =$ the amount invested at 6% and $y =$ the amount invested at 8%. Since Mr. Wilson invested a total of $20,000, we have

$$x+y=20,000$$

The interest he earns comes from 6% of the amount invested at 6% and 8% of the amount invested at 8%.

To find 6% of x, we multiply x by 0.06, which gives us 0.06x. To find 8% of y, we multiply 0.08 times y and get 0.08y.

Interest	+	Interest	=	Total
at 6%		at 8%		Interest
0.06x	+	0.08y	=	1380

Step 3: The system is

$$x+y=20,000$$
$$0.06x+0.08y=1380$$

Step 4: We multiply the first equation by -6 and the second by 100 to eliminate x:

$$x+\quad y=20,000 \quad \text{multiply by } -6 \rightarrow \quad -6x-6y=-120,000$$
$$0.06x+0.08y=1380 \quad \text{multiply by } 100 \rightarrow \quad \underline{6x+8y=138,000}$$
$$2y=18,000$$
$$y=9,000$$

Substituting $y=9000$ into the first equation and solving for x, we get $x=11,000$

Step 5: He invested $11,000 at 6% and $9,000 at 8%.

Step 6: Checking our solutions in the original problem, we have: The sum of $11,000 and $9,000 is $20,000, the total amount he invested. To complete our check, we find the total interest earned from the two accounts:

The interest on $11,000 at 6% is $0.06(11,000) = 660$

The interest on $9,000 at 8% is $\underline{0.08(9,000) = 720}$

$$\$1,380$$

13. Step 1: We know that Ron has 14 coins that are nickels and quarters. We know that a nickel is worth 5 cents and a quarter is worth 25 cents. We do not know the specific number of nickels and quarters he has, but we do know that the total value of the coins is $2.30.

Step 2: Let x = the number of nickels and y = the number of quarters. The total number of coins is 14, so

$$x + y = 14$$

The total amount of money he has is $2.30, which comes from nickels and quarters:

Amount of money in nickels	+	Amount of money in quarters		Total amount of money
$0.05x$	+	$0.25y$	=	2.30

Step 3: The system that represents the situation is:

$$x + y = 14 \qquad \text{The number of coins}$$
$$0.05x + 0.25y = 2.30 \qquad \text{The value of the coins}$$

Step 4: We multiply the first equation by -5 and the second by 100 to eliminate the variable x:

$$x + y = 14 \quad \underline{\text{multiply by } -5} \rightarrow \quad -5x - 5y = -70$$
$$0.05x + 0.25y = 2.30 \quad \underline{\text{multiply by } 100} \rightarrow \quad 5x + 25y = 230$$
$$20y = 160$$
$$y = 8$$

Substituting $y = 8$ into our first equation, we get $x = 6$

Step 5: Ron has 6 nickels and 8 quarters.

Step 6: Six nickels and 8 quarters are 14 coins.

6 nickels are worth $6(0.05)$ = .30
8 quarters are worth $8(0.25)$ = 2.00
The total value is   2.30

17. Step 1: We know there are two solutions that together must total 18 liters. A 50% alcohol solution and a 20% alcohol solution will need to be mixed to form a 30% alcohol solution.

Step 2: Let x = the number of liters of 20% alcohol solution. Let y = the number of liters of 50% alcohol solution. Since the total number of liters we will end up with is 18, the first equation is

$$x + y = 18$$

To obtain our second equation we look at the amount of alcohol in our two original solutions and our final solution. The amount of alcohol in the x gallons of 20% solution is $0.20x$, while the amount of alcohol in y gallons of 50% solution is $0.50y$. The amount of alcohol in the 18 liters of 30% solution is $0.30(18)$. Since the amount of alcohol we start with must equal the amount of alcohol we end up with, our second equation is

$$0.20x + 0.50y = 0.30(18)$$

I have summarized the information in a table so it will be easier to follow.

	20% Solution	50% Solution	Final Solution
Number of liters	x	y	18
Liters of alcohol	$0.20x$	$0.50y$	$0.30(18)$

Step 3: Our system of equations is

$$x + y = 18$$
$$0.20x + 0.50y = 0.30(18)$$

Step 4: We can solve this system by substitution. Solving the first equation for x and substituting the result into the second equation, we have

$$0.20(18 - y) + 0.50y = 0.30(18)$$

Multiplying each side by 10 gives us an equivalent equation that is a little easier to work with:

$$2(18 - y) + 5y = 3(18)$$
$$36 - 2y + 5y = 54$$
$$3y + 36 = 54$$
$$3y = 18$$
$$y = 6$$

If y is 6, then x must be 12 because $x + y = 18$.

Step 5: It takes 12 liters of 20% alcohol solution and 6 liters of 50% alcohol solution to produce 18 liters of 30% alcohol solution.

Step 6: 12 liters of 20% solution 0.20(12)   =   2.40

6 liters of 50% solution   0.50(6)   =   3.00

18 liters of 30% solution 0.30(18)   =   5.40

21. Step 1: We know that 70 tickets were sold for $310 for the Saturday matinee. Adult tickets cost $5.50 and kids under 12 pay only $4.00. We do not know how many of each type of ticket was sold.

Step 2: Let x = the number of adult tickets. Let y = the number of kid's tickets. The total number of tickets is 70, so

$$x + y = 70$$

The total amount of money for the tickets is $310.00

Amount of money in adult tickets	+	Amount of money in kid's tickets	=	Total amount of money
$5.50x$	+	$4.00y$	=	310

Step 3: Our system of equations is

$$x + \quad y = 70$$
$$5.50x + 4.00y = 310$$

Step 4: We can solve this system by substitution. Solving the first equation for y and substituting the result into the second equation, we have

$$5.50x + 4.00(70 - x) = 310$$

Multiplying each side by 10 gives us an equivalent equation that is a little easier to work with:

$$55x + 40(70 - x) = 3100$$
$$55x + 2800 - 40x = 3100$$
$$15x + 2800 = 3100$$
$$15x = 300$$
$$x = 20$$

If x is 20, then y must be 50 because $x + y = 70$.

Step 5: 20 adult tickets and 50 kids under 12 tickets were sold for the Saturday matinee.

Step 6: 20 adult tickets 5.50(20)   =   110.00

<u>50 kid's tickets   4.00(50)   =   200.00</u>

Total amount of money   =   310.00

25. Step 1: We know the gambler had 45 chips. He had \$5 chips and \$25 chips that totaled \$465.00. We do not know how many of each chip the gambler had.

Step 2: Let $x =$ the \$5 chips and $y =$ the \$25 chips. The total number of chips is 45, so

$$x + y = 45$$

The total amount of money for the chips is \$465.

Amount of money	+	Amount of money	=	Total amount
for \$5 chips		for \$25 chips		
5x	+	25y	=	465

Step 3: Our system of equations is

$$x + \quad y = 45$$
$$5x + 25y = 465$$

Step 4: We can solve this system by substitution. Solving the first equation for x and substituting the result into the second equation, we have

$$5(45 - y) + 25y = 465$$
$$225 - 5y + 25y = 465$$
$$225 + 20y = 465$$
$$20y = 240$$
$$y = 12$$

If $y = 12$, then x must be 33 because $x + y = 45$.

Step 5: The gambler had 33 $5 chips and 12 $25 chips.

Step 6:   33 $5 chips   33(5)   =   165

            12 $25 chips   12(25)   =   300

            Total amount of money   =   465

29. $2x - 6 > 5x + 6$

     $-6 > 3x + 6$ Add $-2x$ to each side

     $-12 > 3x$ Add -6 to each side

      $-4 > x$ Divide each side by 3

       $x < -4$

33. $4(2 - x) \geq 12$

     $8 - 4x \geq 12$ Distributive property

      $-4x \geq 4$ Add -8 to each side

        $\downarrow$

      $x \leq -1$ Divide by -4 and reverse the inequality sign

CHATPER 3 REVIEW

Problems 1 - 6 see the graph in the back of the textbook.

7. $3x + y = 6$ $(4, -6)$ $3x + y = 6$ $(0, 6)$

    $3(4) + y = 6$ $3(0) + y = 6$

    $12 + y = 6$ $y = 6$

      $y = -6$

    $3x + y = 6$ $(1, 3)$ $3x + y = 6$ $(2, 0)$

    $3x + 3 = 6$ $3x + 0 = 6$

     $3x = 3$ $3x = 6$

      $x = 1$ $x = 2$

8. $2x - 5y = 20$ $(5, -2)$ $2x - 5y = 20$ $(0, -4)$

    $2(5) - 5y = 20$ $2(0) - 5y = 20$

    $10 - 5y = 20$ $-5y = 20$

     $-5y = 10$ $y = -4$

      $y = -2$

$$2x - 5y = 20 \quad (15, 2)$$
$$2x - 5(2) = 20$$
$$2x - 10 = 20$$
$$2x = 30$$
$$x = 15$$

$$2x - 5y = 20 \quad (10, 0)$$
$$2x - 5(0) = 20$$
$$2x = 20$$
$$x = 10$$

9. $y = 2x - 6 \quad (4, 2)$
   $y = 2(4) - 6$
   $y = 8 - 6$
   $y = 2$

   $y = 2x - 6 \quad (2, -2)$
   $-2 = 2x - 6$
   $4 = 2x$
   $2 = x$

   $y = 2x - 6 \quad \left(\dfrac{9}{2}, 3\right)$
   $3 = 2x - 6$
   $9 = 2x$
   $\dfrac{9}{2} = x$

10. $y = 5x + 3 \quad (2, 13)$
    $y = 5(2) + 3$
    $y = 10 + 3$
    $y = 13$

    $y = 5x + 3 \quad \left(-\dfrac{3}{5}, 0\right)$
    $0 = 5x + 3$
    $-3 = 5x$
    $-\dfrac{3}{5} = x$

    $y = 5x + 3 \quad \left(-\dfrac{6}{5}, -3\right)$
    $-3 = 5x + 3$
    $-6 = 5x$
    $-\dfrac{6}{5} = x$

11. $y = -3 \quad (2, -3) \quad (-1, -3) \quad (-3, -3)$
12. $x = 6 \quad (6, 5) \quad (6, 0) \quad (6, -1)$

13. $\quad 3x - 4y = 12 \quad \left(2, -\dfrac{3}{2}\right)$ is a solution

$\quad 3(2) - 4\left(-\dfrac{3}{2}\right) = 12$

$\qquad 6 + \dfrac{12}{2} = 12$

$\qquad 6 + 6 = 12$

$\qquad 12 = 12 \quad$ true

14. $\quad 2x + 4y = 8 \quad (6, -1)$ is a solution

$\quad 2(6) + 4(-1) = 8$

$\qquad 12 + (-4) = 8$

$\qquad 8 = 8 \quad$ true

15. $y = 3x + 7 \qquad \left(-\dfrac{8}{3}, -1\right)$ is a solution

$\quad -1 = 3\left(-\dfrac{8}{3}\right) + 7$

$\quad -1 = -8 + 7$

$\quad -1 = -1 \qquad$ true

$\qquad y = 3x + 7 \qquad (-3, -2)$ is a solution

$\qquad -2 = 3(-3) + 7$

$\qquad -2 = -9 + 7$

$\qquad -2 = -2 \qquad$ true

16. $y = 2x - 5 \qquad (1, -3)$ is a solution

$\quad -3 = 2(1) - 5$

$\quad -3 = 2 - 5$

$\quad -3 = -3 \qquad$ true

17. $x + y = -2 \quad (-2, 0)$

$\quad x + 0 = -2$

$\qquad x = -2$

$\qquad x + y = -2 \quad (0, -2)$

$\qquad 0 + y = -2$

$\qquad\qquad y = -2$

$\qquad x + y = -2 \quad (1, -3)$

$\qquad 1 + y = -2$

$\qquad\quad y = -3$

See the graph in the back of the textbook.

18. $x - y = 5 \quad (2, -3)$

$\quad 2 - y = 5$

$\quad -y = 3$

$\qquad y = -3$

$\qquad x - y = 5 \quad (3, -2)$

$\qquad x - (-2) = 5$

$\qquad\quad x + 2 = 5$

$\qquad\qquad x = 3$

$\qquad x - y = 5 \quad (0, -5)$

$\qquad 0 - y = 5$

$\qquad -y = 5$

$\qquad\quad y = -5$

See the graph in the back of the textbook.

19. $y = 3x \qquad (-1, -3)$

$\quad y = 3(-1)$

$\quad y = -3$

$\qquad y = 3x \qquad (1, 3)$

$\qquad y = 3(1)$

$\qquad y = 3$

$\qquad y = 3x \qquad (0, 0)$

$\qquad 0 = 3x$

$\qquad 0 = x$

See the graph in the back of the textbook.

20. $y = \dfrac{1}{3}x \quad (3, 1)$ $y = \dfrac{1}{3}x \quad (0, 0)$ $y = \dfrac{1}{3}x \quad (-3, -1)$

$1 = \dfrac{1}{3}x$ $0 = \dfrac{1}{3}x$ $y = \dfrac{1}{3}(-3)$

$3 = x$ $0 = x$ $y = -1$

See the graph in the back of the textbook.

21. $y = 2x - 1 \quad (1, 1)$ $y = 2x - 1 \quad (0, -1)$ $y = 2x - 1 \quad (-1, -3)$

$y = 2(1) - 1$ $y = 2(0) - 1$ $-3 = 2x - 1$

$y = 2 - 1$ $y = -1$ $-2 = 2x$

$y = 1$ $-1 = x$

See the graph in the back of the textbook.

22. $y = -3x + 4 \quad (0, 4)$ $y = -3x + 4 \quad (1, 1)$ $y = -3x + 4 \quad (2, -2)$

$y = -3(0) + 4$ $y = -3(1) + 4$ $-2 = -3x + 4$

$y = 0 + 4$ $y = -3 + 4$ $-6 = -3x$

$y = 4$ $y = 1$ $2 = x$

See the graph in the back of the textbook.

23.   **x-Intercept**   **y-Intercept**

If $y = 0$ If $x = 0$

the equation $3x - y = 3$ the equation $3x - y = 3$

becomes $3x - 0 = 3$ becomes $3(0) - y = 3$

                    $3x = 3$ $-y = 3$

                      $x = 1$ $y = -3$

                    $(1, \ 0)$ $(0, \ -3)$

See the graph in the back of the textbook.

24.   **x-Intercept**   **y-Intercept**

If $y = 0$ If $x = 0$

the equation $3x + 2y = 6$ the equation $3x + 2y = 6$

becomes $3x + 2(0) = 6$ becomes $3(0) + 2y = 6$

                    $3x = 6$ $2y = 6$

                      $x = 2$ $y = 3$

See the graph in the back of the textbook.

25. If $\quad\quad\quad x = 0 \quad\quad\quad$ If $\quad\quad\quad x = 3$

the equation $\quad y = -\dfrac{1}{3}x \quad\quad$ the equation $\quad y = -\dfrac{1}{3}x$

becomes $\quad\quad y = -\dfrac{1}{3}(0) \quad$ becomes $\quad\quad y = -\dfrac{1}{3}(3)$

$\quad\quad\quad\quad\quad\quad y = 0 \quad\quad\quad\quad\quad\quad\quad\quad\quad y = -1$

$\quad\quad\quad\quad\quad\quad (0, 0) \quad\quad\quad\quad\quad\quad\quad\quad\quad (3, -1)$

See the graph in the back of the textbook.

26. If $\quad\quad\quad x = 0 \quad\quad\quad$ If $\quad\quad\quad x = 1$

the equation $\quad y = -3x \quad\quad$ the equation $\quad y = -3x$

becomes $\quad\quad y = -3(0) \quad$ becomes $\quad\quad y = -3(1)$

$\quad\quad\quad\quad\quad\quad y = 0 \quad\quad\quad\quad\quad\quad\quad\quad\quad y = -3$

$\quad\quad\quad\quad\quad\quad (0, 0) \quad\quad\quad\quad\quad\quad\quad\quad\quad (1, -3)$

See the graph in the back of the textbook.

27. If $\quad\quad\quad x = 0 \quad\quad\quad$ If $\quad\quad\quad x = 1$

the equation $\quad y = 2x + 1 \quad\quad$ the equation $\quad y = 2x + 1$

becomes $\quad\quad y = 2(0) + 1 \quad$ becomes $\quad\quad y = 2(1) + 1$

$\quad\quad\quad\quad\quad\quad y = 1 \quad\quad\quad\quad\quad\quad\quad\quad\quad y = 3$

$\quad\quad\quad\quad\quad\quad (0, 1) \quad\quad\quad\quad\quad\quad\quad\quad\quad (1, 3)$

See the graph in the back of the textbook.

28. If $\quad\quad\quad x = 0 \quad\quad\quad$ If $\quad\quad\quad x = 1$

the equation $\quad y = -3x + 2 \quad\quad$ the equation $\quad y = -3x + 2$

becomes $\quad\quad y = -3(0) + 2 \quad$ becomes $\quad\quad y = -3(1) + 2$

$\quad\quad\quad\quad\quad\quad y = 2 \quad\quad\quad\quad\quad\quad\quad\quad\quad y = -3 + 2$

$\quad\quad\quad\quad\quad\quad (0, 2) \quad\quad\quad\quad\quad\quad\quad\quad\quad y = -1$

$\quad\quad\quad\quad\quad\quad\quad\quad\quad\quad\quad\quad\quad\quad\quad\quad\quad\quad (1, -1)$

See the graph in the back of the textbook.

29 - 32. See the graphs in the back of the textbook.

33.   **x-Intercept** = 2   **y-Intercept** = -6

If $y = 0$ If $x = 0$

the equation $3x - y = 6$ the equation $3x - y = 6$

becomes $3x - 0 = 6$ becomes $3(0) - y = 6$

                     $3x = 6$ $-y = 6$

                       $x = 2$ $y = -6$

34.   **x-Intercept** = 5   **y-Intercept** = 2

If $y = 0$ If $x = 0$

the equation $2x + 5y = 10$ the equation $2x + 5y = 10$

becomes $2x + 5(0) = 10$ becomes $2(0) + 5y = 10$

                     $2x = 10$ $5y = 10$

                       $x = 5$ $y = 2$

35.   **x-Intercept** = 3   **y-Intercept** = -3

If $y = 0$ If $x = 0$

the equation $y = x - 3$ the equation $y = x - 3$

becomes $0 = x - 3$ becomes $y = (0) - 3$

                      $3 = x$ $y = -3$

36.   **x-Intercept** = 7   **y-Intercept** = 7

If $y = 0$ If $x = 0$

the equation $y = -x + 7$ the equation $y = -x + 7$

becomes $0 = -x + 7$ becomes $y = -0 + 7$

                    $-7 = -x$ $y = 7$

                      $7 = x$

37.   **x-Intercept** = 2   **y-Intercept** = -6

If $y = 0$ If $x = 0$

the equation $y = 3x - 6$ the equation $y = 3x - 6$

becomes $0 = 3x - 6$ becomes $y = 3(0) - 6$

                      $6 = 3x$ $y = -6$

                      $2 = x$

38.   **x-Intercept** $= 5$ **y-Intercept** $= 10$

If $\qquad y = 0$ If $\qquad x = 0$

the equation $\quad y = -2x + 10$ the equation $\quad y = -2x + 10$

becomes $\qquad 0 = -2x + 10$ becomes $\qquad y = -2(0) + 10$

$\qquad\qquad -10 = -2x$ $\qquad\qquad y = 10$

$\qquad\qquad 5 = x$

39. We can let $(x_1,\ y_1) = (2,\ 3)$ and $(x_2,\ y_2) = (3,\ 5)$ then $m = \frac{y_2 - y_1}{x_2 - x_1} = \frac{5-3}{3-2} = \frac{2}{1} = 2$

The slope is 2.

40. We can let $(x_1,\ y_1) = (3,\ -2)$ and $(x_2,\ y_2) = (1,\ 8)$ then $m = \frac{y_2 - y_1}{x_2 - x_1} = \frac{8-(-2)}{1-3} = \frac{10}{-2} = -5$

The slope is -5.

41. We can let $(x_1,\ y_1) = (-2,\ 3)$ and $(x_2,\ y_2) = (6,\ -5)$ then $m = \frac{y_2 - y_1}{x_2 - x_1} = \frac{-5-3}{6-(-2)} = \frac{-8}{8} = -1$

The slope is -1.

42. We can let $(x_1,\ y_1) = (7,\ 3)$ and $(x_2,\ y_2) = (-2,\ -4)$ then $m = \frac{y_2 - y_1}{x_2 - x_1} = \frac{-4-3}{-2-7} = \frac{-7}{-9} = \frac{7}{9}$

The slope is $\frac{7}{9}$.

43. We can let $(x_1,\ y_1) = (-1,\ -4)$ and $(x_2,\ y_2) = (-3,\ -8)$ then $m = \frac{y_2 - y_1}{x_2 - x_1} = \frac{-8-(-4)}{-3-(-1)} = \frac{-4}{-2} = 2$

The slope is 2.

44. We can let $(x_1,\ y_1) = (-2,\ -5)$ and $(x_2,\ y_2) = (-4,\ -1)$ then $m = \frac{y_2 - y_1}{x_2 - x_1} = \frac{-1-(-5)}{-4-(-2)} = \frac{4}{-2} = -2$

The slope is -2.

45. Substituting $m = 3$ and $b = 2$ into the equation

$y = mx + b$, we have

$$y = 3x + 2$$

46. Substituting $m = -5$ and $b = -1$ into the equation

$y = mx + b$, we have

$$y = -5x - 1$$

47. Substituting $m = -1$ and $b = 6$ into the equation

$y = mx + b$, we have

$$y = -x + 6$$

48. Substituting $m = -4$ and $b = -3$ into the equation
    $y = mx + b$, we have
    $$y = -4x - 3$$

49. Substituting $m = -\dfrac{1}{3}$ and $b = \dfrac{3}{4}$ into the equation
    $y = mx + b$, we have
    $$y = -\dfrac{1}{3}x + \dfrac{3}{4}$$

50. Substituting $m = -\dfrac{2}{5}$ and $b = -\dfrac{2}{3}$ into the equation
    $y = mx + b$, we have
    $$y = -\dfrac{2}{5}x - \dfrac{2}{3}$$

51. The equation is in the slope-intercept form, so the slope must be 4
    and the y-intercept -1. $(m = 4, \quad b = -1)$

52. The equation is in the slope-intercept form, so the slope must be -3
    and the y-intercept 2. $(m = -3, \quad b = 2)$

53. To identify the slope and y-intercept, we must solve for y to get
    the equation in the form $y = mx + b$.
    $$2x + y = -5 \qquad \text{Original equation}$$
    $$y = -2x - 5 \quad \text{Add } -2x \text{ to each side}$$
    The slope must be -2 and the y-intercept -5.

54. To identify the slope and y-intercept, we must solve for y to get
    the equation in the form $y = mx + b$.
    $$3x - y = 4 \qquad \text{Original equation}$$
    $$-y = -3x + 4 \quad \text{Add } -3x \text{ to each side}$$
    $$y = 3x - 4 \quad \text{Multiply each side by } -1$$
    The slope must be 3 and the y-intercept -4.

55. To identify the slope and y-intercept, we must solve for y to get

the equation in the form $y = mx + b$.

$$6x + 3y = 9 \qquad \text{Original equation}$$
$$3y = -6x + 9 \quad \text{Add } -6x \text{ to each side}$$
$$y = -2x + 3 \quad \text{Divide each side by 3}$$

The slope must be -2 and the y-intercept 3.

56. To identify the slope and y-intercept, we must solve for y to get

the equation in the form $y = mx + b$.

$$8x - 4y = -12 \qquad \text{Original equation}$$
$$-4y = -8x - 12 \qquad \text{Add } -8x \text{ to each side}$$
$$y = 2x + 3 \qquad \text{Divide each side by } -4$$

The slope must be 2 and the y-intercept 3.

57. To identify the slope and y-intercept, we must solve for y to get

the equation in the form $y = mx + b$.

$$5x + 2y = 8 \qquad \text{Original equation}$$
$$2y = -5x + 8 \qquad \text{Add } -5x \text{ to each side}$$
$$y = -\tfrac{5}{2}x + 4 \qquad \text{Divide each side by 2}$$

The slope must be $-\tfrac{5}{2}$ and the y-intercept 4.

58. To identify the slope and y-intercept, we must solve for y to get

the equation in the form $y = mx + b$.

$$3x - 4y = -16 \qquad \text{Original equation}$$
$$-4y = -3x - 16 \qquad \text{Add } -3x \text{ to each side}$$
$$y = \tfrac{3}{4}x + 4 \qquad \text{Divide each side by } -4$$

The slope must be $\tfrac{3}{4}$ and the y-intercept 4 .

59. Using $(x_1, \, y_1) = (3, \, 1)$ and $m = 4$

in $\qquad y - y_1 = m(x - x_1) \quad$ Point - slope form

gives us $\quad y - 1 = 4(x - 3)$
$$y - 1 = 4x - 12 \qquad \text{Multiply out right side}$$
$$y = 4x - 11 \qquad \text{Add 1 to each side}$$

122

Chapter 3 Review

60. Using $(x_1, y_1) = (-1, 4)$ and $m = -2$

  in $y - y_1 = m(x - x_1)$ Point - slope form

  gives us $y - 4 = -2(x + 1)$ Note: $x - (-1) = x + 1$

      $y - 4 = -2x - 2$ Multiply out right side

      $y = -2x + 2$ Add 4 to each side

61. Using $(x_1, y_1) = (4, 3)$ and $m = \dfrac{1}{2}$

  in $y - y_1 = m(x - x_1)$ Point - slope form

  gives us $y - 3 = \dfrac{1}{2}(x - 4)$

      $y - 3 = \dfrac{1}{2}x - 2$ Multiply out right side

      $y = \dfrac{1}{2}x + 1$ Add 3 to each side

62. Using $(x_1, y_1) = (-6, 4)$ and $m = \dfrac{2}{3}$

  in $y - y_1 = m(x - x_1)$ Point - slope form

  gives us $y - 4 = \dfrac{2}{3}(x + 6)$ Note: $x - (-6) = x + 6$

      $y - 4 = \dfrac{2}{3}x + 4$ Multiply out right side

      $y = \dfrac{2}{3}x + 8$ Add 4 to each side

63. Using $(x_1, y_1) = (3, -2)$ and $m = -\dfrac{3}{4}$

  in $y - y_1 = m(x - x_1)$ Point - slope form

  gives us $y + 2 = -\dfrac{3}{4}(x - 3)$ Note: $y - (-2) = y + 2$

      $y + 2 = -\dfrac{3}{4}x + \dfrac{9}{4}$ Multiply out right side

      $y = -\dfrac{3}{4}x + \dfrac{1}{4}$ Add -2 to each side

64. Using $(x_1, y_1) = (-2, -1)$ and $m = -\dfrac{2}{5}$

     in $y - y_1 = m(x - x_1)$ Point - slope form

     gives us $y + 1 = -\dfrac{2}{5}(x + 2)$ Note: $y - (-1) = y + 1$

                  $y + 1 = -\dfrac{2}{5}x - \dfrac{4}{5}$ Multiply out right side

                  $y = -\dfrac{2}{5}x - \dfrac{9}{5}$ Add -1 to each side

65. We begin by finding the slope of the line:

$$m = \frac{4 - 6}{2 - 1} = \frac{-2}{1} = -2$$

     Using $(x_1, y_1) = (2, 4)$ and $m = -2$

     in $y - y_1 = m(x - x_1)$

     gives us $y - 4 = -2(x - 2)$

                  $y - 4 = -2x + 4$ Multiply out right side

                  $y = -2x + 8$ Add 4 to each side

66. We begin by finding the slope of the line:

$$m = \frac{5 - 8}{2 - 4} = \frac{-3}{-2} = \frac{3}{2}$$

     Using $(x_1, y_1) = (2, 5)$ and $m = \dfrac{3}{2}$

     in $y - y_1 = m(x - x_1)$

     gives us $y - 5 = \dfrac{3}{2}(x - 2)$

                  $y - 5 = \dfrac{3}{2}x - 3$ Multiply out right side

                  $y = \dfrac{3}{2}x + 2$ Add 5 to each side

67. We begin by finding the slope of the line:

$$m = \frac{2 - 4}{2 - 4} = \frac{-2}{-2} = 1$$

Using $(x_1, y_1) = (2, 2)$ and $m = 1$

in $\qquad y - y_1 = m(x - x_1)$

gives us $\quad y - 2 = 1(x - 2)$

$\qquad\qquad y - 2 = x - 2 \qquad$ Multiply out right side

$\qquad\qquad\quad y = x \qquad\qquad$ Add 2 to each side

68. We begin by finding the slope of the line:

$$m = \frac{3 - 6}{1 - 2} = \frac{-3}{-1} = 3$$

Using $(x_1, y_1) = (1, 3)$ and $m = 3$

in $\qquad y - y_1 = m(x - x_1)$

gives us $\quad y - 3 = 3(x - 1)$

$\qquad\qquad y - 3 = 3x - 3 \qquad$ Multiply out right side

$\qquad\qquad\quad y = 3x \qquad\qquad$ Add 3 to each side

69. We begin by finding the slope of the line:

$$m = \frac{-7 - 1}{3 - (-3)} = \frac{-8}{6} = -\frac{4}{3}$$

Using $(x_1, y_1) = (3, -7)$ and $m = -\dfrac{4}{3}$

in $\qquad y - y_1 = m(x - x_1)$

gives us $\quad y + 7 = -\dfrac{4}{3}(x - 3) \quad$ Note: $y - (-7) = y + 7$

$\qquad\qquad y + 7 = -\dfrac{4}{3}x + 4 \qquad$ Multiply out right side

$\qquad\qquad\quad y = -\dfrac{4}{3}x - 3 \qquad$ Add -7 to each side

70. We begin by finding the slope of the line:

$$m = \frac{5 - 3}{-3 - 3} = \frac{2}{-6} = -\frac{1}{3}$$

Using $(x_1, y_1) = (-3, 5)$ and $m = -\dfrac{1}{3}$

in $\qquad y - y_1 = m(x - x_1)$

gives us $\quad y - 5 = -\dfrac{1}{3}(x + 3) \quad$ Note: $x - (-3) = x + 3$

$\qquad\qquad y - 5 = -\dfrac{1}{3}x - 1 \qquad$ Multiply out right side

$\qquad\qquad\quad y = -\dfrac{1}{3}x + 4 \qquad$ Add 5 to each side

71 - 76. You may graph each line by finding the x-intercept and the y-intercept or by substituting x or y values.
         The following answers are where the two lines intersect.

71. $(4, -2)$ 72. $(-3, 2)$
73. $(3, -2)$ 74. $(2, 0)$
75. $(2, 1)$ 76. $(-1, -2)$

77. $x - y = 4$ Substituting $x = 1$ in the second equation.

$\dfrac{x + y = -2}{2x = 2}$ $1 + y = -2$

$\qquad x = 1$ $y = -3$

The ordered pair is $(1, -3)$.

78. $-x - y = -3$ Substituting $x = -2$ in the first equation.

$\dfrac{2x + y = 1}{x \quad = -2}$ $-(-2) - y = -3$

$\qquad\qquad\qquad\qquad 2 - y = -3$

$\qquad\qquad\qquad\qquad -y = -5$

$\qquad\qquad\qquad\qquad y = 5$

The ordered pair is $(-2, 5)$.

79. $5x - 3y = 2$ $\underline{\text{Multiply by 2} \rightarrow}$ $10x - 6y = 4$

$-10x + 6y = -4$ $\underline{\text{no change} \quad} \rightarrow$ $\dfrac{-10x + 6y = -4}{0 = 0}$

The lines coincide.

80. $2x + 3y = -2$ $\underline{\text{Multiply by 2}} \rightarrow$ $4x + 6y = -4$

$3x - 2y = 10$ $\underline{\text{Multiply by 3}} \rightarrow$ $\dfrac{9x - 6y = 30}{13x = 26}$

$\qquad\qquad\qquad\qquad\qquad\qquad x = 2$

Substituting $x = 2$ in the first equation,

$2(2) + 3y = -2$

$\quad 4 + 3y = -2$

$\qquad 3y = -6$

$\qquad y = -2$

The ordered pair is $(2, -2)$.

81.

$$-3x + 4y = 1 \quad \underline{\text{Multiply by } 4} \quad \rightarrow \quad -12x + 16y = 4$$

$$-4x + 5y = 1 \quad \underline{\text{Multiply by } -3} \rightarrow \quad \frac{12x - 15y = -3}{y = 1}$$

Substituting $y = 1$ in the first equation.

$$-3x + 4(1) = 1$$
$$-3x + 4 = 1$$
$$-3x = -3$$
$$x = 1$$

The ordered pair is $(1, 1)$.

82.

$$-4x - 2y = 3 \quad \underline{\text{Multiply by } 3} \rightarrow \quad -12x - 6y = 9$$

$$6x + 3y = 1 \quad \underline{\text{Multiply by } 2} \rightarrow \quad \frac{12x + 6y = 2}{0 = 11 \text{ false}}$$

The lines are parallel.

83.

$$-2x + 5y = -11 \quad \underline{\text{Multiply by } 7} \rightarrow \quad -14x + 35y = -77$$

$$7x - 3y = -5 \quad \underline{\text{Multiply by } 2} \rightarrow \quad \frac{14x - 6y = -10}{29y = -87}$$

$$y = -3$$

Substituting $y = -3$ in the first equation,

$$-2x + 5(-3) = -11$$
$$-2x - 15 = -11$$
$$-2x = 4$$
$$x = -2$$

The ordered pair is $(-2, -3)$.

84.

$$-2x + 5y = -15 \quad \underline{\text{Multiply by } 3} \rightarrow \quad -6x + 15y = -45$$

$$3x - 4y = 19 \quad \underline{\text{Multiply by } 2} \rightarrow \quad \frac{6x - 8y = 38}{7y = -7}$$

$$y = -1$$

Substituting $y = -1$ in the first equation,

$$-2x + 5(-1) = -15$$
$$-2x - 5 = -15$$
$$-2x = -10$$
$$x = 5$$

The ordered pair is $(5, -1)$.

85. $x + y = 5$ and $y = -3x + 1$

$$x + (-3x + 1) = 5$$
$$-2x + 1 = 5$$
$$-2x = 4$$
$$x = -2$$

Substituting $x = -2$ in the first equation,

$$-2 + y = 5$$
$$y = 7$$

The ordered pair is $(-2, 7)$.

86. $x - y = -2$ and $y = -2x - 10$

$$x - (-2x - 10) = -2$$
$$x + 2x + 10 = -2$$
$$3x = -12$$
$$x = -4$$

Substituting $x = -4$ in the second equation,

$$y = -2(-4) - 10$$
$$y = 8 - 10$$
$$y = -2$$

The ordered pair is $(-4, -2)$.

87. $4x - 3y = -16$ and $y = 3x + 7$

$$4x - 3(3x + 7) = -16$$
$$4x - 9x - 21 = -16$$
$$-5x - 21 = -16$$
$$-5x = 5$$
$$x = -1$$

Substituting $x = -1$ in the second equation,

$$y = 3(-1) + 7$$
$$y = -3 + 7$$
$$y = 4$$

The ordered pair is $(-1, 4)$.

88. $5x + 2y = -2$ and $y = -8x + 10$

$5x + 2(-8x + 10) = -2$

$5x - 16x + 20 = -2$

$-11x + 20 = -2$

$-11x = -22$

$x = 2$

Substituting $x = 2$ in the second equation,

$y = -8(2) + 10$

$y = -16 + 10$

$y = -6$

The ordered pair is $(2, \ -6)$.

89. $-3x + 12y = -8$ and $x = 4y + 2$

$-3(4y + 2) + 12y = -8$

$-12y - 6 + 12y = -8$

$-6 = -8$ False statement

There are no points in common, therefore the lines are parallel.

90. $4x - 2y = 8$ and $y = -3x - 19$

$4x - 2(-3x - 19) = 8$

$4x + 6x + 38 = 8$

$10x + 38 = 8$

$10x = -30$

$x = -3$

Substituting $x = -3$ in the second equation,

$y = -3(-3) - 19$

$y = 9 - 19$

$y = -10$

The ordered pair is $(-3, \ -10)$.

91. $$10x - 5y = 20 \text{ and } x = -6y - 11$$
$$10(-6y - 11) - 5y = 20$$
$$-60y - 110 - 5y = 20$$
$$-65y = 130$$
$$y = -2$$

Substituting $y = -2$ in the second equation,
$$x = -6(-2) - 11$$
$$x = 12 - 11$$
$$x = 1$$

The ordered pair is $(1, -2)$.

92. $$-6x + 2y = -4 \text{ and if } y = 3x - 2 \text{ then}$$
$$-6x + 2(3x - 2) = -4$$
$$-6x + 6x - 4 = -4$$
$$-4 = -4 \quad \text{True}$$

The lines coincide.

93. Step 1: We know that the two numbers have a sum of 18 and twice the smaller is 6 more than the larger.

Step 2: Let $x =$ the smaller number and $y =$ the larger number.

"The sum is 18" translates to
$$x + y = 18$$

"Twice the smaller number is 6 more than the larger number" translates to
$$2x = y + 6$$

Step 3: The system that describes the situation must be
$$x + y = 18$$
$$2x = y + 6 \text{ becomes } 2x - y = 6$$

Step 4: We can solve the system by elimination:
$$x + y = 18$$
$$\underline{2x - y = 6}$$
$$3x = 24$$
$$x = 8$$

Substituting $x = 8$ in the first equation,
$$8 + y = 18$$
$$y = 10$$

Step 5: So 8 and 10 are the numbers for which we are looking.

Step 6: Twice the number 8 is equal to 10 plus six more.

94. Step 1: We know that the difference of two positive numbers is 16 and one number is three times the other.

Step 2: Let x = larger number and y = smaller number.

"The difference of two positive numbers is 16" translates to
$$x - y = 16$$
"One number is three times the other" translates to
$$x = 3y$$

Step 3: The system that describes the situation must be
$$x - y = 16 \text{ and } x = 3y$$

Step 4: We can solve the system by substitution
$$3y - y = 16$$
$$2y = 16$$
$$y = 8$$

Substituting $y = 8$ in the second equation,
$$x = 3(8)$$
$$x = 24$$

Step 5: The numbers are 8, 24.

Step 6: The difference of 24 and 8 is 16 and three times 8 is 24.

95. Step 1: We do not know the specific amounts invested in the two accounts. We do know that their sum is $12,000 and that the interest rates on the two accounts are 4% and 5%. We also know that the total interest earned is $560.

Step 2: Let x = the amount invested at 4%, Let y = the amount invested at 5%. Since the total invested was $12,000, we have
$$x + y = 12,000$$
To find the total interest, we multiply the amount invested by the interest rate and add.

Interest at 4%	+	Interest at 5%	=	Total Interest
$0.04x$	+	$0.05y$	=	560

Step 3: The system is
$$0.04x + 0.05y = 560$$
$$x + y = 12,000$$

Step 4:

$0.04x + 0.05y = 560$	Multiply by 100 $\rightarrow$	$4x + 5y = 56,000$
$x + y = 12,000$	Multiply by -4 $\rightarrow$	$-4x - 4y = -48,000$

$$y = 8,000$$
$$12,000 - y = \$4,000$$

Step 5: They invested $4,000 at 4% and $8,000 at 5%.

Step 6: The interest on $4,000 at 4% is 0.04(4,000) $=$ 160

The interest on $8,000 at 5% is 0.05(8,000) $=$ 400

The total interest is $560

96. Step 1: We do not know the specific amounts invested in the two accounts. We do know that their sum is $14,000 and that the interest rates on the two accounts are 6% and 8%. We also know that the total interest earned is $1,060.

Step 2: Let $x =$ the amount invested at 6% Let $y =$ the amount invested at 5%. Since the total invested was $14,000, we have

$$x + y = 14,000$$

To find the total interest, we multiply the amount invested by the interest rate and add.

Interest	+	Interest	=	Total
at 6%		at 8%		Interest
0.06x	+	0.08y	=	1060

Step 3: The system is

$$x + \quad y = 14,000$$
$$0.06x + 0.08y = 1060$$

Step 4: $x + \quad y = 14,000$ Multiply by $-6 \rightarrow$ $-6x - 6y = -84,000$

$0.06x + 0.08y = 1,060$ Multiply by $100 \rightarrow$ $6x + 8y = 106,000$

$$2y = 22,000$$
$$y = 11,000$$
$$14,000 - y = 3,000$$

Step 5: They invested $3,000 at 6% and $11,000 at 8%.

Step 6: The interest on $3,000 at 6% is 0.06(3,000) $=$ 180

The interest on $11,000 at 8% is 0.08(11,000) $=$ 880

The total interest is $1060

97. Step 1: We know that Barbara has 17 coins that are dimes and nickels. We know a dime is worth 10 cents and a nickel is worth 5 cents. We do not know the specific number of dimes and nickels she has, but we do know that the total value of the coins is $1.35.

Step 2: Let $x =$ number of dimes and $y =$ the number of nickels. The total number of coins is 17, so

$$x + y = 17$$

The total amount of money she has is $1.35, which comes from nickels and dimes:

Amount of money	+	Amount of money	=	Total amount
in dimes		in nickels		of money
0.10x	+	0.05y		1.35

Step 3: The system that represents the situation is

$$x + \quad y = 17 \qquad \text{The number of coins}$$
$$0.10x + 0.05y = 1.35 \qquad \text{The value of the coins}$$

Step 4: $x + \quad y = 17$ Multiply by $-10 \rightarrow$ $-10x - 10y = -170$

$0.10x + 0.05y = 1.35$ Multiply by $100 \rightarrow$ $\underline{10x + 5y = 135}$

$$-5y = -35$$
$$y = 7$$

Step 5: Barbara has 10 dimes and 7 nickels.

Step 6: 10 dimes are worth $10(0.10)$ =   1.00

$\underline{7 \text{ nickels are worth } 7(0.05) \quad = \quad 0.35}$

The total value is $1.35

98. Step 1: We know that Tom has 15 coins that are dimes and quarters. We know a dime is worth 10 cents and a quarter is worth 25 cents. We do not know the specific number of dimes and quarters, but we do know that the total value of the coins is $2.40.

Step 2: Let x = number of dimes and y = number of quarters. The total number of coins is 15, so

$$x + y = 15$$

The total amount of money he has is $2.40, which comes from dimes and quarters:

Amount of money in dimes	+	Amount of money in quarters	=	Total amount of money
$0.10x$		$0.25y$		2.40

Step 3: The system that represents the situation is:

$$x + \quad y = 15 \qquad \text{The number of coins}$$
$$0.10x + 0.25y = 2.40 \qquad \text{The value of the coins}$$

Step 4: $x + \quad y = 15$ Multiply by $-10 \rightarrow$ $-10x - 10y = -150$

$0.10x + 0.25y = 2.40$ Multiply by $100 \rightarrow$ $\underline{10x + 25y = 240}$

$$15y = 90$$
$$y = 6$$
$$15 - y = 9$$

Step 5: Tom has 9 dimes and 6 quarters.

Step 6: 9 dimes are worth $9(0.10)$ =   0.90

$\underline{6 \text{ quarters are worth } 6(0.25) \quad = \quad 1.50}$

The total value is $2.40

99. Step 1: We know there are two solutions that together must total 50 liters. The solutions of alcohol are 20% and 10% and the final solution is to be 12%.   We do not know how many liters of each individual solution we need.

Step 2: Let x = number of liters of 20% alcohol solution needed and y = number of liters of 10% alcohol solution needed. Since the total number of liters is 50, we have

$$x + y = 50$$

To obtain our second equation, we add 20% of x liters to 10% of y liters to get 12% of 50 liters (the amount of alcohol in the final solution).

$$0.20x + 0.10y = 0.12(50)$$

The information has been put in a table for you:

	20% Solution	10% Solution	Final Solution
Number of liters	x	y	50
Liters of alcohol	0.20x	0.10y	0.12(50)

Step 3: Our system of equations is:

$$x + \quad y = 50$$
$$0.20x + 0.10y = 0.12(50)$$

Step 4: Solving by substitution, we have:

$$0.20(50 - y) + 0.10y = 0.12(50)$$

Multiply each side by 100:

$$20(50 - y) + 10y = 12(50)$$
$$1000 - 20y + 10y = 600$$
$$1000 - 10y = 600$$
$$-10y = -400$$
$$y = 40$$
$$50 - y = 10$$

Step 5: We have 10 liters of 20% solution and 40 liters of 10% solution.

Step 6: $0.20(10) + 0.10(40) \overset{?}{=} 0.12(50)$

$$2 + 4 \overset{?}{=} 6$$
$$6 = 6 \quad \text{It checks}$$

100. Step 1: We know that there are two solutions that together must total 40 liters. The two solutions of alcohol are 25% and 15% and the final solution is to be 20%.   We do not know how many liters of each individual solution we need.

Step 2: Let x = number of liters of 25% alcohol solution and y = number of liters of 15% alcohol solution needed. Since the total number of liters is 40, we have

$$x + y = 40$$

To obtain our second equation, we add 25% of x liters to 15% of y liters to get 20% of 40 liters (the amount of alcohol in the final solution).

$$0.25x + 0.15y = 0.20(40)$$

The information has been put in a table for you:

	25% Solution	15% Solution	Final Solution
Number of liters	x	y	40
Liters of alcohol	$0.25x$	$0.15y$	$0.20(40)$

Step 3: Our system of equations is:

$$x + \quad y = 40$$
$$0.25x + 0.15y = 0.20(40)$$

Step 4: Solving by substitution, we have:

$$0.25(40 - y) + 0.15y = 0.20(40)$$

Multiply each side by 100

$$25(40 - y) + 15y = 20(40)$$
$$1000 - 25y + 15y = 800$$
$$1000 - 10y = 800$$
$$-10y = -200$$
$$y = 20$$
$$40 - y = 20$$

Step 5: We have 20 liters of 25% solution and 20 liters of 15% solution.

Step 6: $0.25(20) + 0.15(20) \overset{?}{=} 0.20(40)$

$$5 + 3 \overset{?}{=} 8$$
$$8 = 8 \quad \text{It checks}$$

101 - 102. See the graphs in the back of the textbook.

CHAPTER 3 TEST

1. Given the equation $2x - 5y = 10$

Given $(0, \ )$, substitute $x = 0$ Given $(\ , 0)$, substitute $y = 0$

$$2(0) - 5y = 10 \qquad\qquad 2x - 5(0) = 10$$
$$-5y = 10 \qquad\qquad\qquad 2x = 10$$
$$y = -2 \qquad\qquad\qquad\quad x = 5$$

The ordered pair is $(0, -2)$.   The ordered pair is $(5, 0)$.

Given (10,), substitute $x = 10$

$$2(10) - 5y = 10$$
$$20 - 5y = 10$$
$$-5y = -10$$
$$y = 2$$

The ordered pair is (10, 2).

Given (, -3), substitute $y = -3$

$$2x - 5(-3) = 10$$
$$2x + 15 = 10$$
$$2x = -5$$
$$x = -\frac{5}{2}$$

The ordered pair is $\left(-\frac{5}{2},\ -3\right)$.

2. $y = 4x - 3$ (2, 5)

$$5 = 4(2) - 3 \qquad \text{a solution}$$
$$5 = 8 - 3$$
$$5 = 5 \qquad\qquad \text{True}$$

$y = 4x - 3$ (0, -3)

$$-3 = 4(0) - 3 \qquad \text{a solution}$$
$$-3 = -3 \qquad\qquad \text{True}$$

3. $y = 3x - 2$

x-Intercept		**y-Intercept**	
When	$y = 0$	When	$x = 0$
the equation	$y = 3x - 2$	the equation	$y = 3x - 2$
becomes	$0 = 3x - 2$	becomes	$y = 3(0) - 2$
	$2 = 3x$		$y = -2$
	$\dfrac{2}{3} = x$		$(0, -2)$
	$\left(\dfrac{2}{3}, 0\right)$		

See the graph in the back of the textbook.

4. $x = -2$ y is all real numbers.

See the graph in the back of the textbook.

5. We know from Section 3.6 that the equation of the line with slope m and y-intercept b is

$$y = mx + b$$

Rewriting $3x - 2y = 6$ in this form, we have

$$3x - 2y = 6 \qquad \text{Original equation}$$
$$-2y = -3x + 6 \qquad \text{Add } -3x \text{ to each side}$$
$$y = \tfrac{3}{2}x - 3 \qquad \text{Divide each side by } -2$$

We now find the slope $= \frac{3}{2}$ and the y-intercept is -3. To find the **x-intercept**

$$\text{When} \qquad\qquad y = 0$$
$$\text{the equation} \quad 3x - 2y = 6$$
$$\text{becomes} \qquad 3x - 2(0) = 6$$
$$3x = 6$$
$$x = 2$$

The x-intercept is 2. See the graph in the back of the textbook.

6. We know from Section 3.6 that the equation of the line with slope m and y-intercept b is

$$y = mx + b$$

Rewriting $2x - y = 5$ in this form, we have

$$2x - y = 5 \qquad\qquad \text{Original equation}$$
$$-y = -2x + 5 \qquad\quad \text{Add } -2x \text{ to each side}$$
$$y = 2x - 5 \qquad\qquad \text{Divide each side by } -1$$

We now find the slope is 2 and the y-intercept is -5. To find the **x-intercept**

$$\text{When} \qquad\qquad y = 0$$
$$\text{the equation} \quad 2x - y = 5$$
$$\text{becomes} \qquad\quad 2x = 5$$
$$x = \frac{5}{2}$$

The x-intercept is $\frac{5}{2}$. See the graph in the back of the textbook.

7. Substituting $m = 4$ and y–intercept 8, we have

$$y = 4x + 8$$

8. We begin by finding the slope of the line:

$$m = \frac{1-4}{-3-(-2)} = \frac{-3}{-1} = 3$$

Using $(x_1,\, y_1) = (-3,\ 1)$ and $m = 3$ in $y - y_1 = m(x - x_1)$ yields

$$y - 1 = 3(x + 3) \quad \text{Note: } x - (-3) = x + 3$$
$$y - 1 = 3x + 9 \qquad \text{Multiply out right side}$$
$$y = 3x + 10 \qquad \text{Add } 1 \text{ to each side}$$

9. $x + 2y = 5$

x-Intercept		**y-Intercept**	
When	$y = 0$	When	$x = 0$
the equation	$x + 2y = 5$	the equation	$x + 2y = 5$
becomes	$x + 2(0) = 5$	becomes	$0 + 2y = 5$
	$x = 5$		$y = \dfrac{5}{2}$
	$(5, 0)$		$\left(0, \dfrac{5}{2}\right)$

$y = 2x$

x-Intercept		**y-Intercept**	
When	$y = 0$	When	$x = 1$
the equation	$y = 2x$	the equation	$y = 2x$
becomes	$0 = 2x$	becomes	$y = 2(1)$
	$0 = x$		$y = 2$
	$(0, 0)$		$(1, 2)$

See the graph in the back of the textbook. The lines intersect at the point $(1, 2)$.

10. $x - y = 1$ <u>Multiply by -2</u> $\rightarrow$ $\quad -2x + 2y = -2$

$\quad 2x - 3y = 6$ <u>No change</u> $\quad \rightarrow$ $\quad \dfrac{2x - 3y = 6}{}$

$$-y = 4$$
$$y = -4$$

Substituting $y = -4$ in the first equation,

$$x - (-4) = 1$$
$$x + 4 = 1$$
$$x = -3$$

The solution to the system is $(-3, -4)$.

11. $2x + y = 7$ <u>Multiply by 3</u> $\rightarrow$ $\quad 6x + 3y = 21$

$\quad 3x + 7y = -6$ <u>Multiply by -2</u> $\rightarrow$ $\dfrac{-6x - 14y = 12}{}$

$$-11y = 33$$
$$y = -3$$

Substituting $y = -3$ in the first equation,

$$2x + (-3) = 7$$
$$2x = 10$$
$$x = 5$$

The solution to the system is $(5, -3)$.

12. $7x + 8y = -2$ <u>No change</u> $\rightarrow$ $7x + 8y = -2$

    $3x - 2y = 10$ <u>Multiply by 4</u> $\rightarrow$ $12x - 8y = 40$

$$19x \qquad = 38$$
$$x \qquad = 2$$

$7x + 8y = -2$ when $x = 2$

$7(2) + 8y = -2$

$14 + 8y = -2$

$8y = -16$

$y = -2$

The solution to the system is $(2, -2)$.

13. $6x - 10y = 6$ <u>Multiply by 3</u> $\rightarrow$ $18x - 30y = 18$

    $9x - 15y = 9$ <u>Multiply by -2</u> $\rightarrow$ $-18x + 30y = -18$

$$0 = 0$$

The lines coincide and there is an infinite number of solutions to the system.

14. $3x + 2y = 20$

       $y = 2x + 3$

Substituting $2x + 3$ for y in the first equation.

$$3x + 2(2x + 3) = 20$$
$$3x + 4x + 6 = 20$$
$$7x + 6 = 20$$
$$7x = 14$$
$$x = 2$$
$$y = 2x + 3 \quad \text{when } x = 2$$
$$y = 2(2) + 3$$
$$y = 4 + 3$$
$$y = 7$$

The solution to the system is $(2, 7)$.

15. $3x - 6y = -6$ and $x = 3$

$$3(3) - 6y = -6$$
$$9 - 6y = -6$$
$$-6y = -15$$
$$y = \frac{15}{6}$$
$$y = \frac{5}{2}$$

The solution to the system is $\left(3, \frac{5}{2}\right)$.

16. $2x - 7y = 2$ and $x = 4y$

$$2(4y) - 7y = 2$$
$$8y - 7y = 2$$
$$y = 2$$

Substituting $y = 2$ in the second equation,

$$x = 4(2)$$
$$x = 8$$

The solution to the system is $(8, 2)$.

17. Step 1: We know that two numbers have a sum of 12. Their difference is 2. We do not know the two numbers.

Step 2: Let $x =$ one number and $y =$ the other number. "The sum of the two numbers is 12" translates to

$$x + y = 12$$

"Their difference is 2" translates to

$$x - y = 2$$

Step 3: The system that describes the situation must be

$$x + y = 12$$
$$x - y = 2$$

Step 4: We can solve this system by the elimination method.

$$x + y = 12$$
$$\underline{x - y = 2}$$
$$2x \quad = 14$$
$$x \quad = 7$$

$$x + y = 12 \quad \text{when } x = 7$$
$$7 + y = 12$$
$$y = 5$$

Step 5: The numbers are 7 and 5.

Step 6: The sum of 7 and 5 is 12 and the difference of 7 and 5 is 2.

18. Step 1: We know the sum of two numbers is 15. We also know one number is six more than twice the other. We do not know the two numbers.

Step 2: Let x = one number and y = the other number. "The sum of two numbers is 15" translates to

$$x + y = 15$$

"One number is six more than twice the other" translates to

$$x = 2y + 6$$

Step 3: The system that describes the situation must be

$$x + y = 15$$
$$x = 2y + 6$$

Step 4: Substituting $2y + 6$ for x in the first equation.

$$2y + 6 + y = 15$$
$$3y + 6 = 15$$
$$3y = 9$$
$$y = 3$$

$x = 2y + 6$ when $y = 3$
$x = 2(3) + 6$
$x = 6 + 6$
$x = 12$

Step 5: The numbers are 12 and 3.

Step 6: The sum of 12 and 3 is 15. 12 is six more than twice 3.

19. Step 1: We do not know the specific amounts in the accounts. We do know that their sum is $10,000 and that the interest rates on the two accounts are 9% and 11%. We also know that the total interest earned is $980.

Step 2: Let x = amount invested at 9% and y = amount invested at 11%. "Total invested is $10,000" translates to $x + y = 10,000$. To find the total interest, we multiply the amount invested by the interest rate and add.

Interest at 9%	+	Interest at 11%	=	Total interest
$0.09x$		$0.11y$		$980

Step 3: The system is

$$x + y = 10,000$$
$$0.09x + 0.11y = 980$$

Step 4:

$x + y = 10,000$ <u>Multiply by -9</u> $\rightarrow$ $-9x - 9y = -90,000$

$0.09x + 0.11y = 980$ <u>Multiply by 100</u> $\rightarrow$ <u>$9x + 11y = 98,000$</u>

$$2y = 8,000$$
$$y = 4,000$$

$$x + y = 10,000 \quad \text{when } y = 4000$$
$$x + 4000 = 10,000$$
$$x = 6000$$

Step 5. Dr. Stork invested $6,000 at 9% and $4,000 at 11%.

Step 6: $6,000 at 9% added to $4,000 at 11% equals $10,000.

The interest on $6,000 at 9% is 0.09(6000)   = 540

<u>The interest on $4,000 at 11% is 0.11(4000) = 440</u>

The total interest is $980

20. Step 1: We know that Diane has 12 coins that are nickels and quarters. We know that a nickel is worth 5 cents and a quarter is worth 25 cents. We do not know the specific number of nickels and quarters, but we do know that the total value of the coins is $1.60.

Step 2: Let x = nickels and y = quarters. Total coins are 12 translates to

$$x + y = 12$$

Amount of money	+	Amount of money	=	Total amount
in nickels		in quarters		of money
$0.05x$	+	$0.25y$	=	1.60

Step 3: The system is

$$x + y = 12$$
$$0.05x + 0.25y = 1.60$$

Step 4: $x + y = 12$ <u>Multiply by $-5 \rightarrow$</u> $-5x - 5y = -60$

$0.05x + 0.25y = 1.60$ <u>Multiply by $100 \rightarrow$</u> $5x + 25y = 160$

$$20y = 100$$
$$y = 5$$

$$x + y = 12 \quad \text{when } y = 5$$
$$x + 5 = 12$$
$$x = 7$$

Step 5: Diane has 7 nickels and 5 quarters.

Step 6: 7 nickels and 5 quarters equal 12 coins.

7 nickels are worth 7(0.05)   = 0.35

<u>5 quarters are worth 5(0.25) = 1.25</u>

The total value is $1.60

21. See the graph in the back of the textbook.

CHAPTER 4

SECTION 4.1

1. $4^2 = 4 \cdot 4 = 16$ Base 4, exponent 2

5. $4^3 = 4 \cdot 4 \cdot 4 = 16 \cdot 4 = 64$ Base 4, exponent 3

9. $-2^3 = -2 \cdot 2 \cdot 2 = -4 \cdot 2 = -8$ Base 2, exponent 3

13. $\left(\frac{2}{3}\right)^2 = \frac{2}{3} \cdot \frac{2}{3} = \frac{4}{9}$ Base $\frac{2}{3}$, exponent 2

17. $x^4 \cdot x^5 = x^{4+5} = x^9$

21. $2^5 \cdot 2^4 \cdot 2^3 = 2^{5+4+3} = 2^{12}$

25. $(x^2)^5 = x^{2 \cdot 5} = x^{10}$

29. $(2^5)^{10} = 2^{5 \cdot 10} = 2^{50}$

33. $(b^x)^y = b^{xy}$

37. $(-3x)^4 = (-3)^4 x^4 = 81x^4$

41. $(4xyz)^3 = 4^3 x^3 y^3 z^3 = 64x^3 y^3 z^3$

45. $(x^2)^3 (x^4)^2 = x^{2 \cdot 3} x^{4 \cdot 2}$

$\qquad\qquad\quad = x^6 x^8$

$\qquad\qquad\quad = x^{6+8}$

$\qquad\qquad\quad = x^{14}$

49. $(3x^2)^3 (2x)^4 = 3^3 x^{2 \cdot 3} 2^4 x^4$

$\qquad\qquad\qquad = 27 \cdot 16 x^6 x^4$

$\qquad\qquad\qquad = 432 x^{6+4}$

$\qquad\qquad\qquad = 432 x^{10}$

53. $\left(\frac{2}{3} a^4 b^5\right)^3 = \left(\frac{2}{3}\right)^3 a^{4 \cdot 3} b^{5 \cdot 3}$

$\qquad\qquad\quad = \frac{8}{27} a^{12} b^{15}$

57. $(4x^2 y)^3 \left(\frac{1}{8} xy\right)^2$

$\qquad = 4^3 x^{2 \cdot 3} y^3 \left(\frac{1}{8}\right)^2 x^2 y^2$

$\qquad = 64 \left(\frac{1}{64}\right) x^6 x^2 y^3 y^2$

$\qquad = 1 x^{6+2} y^{3+2}$

$\qquad = x^8 y^5$

61. $570 = 5.7 \times 10^2$

65. $2.49 \times 10^3 = 2.49 \times 1000 = 2490$

69. $2.8 \times 10^4 = 2.8 \times 10,000 = 28,000$

73. $V = s^3$ when $s = 2.5$ inches

$\qquad V = (2.5)^3$

$\qquad V = 15.6$ inches^3

81. $7.4 \times 10^5 = 7.4 \times 100,000 = \$740,000$

85. 100,000;   the exponent and the number of zeros are the same.

89. (b) 4 because $16 = 2^4$ and $4 = 2^2$

93. $4 - (-7) = 4 + 7 = 11$

97. $-15 - (-20) = -15 + 20 = 5$

SECTION 4.2

1. $3^{-2} = \dfrac{1}{3^2}$ Definition of negative exponents

$= \dfrac{1}{9}$

5. $5^{-3} = \dfrac{1}{5^3}$ Definition of negative exponents

$= \dfrac{1}{125}$

9. $(2x)^{-3} = \dfrac{1}{(2x)^3}$ Definition of negative exponents

$= \dfrac{1}{8x^3}$ Property 3

13. $\dfrac{5^3}{5^1} = 5^{3-1}$ Property 4

$= 5^2$ Subtract

$= 25$

17. $\dfrac{x^{10}}{x^4} = x^{10-4}$ Property 4

$= x^6$

21. $\dfrac{6}{6^{11}} = \dfrac{6^1}{6^{11}}$ Definition

$= 6^{1-11}$ Property 4

$= 6^{-10}$ Subtract

$= \dfrac{1}{6^{10}}$ Definition of negative exponents

25. $\dfrac{2^5}{2^{-3}} = 2^{5-(-3)}$ Property 4

$= 2^8$ Subtract

29. $10^0 = 1$ Definition

33. $\left(7y^3\right)^{-2} = \dfrac{1}{\left(7y^3\right)^2}$ Definition of negative exponents

$= \dfrac{1}{7^2 y^6}$ Property 2 and 3

$= \dfrac{1}{49y^6}$

37. $y^7 \cdot y^{-10} = y^{7+(-10)}$ Property 1

$= y^{-3}$ Add

$= \dfrac{1}{y^3}$ Definition of neg exponents

41. $\dfrac{\left(a^4\right)^3}{\left(a^3\right)^2} = \dfrac{a^{12}}{a^6}$ Property 2

$= a^{12-6}$ Property 4

$= a^6$ Subtract

45. $\dfrac{\left(x^{-2}\right)^3}{x^{-5}} = \dfrac{x^{-6}}{x^{-5}}$ Property 2

$= x^{-6-(-5)}$ Property 4

$= x^{-1}$ Subtract

$= \dfrac{1}{x}$ Definition of neg exponents

49. $\dfrac{\left(a^3\right)^2\left(a^4\right)^5}{\left(a^5\right)^2} = \dfrac{a^6 a^{20}}{a^{10}}$ Property 2

$\quad\quad\quad\quad = \dfrac{a^{26}}{a^{10}}$ Property 1

$\quad\quad\quad\quad = a^{26-10}$ Property 4

$\quad\quad\quad\quad = a^{16}$ Subtract

53. $\dfrac{\left(x^{-7}\right)^3\left(x^4\right)^5}{\left(x^3\right)^2\left(x^{-1}\right)^8} = \dfrac{x^{-21} x^{20}}{x^6 x^{-8}}$ Property 2

$\quad\quad\quad\quad = \dfrac{x^{-1}}{x^{-2}}$ Property 1

$\quad\quad\quad\quad = x^{-1-(-2)}$ Property 4

$\quad\quad\quad\quad = x^1$ Subtract

$\quad\quad\quad\quad = x$ Definition

57. $35,700 = 3.57 \times 10^4$

61. $0.000009 = 9 \times 10^{-6}$

65. $5.6 \times 10^4 = 56,000$

69. $7.89 \times 10^1 = 78.9$

73. $0.006 = 6 \times 10^{-3}$

77. $23.5 \times 10^4 = 2.35 \times 10^1 \times 10^4$

$\quad\quad\quad\quad\quad = 2.35 \times 10^5$

81. Smaller square $A = s^2$

 Larger square $A = (2s)^2$

$\quad\quad A = s^2$ when $s = 10$ inches

$\quad\quad A = 10^2$

$\quad\quad\quad = 100$ square inches

$\quad\quad A = (2s)^2$ when $s = 10$

$\quad\quad A = [2(10)]^2$

$\quad\quad\quad = (20)^2$

$\quad\quad\quad = 400$ square inches

$400 \div 100 = 4$ squares to cover the larger square's area by the smaller square's area.

85. First box $V = x^3$

 Second box $V = (2x)^3$

$\quad\quad V = x^3$ when $x = 6$ inches

$\quad\quad V = 6^3$

$\quad\quad\quad = 216$ inches^3

$\quad\quad V = (2x)^3$ when $x = 6$ inches

$\quad\quad V = [2(6)]^3$

$\quad\quad\quad = (12)^3$

$\quad\quad\quad = 1728$ inches^3

$1728 \div 216 = 8$ smaller boxes that will fit inside the larger box.

89. $4x + 3x = (4+3)x = 7x$

93. $4y + 5y + y = (4+5+1)y = 10y$

97. x-intercept $(2, 0)$ y-intercept $(0, 4)$

$\quad 2x + y = 4$ let $y = 0$ $2x + y = 4$ let $x = 0$

$\quad 2x + 0 = 4$ $2(0) + y = 4$

$\quad\quad 2x = 4$ $y = 4$

$\quad\quad\; x = 2$

See the graph in the back of the textbook.

SECTION 4.3

1. $(3x^4)(4x^3) = (3 \cdot 4)(x^4 \cdot x^3)$ Commutative and associative properties

         $= 12x^7$ Multiply coefficients, add exponents

5. $(8x)(4x) = (8 \cdot 4)(x \cdot x)$ Commutative and associative properties

         $= 32x^2$ Multiply coefficients, add exponents

9. $(6ab^2)(-4a^2b) = 6(-4)(aa^2)(b^2b)$ Commutative and associative properties

            $= -24a^3b^3$ Multiply coefficients, add exponents

13. $\frac{15x^3}{5x^2} = \frac{15}{5} \cdot \frac{x^3}{x^2}$ Write as separate fractions

       $= 3x$ Divide coefficients, subtract exponents

17. $\frac{32a^3}{64a^4} = \frac{32}{64} \cdot \frac{a^3}{a^4}$ Write as separate fractions

       $= \frac{1}{2} \cdot \frac{1}{a}$ Divide coefficients, subtract exponents

       $= \frac{1}{2a}$

21. $\frac{3x^3y^2z}{27xy^2z^3} = \frac{3}{27} \cdot \frac{x^3}{x} \cdot \frac{y^2}{y^2} \cdot \frac{z}{z^3}$ Write as separate fractions

       $= \frac{1}{9} \cdot \frac{x^2}{1} \cdot \frac{1}{z^2}$ Divide coefficients, subtract exponents

       $= \frac{x^2}{9z^2}$

25. $(3 \times 10^3)(2 \times 10^5) = (3 \times 2)(10^3 \times 10^5)$

                 $= 6 \times 10^8$ Add exponents

29. $(5.5 \times 10^{-3})(2.2 \times 10^{-4}) = (5.5 \times 2.2)(10^{-3} \times 10^{-4})$

                      $= 12.1 \times 10^{-7}$ Add exponents

                      $= 1.21 \times 10^1 \times 10^{-7}$ Change to scientific notation

                      $= 1.21 \times 10^{-6}$

33. $\frac{6 \times 10^8}{2 \times 10^{-2}} = \frac{6}{2} \times \frac{10^8}{10^{-2}}$ Write as separate fractions

       $= 3 \times 10^{8-(-2)}$ Subtract exponents

       $= 3 \times 10^{10}$

37. $3x^2 + 5x^2 = (3 + 5)x^2 = 8x^2$

41. $2a + a - 3a = (2 + 1 - 3)a = 0a = 0$

45. $20ab^2 - 19ab^2 + 30ab^2 = (20 - 19 + 30)ab^2$
$$= 31ab^2$$

49. $\dfrac{(3x^2)(8x^5)}{6x^4} = \left(\dfrac{3 \cdot 8}{6}\right)\dfrac{x^2 \cdot x^5}{x^4}$ Write as separate fractions

$\qquad\qquad = 4\dfrac{x^7}{x^4}$ Multiply and divide coefficients, add exponents

$\qquad\qquad = 4x^{7-4}$ Subtract exponents

$\qquad\qquad = 4x^3$

53. $\dfrac{(4x^3y^2)(9x^4y^{10})}{(3x^5y)(2x^6y)} = \dfrac{4 \cdot 9}{3 \cdot 2} \cdot \dfrac{x^3x^4}{x^5x^6} \cdot \dfrac{y^2y^{10}}{yy}$ Write as separate fractions

$\qquad\qquad = \dfrac{36}{6} \cdot \dfrac{x^7}{x^{11}} \cdot \dfrac{y^{12}}{y^2}$ Multiply ciefficients, add exponents

$\qquad\qquad = \dfrac{6y^{10}}{x^4}$ Divide coefficients, subtract exponents

57. $\dfrac{(5 \times 10^3)(4 \times 10^{-5})}{2 \times 10^{-2}} = \dfrac{5 \cdot 4}{2} \cdot \dfrac{10^3 \cdot 10^{-5}}{10^{-2}}$ Write as separate fractions

$\qquad\qquad = 10 \cdot \dfrac{10^{-2}}{10^{-2}}$ Multiply and divide coefficients, add exponents

$\qquad\qquad = 10 \cdot 10^0$ Subtract exponents

$\qquad\qquad = 10$ Definition $a^0 = 1\,(a \neq 0)$

$\qquad\qquad = 1 \times 10^1$ Scientific notation

61. $\dfrac{18x^4}{3x} + \dfrac{21x^7}{7x^4} = 6x^3 + 3x^3$ Divide coefficients, subtract exponents

$\qquad\qquad = 9x^3$ Combine like terms

65. $\dfrac{6x^7y^4}{3x^2y^2} + \dfrac{8x^5y^8}{2y^6} = 2x^5y^2 + 4x^5y^2$ Divide coefficients, subtract exponents

$\qquad\qquad = 6x^5y^2$ Combine like terms

69. $(7^3)^x = 7^{12}$

$\qquad 7^{3x} = 7^{12}$

$\qquad 3x = 12$

$\qquad\ x = 4$

73. Substituting $a = 3$ and $b = 4$ in the expressions $(a+b)^2$ and $a^2 + 2ab + b^2$, we have

$$(a+b)^2 = a^2 + 2ab + b^2$$
$$(3+4)^2 = 3^2 + 2(3)(4) + 4^2$$
$$7^2 = 9 + 24 + 16$$
$$49 = 49$$

The expressions $(a+b)^2$ and $a^2 + 2ab + b^2$ are equal.

77. $V = LWH$ when width $= x$, length $= 2x$, and height $= 4$ inches

$$V = (2x)x(4)$$
$$V = 8x^2 \text{ inches}^3$$

81. $y = x^2$ when $(-4,)$ $y = x^2$ when $(-2,)$ $y = x^2$ when $(-1,)$ $y = x^2$ when $(0,)$

$y = (-4)^2$ $y = (-2)^2$ $y = (-1)^2$ $y = 0^2$

$y = 16$ $(-4, 16)$ $y = 4$ $(-2, 4)$ $y = 1$ $(-1, 1)$ $y = 0$ $(0, 0)$

$y = x^2$ when $(1,)$ $y = x^2$ when $(2,)$ $y = x^2$ when $(4,)$

$y = 1^2$ $y = 2^2$ $y = 4^2$

$y = 1$ $(1, 1)$ $y = 4$ $(2, 4)$ $y = 16$ $(4, 16)$

See the graph in the back of the textbook.

85. $y = \dfrac{1}{2}x^2$ when $(-4,)$ $y = \dfrac{1}{2}x^2$ when $(-2,)$ $y = \dfrac{1}{2}x^2$ when $(-1,)$

$y = \dfrac{1}{2}(-4)^2$ $y = \dfrac{1}{2}(-2)^2$ $y = \dfrac{1}{2}(-1)^2$

$y = \dfrac{1}{2}(16)$ $y = \dfrac{1}{2}(4)$ $y = \dfrac{1}{2}(1)$

$y = 8$ $(-4, 8)$ $y = 2$ $(-2, 2)$ $y = \dfrac{1}{2}$ $\left(-1, \dfrac{1}{2}\right)$

$y = \dfrac{1}{2}x^2$ when $(0,)$ $y = \dfrac{1}{2}x^2$ when $(1,)$ $y = \dfrac{1}{2}x^2$ when $(2,)$

$y = \dfrac{1}{2}(0)^2$ $y = \dfrac{1}{2}(1)^2$ $y = \dfrac{1}{2}(2)^2$

$y = \dfrac{1}{2}(0)$ $y = \dfrac{1}{2}(1)$ $y = \dfrac{1}{2}(4)$

$y = 0$ $(0, 0)$ $y = \dfrac{1}{2}$ $\left(1, \dfrac{1}{2}\right)$ $y = 2$ $(2, 2)$

$y = \frac{1}{2}x^2$ when $(4, \ )$

$y = \frac{1}{2}(4)^2$

$y = \frac{1}{2}(16)$

$y = 8 \qquad (4, 8)$

See the graph in the back of the textbook.

89. $4x$ when $x = -2$

$\qquad 4(-2) = -8$

93. $x^2 + 5x + 6$ when $x = -2$

$(-2)^2 + 5(-2) + 6 = 4 + (-10) + 6 = 0$

97. $y = \frac{1}{3}x + 1$ when $(-3, \ )$ $\qquad y = \frac{1}{3}x + 1$ when $(0, \ )$

$y = \frac{1}{3}(-3) + 1 \qquad\qquad\qquad y = \frac{1}{3}(0) + 1$

$y = -1 + 1 \qquad\qquad\qquad\qquad y = 1 \qquad (0, 1)$

$y = 0 \qquad (-3, 0)$

$y = \frac{1}{3}x + 1$ when $(3, \ )$

$y = \frac{1}{3}(3) + 1$

$y = 1 + 1$

$y = 2 \qquad (3, 2)$

See the graph in the back of the textbook.

SECTION 4.4

1. $2x^3 - 3x^2 + 1$ $\qquad\qquad$ Degree 3 $\qquad\qquad$ 5. $2x - 1$ $\qquad$ Degree 1

   3 terms is a trinomial $\qquad\qquad\qquad\qquad$ 2 terms is a binomial

9. $7a^2$ $\qquad$ Degree 2

   1 term is a monomial

13. $(2x^2 + 3x + 4) + (3x^2 + 2x + 5) = (2x^2 + 3x^2) + (3x + 2x) + (4 + 5)$

$\qquad\qquad\qquad\qquad\qquad\qquad = 5x^2 + 5x + 9$

17. $x^2 + 4x + 2x + 8 = x^2 + 6x + 8$ $\qquad\qquad\qquad$ 21. $x^2 - 3x + 3x - 9 = x^2 - 9$

25. $(6x^3 - 4x^2 + 2x) + (9x^2 - 6x + 3) = 6x^3 + (-4x^2 + 9x^2) + (2x - 6x) + 3$
$$= 6x^3 + 5x^2 - 4x + 3$$

29. $(a^2 - a - 1) - (-a^2 + a + 1) = a^2 - a - 1 + a^2 - a - 1$
$$= (a^2 + a^2) + (-a - a) + (-1 - 1)$$
$$= 2a^2 - 2a - 2$$

33. $(4y^2 - 3y + 2) + (5y^2 + 12y - 4) - (13y^2 - 6y + 20)$
$$= 4y^2 - 3y + 2 + 5y^2 + 12y - 4 - 13y^2 + 6y - 20$$
$$= (4y^2 + 5y^2 - 13y^2) + (-3y + 12y + 6y) + (2 - 4 - 20)$$
$$= -4y^2 + 15y - 22$$

37. $(11y^2 + 11y + 11) - (3y^2 + 7y - 15) = 11y^2 + 11y + 11 - 3y^2 - 7y + 15$
$$= (11y^2 - 3y^2) + (11y - 7y) + (11 + 15)$$
$$= 8y^2 + 4y + 26$$

41. $\left[(3x - 2) + (11x + 5)\right] - (2x + 1) = \left[(3x + 11x) + (-2 + 5)\right] - (2x + 1)$
$$= 14x + 3 - 2x - 1$$
$$= (14x - 2x) + (3 - 1)$$
$$= 12x + 2$$

45. When $\quad y = 10, \quad (y - 5)^2$
   becomes $\quad (10 - 5)^2 = 5^2 = 25$

49. $2x(5x) = (2 \cdot 5)(x \cdot x) = 10x^2$ 53. $2x(3x^2) = (2 \cdot 3)(x \cdot x^2) = 6x^3$

SECTION 4.5

1. $2x(3x + 1) = 2x(3x) + 2x(1)$
$$= 6x^2 + 2x$$

5. $2ab(a^2 - ab + 1) = 2ab(a^2) + 2ab(-ab) + 2ab(1)$
$$= 2a^3b - 2a^2b^2 + 2ab$$

9. $4x^2y(2x^3y + 3x^2y^2 + 8y^3) = 4x^2y(2x^3y) + 4x^2y(3x^2y^2) + 4x^2y(8y^3)$
$$= 8x^5y^2 + 12x^4y^3 + 32x^2y^4$$

13. $(x+6)(x+1) = x(x) + x(1) + 6(x) + 6(1)$
$\qquad\qquad\qquad$ **F** $\quad$ **O** $\quad$ **I** $\quad$ **L**
$$= x^2 + x + 6x + 6$$
$$= x^2 + 7x + 6$$

17. $(a+5)(a-3) = a(a) + a(-3) + 5(a) + 5(-3)$
$\qquad\qquad\qquad$ **F** $\quad$ **O** $\quad$ **I** $\quad$ **L**
$$= a^2 - 3a + 5a - 15$$
$$= a^2 + 2a - 15$$

21. $(x+6)(x-6) = x(x) + x(-6) + 6(x) + 6(-6)$
$\qquad\qquad\qquad$ **F** $\quad$ **O** $\quad$ **I** $\quad$ **L**
$$= x^2 - 6x + 6x - 36$$
$$= x^2 - 36$$

25. $(2x-3)(x-4) = 2x(x) + 2x(-4) + (-3)x + (-3)(-4)$
$\qquad\qquad\qquad$ **F** $\quad$ **O** $\quad$ **I** $\quad$ **L**
$$= 2x^2 - 8x - 3x + 12$$
$$= 2x^2 - 11x + 12$$

29. $(2x-5)(3x-2) = 2x(3x) + 2x(-2) + (-5)3x + (-5)(-2)$
$\qquad\qquad\qquad$ **F** $\quad$ **O** $\quad$ **I** $\quad$ **L**
$$= 6x^2 - 4x - 15x + 10$$
$$= 6x^2 - 19x + 10$$

33. $(5x-4)(5x+4) = 5x(5x) + 5x(4) + (-4)5x + (-4)4$
$\qquad\qquad\qquad$ **F** $\quad$ **O** $\quad$ **I** $\quad$ **L**
$$= 25x^2 + 20x - 20x - 16$$
$$= 25x^2 - 16$$

37. $(1-2a)(3-4a) = 1(3) + 1(-4a) + (-2a)3 + (-2a)(-4a)$
$\qquad\qquad\qquad$ **F** $\quad$ **O** $\quad$ **I** $\quad$ **L**
$$= 3 - 4a - 6a + 8a^2$$
$$= 3 - 10a + 8a^2$$

41.
$$\begin{array}{r} x^2 + 3x - 4 \\ x + 1 \\ \hline x^3 + 3x^2 - 4x \\ x^2 + 3x - 4 \\ \hline x^3 + 4x^2 - x - 4 \end{array}$$

45.
$$\begin{array}{r} x^2 - 2x + 4 \\ x + 2 \\ \hline x^3 - 2x^2 + 4x \\ 2x^2 - 4x + 8 \\ \hline x^3 \qquad\qquad + 8 \end{array}$$

49.
$$\begin{array}{r} 5x^2 + 2x + 1 \\ x^2 - 3x + 5 \\ \hline 5x^4 + 2x^3 + x^2 \\ -15x^3 - 6x^2 - 3x \\ 25x^2 + 10x + 5 \\ \hline 5x^4 - 13x^3 + 20x^2 + 7x + 5 \end{array}$$

53. $(3a^4 + 2)(2a^2 + 5) = 3a^4(2a^2) + 3a^4(5) + 2(2a^2) + 2(5)$
$\qquad\qquad\qquad$ **F** $\quad$ **O** $\quad$ **I** $\quad$ **L**
$$= 6a^6 + 15a^4 + 4a^2 + 10$$

57. If we let $x =$ the width, then the length is $2x + 5$. The area is

$$A = L \times W$$
$$A = (2x + 5)x$$
$$A = 2x^2 + 5x$$

61. a, b. See the back of the textbook for the complete table.

65. $R = xp$: when $x = 1700 - 100p$
$$R = (1700 - 100p)p$$

69. Using $(x_1, y_1) = (-2, -6)$ and $m = 3$ in $y - y_1 = m(x - x_1)$ gives us

$$y - (-6) = 3[x - (-2)]$$ Substitution
$$y + 6 = 3(x + 2)$$
$$y + 6 = 3x + 6$$ Add -6 to each side
$$y = 3x$$

SECTION 4.6

1. $(x - 2)^2 = x^2 + 2(x)(-2) + 4$
$$= x^2 - 4x + 4$$

5. $(x - 5)^2 = x^2 + 2(x)(-5) + 25$
$$= x^2 - 10x + 25$$

9. $(x + 10)^2 = x^2 + 2(x)(10) + 100$
$$= x^2 + 20x + 100$$

13. $(2x - 1)^2 = 4x^2 + 2(2x)(-1) + 1$
$$= 4x^2 - 4x + 1$$

17. $(3x - 2)^2 = 9x^2 + 2(3x)(-2) + 4$
$$= 9x^2 - 12x + 4$$

21. $(4x - 5y)^2 = 16x^2 + 2(4x)(-5y) + 25y^2$
$$= 16x^2 - 40xy + 25y^2$$

25. $(6x - 10y)^2 = 36x^2 + 2(6x)(-10y) + 100y^2$
$$= 36x^2 - 120xy + 100y^2$$

29. $(a^2 + 1)^2 = a^4 + 2(a^2)(1) + 1^2$
$$= a^4 + 2a^2 + 1$$

33. $(x - 3)(x + 3) = x^2 - 9$

37. $(y - 1)(y + 1) = y^2 - 1$

41. $(2x + 5)(2x - 5) = 4x^2 - 25$

45. $(2a + 7)(2a - 7) = 4a^2 - 49$

49. $(x^2 + 3)(x^2 - 3) = x^4 - 9$

53. $(5y^4 - 8)(5y^4 + 8) = 25y^8 - 64$

57. $(2x+3)^2 - (4x-1)^2$

$= (2x)^2 + 2(2x)(3) + 3^2 - \left[(4x)^2 + 2(4x)(-1) + (-1)^2 \right]$

$= 4x^2 + 12x + 9 - (16x^2 - 8x + 1)$

$= 4x^2 + 12x + 9 - 16x^2 + 8x - 1$

$= -12x^2 + 20x + 8$

61. $(2x+3)^3 = (2x+3)^2(2x+3)$

$= \left[4x^2 + 2(2x)(3) + 9 \right](2x+3)$

$= (4x^2 + 12x + 9)(2x+3)$

$$
\begin{array}{r}
4x^2 + 12x + 9 \\
2x + 3 \\
\hline
8x^3 + 24x^2 + 18x \\
12x^2 + 36x + 27 \\
\hline
8x^3 + 36x^2 + 54x + 27
\end{array}
$$

$(2x+3)^3 = 8x^3 + 36x^2 + 54x + 27$

65. $(x+3)^2$ \qquad Substituting $x=2$

$(2+3)^2 = 5^2 = 25$

$x^2 + 6x + 9$ \qquad Substituting $x=2$

$x^2 + 6x + 9 = 2^2 + 6(2) + 9$

$= 4 + 12 + 9$

$= 16 + 9$

$= 25$ \quad (Both equal 25)

69. Let $x =$ first consecutive integer

Let $x+1 =$ second consecutive integer

Let $x+2 =$ third consecutive integer

$x^2 + (x+1)^2 + (x+2)^2 = x^2 + (x^2 + 2x + 1) + (x^2 + 4x + 4)$

$= 3x^2 + 6x + 5$

73. $\dfrac{10x^3}{5x} = 2x^{3-1}$

$= 2x^2$

77. $\dfrac{35a^6 b^8}{70a^2 b^{10}} = \dfrac{a^{6-2}b^{8-10}}{2}$ \qquad Divide coefficients, subtract exponents

$= \dfrac{a^4 b^{-2}}{2}$

$= \dfrac{a^4}{2b^2}$ \qquad Definition of negative exponents

81. $y = 2x + 3$

x-Intercept		**y-Intercept**	
When	$y = 0$	When	$x = 0$
the equation	$y = 2x + 3$	the equation	$y = 2x + 3$
becomes	$0 = 2x + 3$	becomes	$y = 2(0) + 3$
	$-3 = 2x$		$y = 3$
	$-\dfrac{3}{2} = x$		$(0, 3)$
	$\left(-\dfrac{3}{2}, 0\right)$		

$y = -2x - 1$

x-Intercept		**y-Intercept**	
When	$y = 0$	When	$x = 0$
the equation	$y = -2x - 1$	the equation	$y = -2x - 1$
becomes	$0 = -2x - 1$	becomes	$y = -2(0) - 1$
	$1 = -2x$		$y = -1$
	$-\dfrac{1}{2} = x$		$(0, -1)$
	$\left(-\dfrac{1}{2}, 0\right)$		

See the graph in the back of the textbook. The lines intersect at the point $(-1, \ 1)$.

SECTION 4.7

1. $\dfrac{5x^2 - 10x}{5x} = \dfrac{5x^2}{5x} - \dfrac{10x}{5x}$ Divide each term by $5x$

   $= x - 2$

5. $\dfrac{25x^2 y - 10xy}{5x} = \dfrac{25x^2 y}{5x} - \dfrac{10xy}{5x}$ Divide each term by $5x$

   $= 5xy - 2y$

9. $\dfrac{50x^5 - 25x^3 + 5x}{5x} = \dfrac{50x^5}{5x} - \dfrac{25x^3}{5x} + \dfrac{5x}{5x}$ Divide each term by $5x$

   $= 10x^4 - 5x^2 + 1$

13. $\dfrac{16a^5 + 24a^4}{-2a} = \dfrac{16a^5}{-2a} + \dfrac{24a^4}{-2a}$ Divide each term by $-2a$

   $= -8a^4 - 12a^3$

17. $\dfrac{12a^3b - 6a^2b^2 + 14ab^3}{-2a} = \dfrac{12a^3b}{-2a} - \dfrac{6a^2b^2}{-2a} + \dfrac{14ab^3}{-2a}$ Divide each term by $-2a$

$\qquad\qquad\qquad\qquad\quad = -6a^2b + 3ab^2 - 7b^3$

21. $\dfrac{6x + 8y}{2} = \dfrac{6x}{2} + \dfrac{8y}{2}$ Divide each term by 2

$\qquad\qquad\quad = 3x + 4y$

25. $\dfrac{10xy - 8x}{2x} = \dfrac{10xy}{2x} - \dfrac{8x}{2x}$ Divide each term by $2x$

$\qquad\qquad\qquad = 5y - 4$

29. $\dfrac{x^2y - x^3y^2}{-x^2y} = \dfrac{x^2y}{-x^2y} - \dfrac{x^3y^2}{-x^2y}$ Divide each term by $-x^2y$

$\qquad\qquad\qquad = -1 + xy$

33. $\dfrac{x^3 - 3x^2y + xy^2}{x} = \dfrac{x^3}{x} - \dfrac{3x^2y}{x} + \dfrac{xy^2}{x}$ Divide each term by x

$\qquad\qquad\qquad\quad = x^2 - 3xy + y^2$

37. $\dfrac{26x^2y^2 - 13xy}{-13xy} = \dfrac{26x^2y^2}{-13xy} - \dfrac{13xy}{-13xy}$ Divide each term by $-13xy$

$\qquad\qquad\qquad\quad = -2xy + 1$

41. $\dfrac{5a^2x - 10ax^2 + 15a^2x^2}{20a^2x^2} = \dfrac{5a^2x}{20a^2x^2} - \dfrac{10ax^2}{20a^2x^2} + \dfrac{15a^2x^2}{20a^2x^2}$ Divide each term by $20a^2x^2$

$\qquad\qquad\qquad\qquad\quad = \dfrac{1}{4x} - \dfrac{1}{2a} + \dfrac{3}{4}$

45. $\dfrac{9a^{5m} - 27a^{3m}}{3a^{2m}} = \dfrac{9a^{5m}}{3a^{2m}} - \dfrac{27a^{3m}}{3a^{2m}}$ Divide each term by $3a^{2m}$

$\qquad\qquad\qquad = 3a^{5m-2m} - 9a^{3m-2m}$

$\qquad\qquad\qquad = 3a^{3m} - 9a^m$

49. $\dfrac{2x^3(3x+2)-3x^2(2x-4)}{2x^2}$

$= \dfrac{6x^4+4x^3-6x^3+12x^2}{2x^2}$ Distributive property

$= \dfrac{6x^4}{2x^2}+\dfrac{4x^3}{2x^2}-\dfrac{6x^3}{2x^2}+\dfrac{12x^2}{2x^2}$ Divide each term by $2x^2$

$= 3x^2+2x-3x+6$ Simplify

$= 3x^2-x+6$ Simplify

53. $\dfrac{(x+5)^2+(x+5)(x-5)}{2x}$

$= \dfrac{x^2+10x+25+x^2-25}{2x}$ Multiply

$= \dfrac{2x^2+10x}{2x}$ Simplify

$= \dfrac{2x^2}{2x}+\dfrac{10x}{2x}$ Divide each term by $2x$

$= x+5$

57. When $x=10$ When $x=10$

the expression $\dfrac{3x+8}{2}$ the expression $3x+4$

becomes $\dfrac{3(10)+8}{2}=\dfrac{30+8}{2}$ becomes $3(10)+4=30+4$

$= \dfrac{38}{2}$ $= 34$

$= 19$

61. $2x-3y=-5$ $\underline{\text{No change}} \rightarrow$ $2x-3y=-5$

  $x+y=5$ $\underline{\text{Multiply by } -2} \rightarrow$ $-2x-2y=-10$

                                                          $-5y=-15$

                                                          $y=3$

Substituting $y=3$ into $x+y=5$ gives us

  $x+3=5$

  $x=2$

The solution to the system is $(2,\ 3)$.

65. Substituting $y=-2x+4$ for y from the second equation into the first equation gives us

$$4x+2(-2x+4)=8$$
$$4x-4x+8=8$$
$$8=8$$

A true statement, therefore the lines coincide.

SECTION 4.8

1.
$$\begin{array}{r} x-2 \\ x-3{\overline{\smash{\big)}\,x^2-5x+6}} \\ \text{$-$ $+$} \\ \underline{x^2 \not{-} 3x} \\ -2x+6 \\ \text{$+$ $-$} \\ \underline{\not{-}2x \not{-} 6} \\ 0 \end{array}$$

Our answer: $x-2$.

5.
$$\begin{array}{r} x-3 \\ x-3{\overline{\smash{\big)}\,x^2-6x+9}} \\ \text{$-$ $+$} \\ \underline{x^2 \not{-} 3x} \\ -3x+9 \\ \text{$+$ $-$} \\ \underline{\not{-}3x \not{-} 9} \\ 0 \end{array}$$

Our answer: $x-3$.

9.
$$\begin{array}{r} a-5 \\ 2a+1{\overline{\smash{\big)}\,2a^2-9a-5}} \\ \text{$-$ $-$} \\ \underline{2a^2 \not{-} a} \\ -10a-5 \\ \text{$+$ $+$} \\ \underline{\not{-}10a \not{-} 5} \\ 0 \end{array}$$

Our answer: $a-5$.

13.
$$\begin{array}{r} a-2 \\ a+5{\overline{\smash{\big)}\,a^2+3a+2}} \\ \text{$-$ $-$} \\ \underline{a^2 \not{-} 5a} \\ -2a+2 \\ \text{$+$ $+$} \\ \underline{\not{-}2a \not{-} 10} \\ 12 \end{array}$$

Our answer: $a-2+\frac{12}{a+5}$

17.
$$\begin{array}{r} x+4 \\ x+1{\overline{\smash{\big)}\,x^2+5x-6}} \\ \text{$-$ $-$} \\ \underline{x^2 \not{-} x} \\ 4x-6 \\ \text{$-$ $-$} \\ \underline{4x \not{-} 4} \\ -10 \end{array}$$

Our answer: $x+4+\frac{-10}{x+1}$

21.
$$\begin{array}{r} x-3 \\ 2x+4{\overline{\smash{\big)}\,2x^2-2x+5}} \\ \text{$-$ $-$} \\ \underline{2x^2 \not{-} 4x} \\ -6x+5 \\ \text{$+$ $+$} \\ \underline{\not{-}6x \not{-} 12} \\ 17 \end{array}$$

Our answer: $x-3+\frac{17}{2x+4}$

25.
$$
\begin{array}{r}
2a^2 - a - 3 \\
3a - 5 \overline{\smash{\big)}\ 6a^3 - 13a^2 - 4a + 15} \\
\underline{-\quad +} \\
6a^3 \cancel{-} 10a^2 \\
-3a^2 - 4a \\
\underline{+\quad -} \\
\cancel{-}3a^2 \cancel{-} 5a \\
-9a + 15 \\
\underline{+\quad -} \\
\cancel{-}9a \cancel{-} 15 \\
0
\end{array}
$$

29.
$$
\begin{array}{r}
x^2 + x + 1 \\
x - 1 \overline{\smash{\big)}\ x^3 + 0x^2 + 0x - 1} \\
\underline{-\quad +} \\
x^3 \cancel{-} x^2 \\
x^2 + 0x \\
\underline{-\quad +} \\
x^2 \cancel{-} x \\
x - 1 \\
\underline{-\quad +} \\
x \cancel{-} 1 \\
0
\end{array}
$$

Our answer: $2a^2 - a - 3$ Our answer: $x^2 + x + 1$

33. Step 1: The sum of two numbers is 25. One of the numbers is four times the other. We do not know the two numbers.

Step 2: One number is four times the other translates to $x = 4y$. The sum of two numbers is 25 translates to $x + y = 25$.

Step 3: The system that describes this situation is

$$x = 4y$$
$$x + y = 25$$

Step 4: Substituting $x = 4y$ into the second equation, we have

$$4y + y = 25$$
$$5y = 25$$
$$y = 5$$

$x = 4y$ when $y = 5$
$x = 4 \cdot 5 = 20$

Step 5: The two numbers are 5 and 20.

Step 6: 20 is four times 5 and 5 + 20 = 25.

37. Step 1: We have a total of $160 in $5 bills and $10 bills. You have four more $10 bills than $5 bills. We do not know how many of each bill.

Step 2: Let x = number of $5 bills and y = number of $10 bills. The total of $5 and $10 bills is $160 translates to $5x + 10y = 160$. You have four more $10 bills than $5 bills translates to $y = x + 4$.

Step 3: The system that describes this situation is

$$5x + 10y = 160$$
$$y = x + 4$$

Step 4: Substituting $y = x + 4$ into the first equation, we have

$$5x + 10(x + 4) = 160$$
$$5x + 10x + 40 = 160$$
$$15x + 40 = 160$$
$$15x = 120$$
$$x = 8$$

$y = x + 4$ when $x = 8$

$y = 8 + 4 = 12$

Step 5: You will have eight $5 bills and twelve $10 bills.

Step 6: Twelve $10 bills is four more than eight $5 bills.

12($10 bills)   +   8($5 bills)   =   160

$120 $40   = $160

CHAPTER 4 REVIEW

1. $(-1)^3 = (-1)(-1)(-1) = -1$

2. $(-4)^3 = (-4)(-4)(-4) = -64$

3. $-8^2 = -8(8) = -64$

4. $-2^4 = -2(2)(2)(2) = -16$

5. $\left(\frac{3}{7}\right)^2 = \left(\frac{3}{7}\right)\left(\frac{3}{7}\right) = \frac{9}{49}$

6. $\left(\frac{1}{3}\right)^3 = \left(\frac{1}{3}\right)\left(\frac{1}{3}\right)\left(\frac{1}{3}\right) = \frac{1}{27}$

7. $y^3 \cdot y^9 = y^{3+9} = y^{12}$

8. $y^{10} \cdot y^6 = y^{10+6} = y^{16}$

9. $x^{15} \cdot x^7 \cdot x^5 \cdot x^3 = x^{15+7+5+3} = x^{30}$

10. $x^{21} \cdot x^{13} \cdot x^9 \cdot x^7 = x^{21+13+9+7} = x^{50}$

11. $(x^7)^5 = x^{7\cdot5} = x^{35}$

12. $(x^4)^{10} = x^{4\cdot10} = x^{40}$

13. $(2^6)^4 = 2^{6\cdot4} = 2^{24}$

14. $(3^8)^5 = 3^{8\cdot5} = 3^{40}$

15. $(3y)^3 = 3^3 y^3 = 27y^3$

16. $(2y)^4 = 2^4 y^4 = 16y^4$

17. $(-2xyz)^3 = (-2)^3 x^3 y^3 z^3 = -8x^3 y^3 z^3$

18. $\left(-\frac{2}{3}a^2 b^3 c^4\right)^2 = \left(-\frac{2}{3}\right)^2 (a^2)^2 (b^3)^2 (c^4)^2$
$$= \frac{4}{9}a^4 b^6 c^8$$

19. $7^{-2} = \frac{1}{7^2} = \frac{1}{49}$

20. $9^{-2} = \frac{1}{9^2} = \frac{1}{81}$

21. $4x^{-5} = 4\left(\frac{1}{x^5}\right) = \frac{4}{x^5}$

22. $7x^{-6} = 7\left(\frac{1}{x^6}\right) = \frac{7}{x^6}$

23. $(3y)^{-3} = \frac{1}{(3y)^3} = \frac{1}{27y^3}$

24. $(6y)^{-2} = \frac{1}{(6y)^2} = \frac{1}{36y^2}$

25. $\frac{a^9}{a^3} = a^{9-3} = a^6$

26. $\frac{a^{10}}{a^2} = a^{10-2} = a^8$

27. $\left(\dfrac{x^3}{x^5}\right)^2 = \dfrac{x^6}{x^{10}}$ or $\left(\dfrac{x^3}{x^5}\right)^2 = \left(x^{3-5}\right)^2$

$\qquad\qquad = x^{6-10} \qquad\qquad\quad = \left(x^{-2}\right)^2$

$\qquad\qquad = x^{-4} \qquad\qquad\qquad = x^{-4}$

$\qquad\qquad = \dfrac{1}{x^4} \qquad\qquad\qquad = \dfrac{1}{x^4}$

28. $\left(\dfrac{x^5}{x^3}\right)^{-2} = \left(x^{5-3}\right)^{-2}$

$\qquad\qquad = \left(x^2\right)^{-2}$

$\qquad\qquad = x^{-4}$

$\qquad\qquad = \dfrac{1}{x^4}$

29. $\dfrac{x^9}{x^{-6}} = x^{9-(-6)} = x^{15}$

30. $\dfrac{x^4}{x^{-8}} = x^{4-(-8)} = x^{12}$

31. $\dfrac{x^{-7}}{x^{-2}} = x^{-7-(-2)} = x^{-5} = \dfrac{1}{x^5}$

32. $\dfrac{x^{-9}}{x^{-13}} = x^{-9-(-13)} = x^4$

33. $\left(-3xy\right)^0 = 1$

34. $\left(9xy\right)^0 = 1$

35. $3^0 - 5^1 + 5^0 = 1 - 5 + 1 = -3$

36. $4^1 + 9^0 + (-7)^0 = 4 + 1 + 1 = 6$

37. $(3x^3y^2)^2 = 3^2(x^3)^2(y^2)^2 = 9x^{3(2)}y^{2(2)} = 9x^6y^4$

38. $(-2x^4y^3)^3 = (-2)^3(x^4)^3(y^3)^3 = -8x^{12}y^9$

39. $\left(2a^3b^2\right)^4\left(2a^5b^6\right)^2 = 2^4 a^{12}b^8 2^2 a^{10}b^{12}$

$\qquad\qquad\qquad\qquad = 16(4)a^{12+10}b^{8+12}$

$\qquad\qquad\qquad\qquad = 64a^{22}b^{20}$

40. $\left(3a^2b^5\right)^3\left(2a^6b^7\right)^2 = 3^3 a^6 b^{15} 2^2 a^{12} b^{14}$

$\qquad\qquad\qquad\qquad = 27(4)a^{6+12}b^{15+14}$

$\qquad\qquad\qquad\qquad = 108a^{18}b^{29}$

41. $\left(-3xy^2\right)^{-3} = \dfrac{1}{\left(-3xy^2\right)^3}$

$\qquad\qquad\quad = \dfrac{1}{(-3)^3 x^3 \left(y^2\right)^3}$

$\qquad\qquad\quad = \dfrac{1}{-27x^3 y^{2(3)}} \qquad (-3)^3 = (-3)(-3)(-3) = -27$

$\qquad\qquad\quad = \dfrac{-1}{27x^3 y^6}$

42. $\left(-2x^3y^5\right)^{-4} = \dfrac{1}{\left(-2x^3y^5\right)^4}$

$\qquad\qquad = \dfrac{1}{16x^{12}y^{20}}$

43. $\dfrac{\left(b^3\right)^4\left(b^2\right)^5}{\left(b^7\right)^3} = \dfrac{b^{12}b^{10}}{b^{21}}$

$\qquad\qquad = \dfrac{b^{22}}{b^{21}}$

$\qquad\qquad = b$

44. $\dfrac{\left(b^6\right)^2\left(b^3\right)^4}{\left(b^{10}\right)^3} = \dfrac{b^{12}b^{12}}{b^{30}}$

$\qquad\qquad = \dfrac{b^{24}}{b^{30}}$

$\qquad\qquad = b^{-6}$

$\qquad\qquad = \dfrac{1}{b^6}$

45. $\dfrac{\left(x^{-3}\right)^3\left(x^6\right)^{-1}}{\left(x^{-5}\right)^{-4}} = \dfrac{x^{-3(3)}x^{6(-1)}}{x^{-5(-4)}}$

$\qquad\qquad = \dfrac{x^{-9}x^{-6}}{x^{20}}$

$\qquad\qquad = \dfrac{x^{-15}}{x^{20}}$

$\qquad\qquad = \dfrac{1}{x^{20}x^{15}}$

$\qquad\qquad = \dfrac{1}{x^{35}}$

46. $\dfrac{\left(x^{-4}\right)^{-3}\left(x^{-3}\right)^4}{x^0} = \dfrac{x^{12}x^{-12}}{1}$

$\qquad\qquad = x^0$

$\qquad\qquad = 1$

47. $\dfrac{\left(2x^4\right)\left(15x^9\right)}{\left(6x^6\right)} = \dfrac{30x^{13}}{6x^6}$

$\qquad\qquad = 5x^7$

48. $\dfrac{\left(3x^5\right)\left(20x^3\right)}{15x^{10}} = \dfrac{60x^8}{15x^{10}}$

$\qquad\qquad = 4x^{-2}$

$\qquad\qquad = \dfrac{4}{x^2}$

49. $\dfrac{\left(10x^3y^5\right)\left(21x^2y^6\right)}{\left(7xy^3\right)\left(5x^9y\right)} = \dfrac{10\cdot21x^3x^2y^5y^6}{7\cdot5xx^9y^3y}$

$\qquad\qquad = \dfrac{210x^{3+2}y^{5+6}}{35x^{1+9}y^{3+1}}$

$\qquad\qquad = \dfrac{6x^5y^{11}}{x^{10}y^4}$

$\qquad\qquad = 6x^{5-10}y^{11-4}$

$\qquad\qquad = 6x^{-5}y^7$

$\qquad\qquad = \dfrac{6y^7}{x^5}$

50. $\dfrac{(12xy^5)(16x^2y^2)}{(8x^3y^3)(3x^5y)} = \dfrac{192x^3y^7}{24x^8y^4}$

$= 8x^{-5}y^3$

$= \dfrac{8y^3}{x^5}$

51. $\dfrac{21a^{10}}{3a^4} - \dfrac{18a^{17}}{6a^{11}} = 7a^6 - 3a^6$

$= 4a^6$

52. $\dfrac{24a^{12}}{6a^3} + \dfrac{30a^{24}}{10a^{15}} = 4a^9 + 3a^9$

$= 7a^9$

53. $\dfrac{8x^8y^3}{2x^3y} - \dfrac{10x^6y^9}{5xy^7} = 4x^{8-3}y^{3-1} - 2x^{6-1}y^{9-7}$

$= 4x^5y^2 - 2x^5y^2$

$= (4-2)x^5y^2$

$= 2x^5y^2$

54. $\dfrac{50x^8y^8}{25x^4y^2} + \dfrac{28x^7y^7}{14x^3y} = 2x^4y^6 + 2x^4y^6$

$= 4x^4y^6$

55. $(3.2 \times 10^3)(2 \times 10^4) = (3.2 \times 2)(10^3 \times 10^4)$

$= 6.4 \times 10^7$

56. $(5 \times 10^{-5})(2.1 \times 10^3) = (5 \times 2.1)(10^{-5} \times 10^3)$

$= 10.5 \times 10^{-2}$

$= 1.05 \times 10^1 \times 10^{-2}$

$= 1.05 \times 10^{-1}$

57. $\dfrac{4.6 \times 10^5}{2 \times 10^{-3}} = 2.3 \times 10^{5-(-3)}$

$= 2.3 \times 10^8$

58. $\dfrac{3.5 \times 10^{-7}}{7 \times 10^{-3}} = 0.5 \times 10^{-4}$

$= 5 \times 10^{-1} \times 10^{-4}$

$= 5 \times 10^{-5}$

59. $\dfrac{(4 \times 10^6)(6 \times 10^5)}{3 \times 10^8} = \dfrac{24 \times 10^{11}}{3 \times 10^8}$

$= 8 \times 10^3$

60. $\dfrac{(6 \times 10^5)(6 \times 10^{-3})}{9 \times 10^{-4}} = \dfrac{36 \times 10^2}{9 \times 10^{-4}}$

$= 4 \times 10^2 \times 10^4$

$= 4 \times 10^6$

61. $(3a^2 - 5a + 5) + (5a^2 - 7a - 8) = 3a^2 + 5a^2 - 5a - 7a + 5 - 8$

$= 8a^2 - 12a - 3$

62. $\left(4a^3 - 10a^2 + 6\right) - \left(6a^3 + 5a - 7\right) = 4a^3 - 10a^2 + 6 - 6a^3 - 5a + 7$

$$= -2a^3 - 10a^2 - 5a + 13$$

63. $\left(-7x^2 + 3x - 6\right) - \left(8x^2 - 4x + 7\right) + \left(3x^2 - 2x - 1\right)$

$= -7x^2 + 3x - 6 - 8x^2 + 4x - 7 + 3x^2 - 2x - 1$

$= -12x^2 + 5x - 14$

64. $\left(-4x^2 - 5x + 2\right) + \left(3x^2 - 6x + 1\right) - \left(-x^2 + 2x - 7\right)$

$= -4x^2 - 5x + 2 + 3x^2 - 6x + 1 + x^2 - 2x + 7$

$= -13x + 10$

65. $\left(4x^2 - 3x - 2\right) - \left(8x^2 + 3x - 2\right) = 4x^2 - 3x - 2 - 8x^2 - 3x + 2$

$$= -4x^2 - 6x$$

66. $\left[(2x - 7) + (9x + 5)\right] - (4x + 3) = (11x - 2) - (4x + 3)$

$$= 11x - 2 - 4x - 3$$

$$= 7x - 5$$

67. $2x^2 - 3x + 5$ when $x = 3$

$2(3)^2 - 3(3) + 5 = 18 - 9 + 5$

$$= 14$$

68. $(y - 3)^2$ when $y = -8$

$(-8 - 3) = (-11)^2 = 121$

69. $3x(4x - 7) = 3x(4x) + 3x(-7) = 12x^2 - 21x$

70. $5x(8x + 3) = 5x(8x) + 5x(3) = 40x^2 + 15x$

71. $8x^3y\left(3x^2y - 5xy^2 + 4y^3\right)$

$= 8x^3y\left(3x^2y\right) + 8x^3y\left(-5xy^2\right) + 8x^3y\left(4y^3\right)$

$= 24x^5y^2 - 40x^4y^3 + 32x^3y^4$

72. $5xy^2\left(3x^3 + 4x^2y + 10xy^2\right)$

$= 5xy^2\left(3x^3\right) + 5xy^2\left(4x^2y\right) + 5xy^2\left(10xy^2\right)$

$= 15x^4y^2 + 20x^3y^3 + 50x^2y^4$

73. $a^2 + 5a - 4$
       $\underline{a + 1}$
     $a^3 + 5a^2 - 4a$
        $\underline{a^2 + 5a - 4}$
     $a^3 + 6a^2 + a - 4$

74. $a^2 - 6a + 7$
       $\underline{a - 2}$
     $a^3 - 6a^2 + 7a$
       $\underline{-2a^2 + 12a - 14}$
     $a^3 - 8a^2 + 19a - 14$

75. $x^2 - 5x + 25$
        $\underline{x + 5}$
     $x^3 - 5x^2 + 25x$
         $\underline{5x^2 - 25x + 125}$
     $x^3 \qquad\qquad + 125$

76. $x^2 + x + 1$
        $\underline{x - 1}$
     $x^3 + x^2 + x$
        $\underline{-x^2 - x - 1}$
     $x^3 \qquad\quad -1$

77.
$$\overset{\mathbf{F}\qquad\quad\mathbf{O}\qquad\quad\mathbf{I}\qquad\quad\mathbf{L}}{(3x-7)(2x-5)=3x(2x)+3x(-5)+(-7)(2x)+(-7)(-5)}$$
$$=6x^2-15x-14x+35$$
$$=6x^2-29x+35$$

78.
$$\overset{\mathbf{F}\qquad\quad\mathbf{O}\qquad\quad\mathbf{I}\qquad\quad\mathbf{L}}{(5x-2)(3x+4)=5x(3x)+5x(4)+(-2)(3x)+(-2)4}$$
$$=15x^2+20x-6x-8$$
$$=15x^2+14x-8$$

79. $\left(5y+\dfrac{1}{5}\right)\left(5y-\dfrac{1}{5}\right)=(5y)^2-\left(\dfrac{1}{5}\right)^2=25y^2-\dfrac{1}{25}$

80. $\left(\dfrac{1}{2}y+2\right)\left(\dfrac{1}{2}y-2\right)=\left(\dfrac{1}{2}y\right)^2-(2)^2=\dfrac{1}{4}y^2-4$

81. $(a^2-3)(a^2+3)=(a^2)^2-(3)^2=a^4-9$

82. $(a^2+7)(a^2-7)=(a^2)^2-7^2=a^4-49$

83. $(a-5)^2=a^2+2(a)(-5)+(-5)^2=a^2-10a+25$

84. $(a+6)^2=a^2+2(a)(6)+6^2=a^2+12a+36$

85. $(3x+4)^2=(3x)^2+2(3x)(4)+4^2=9x^2+24x+16$

86. $(5x-1)^2=(5x)^2+2(5x)(-1)+(-1)^2=25x^2-10x+1$

87. $(y^2+3)^2=(y^2)^2+2(y^2)(3)+3^2=y^4+6y^2+9$

88. $(y^2-2)^2=(y^2)^2+2(y^2)(-2)+(-2)^2=y^4-4y^2+4$

89. $\dfrac{10ab}{-5a}+\dfrac{20a^2}{-5a}=-2b-4a$

90. $\dfrac{15a^3b}{5ab}-\dfrac{10a^2b^2}{5ab}-\dfrac{20ab^3}{5ab}=3a^2-2ab-4b^2$

91. $\dfrac{40x^5y^4}{-8xy}-\dfrac{32x^3y^3}{-8xy}-\dfrac{16x^2y}{-8xy}=-5x^4y^3+4x^2y^2+2x$

92. $\dfrac{28x^4y^4}{-7xy^2}-\dfrac{14x^2y^3}{-7xy^2}+\dfrac{21xy^2}{-7xy^2}=-4x^3y^2+2xy-3$

93. $\dfrac{16xy^2-10xy}{-2xy}=\dfrac{16xy^2}{-2xy}-\dfrac{10xy}{-2xy}=-8y+5$

94. $\dfrac{40x^3y^2+20x^2y^2}{-10xy^2}=\dfrac{40x^3y^2}{-10xy^2}+\dfrac{20x^2y^2}{-10xy^2}=-4x^2-2x$

The answer: $-4x^2-2x$

95.

$$\frac{20a^2 - 16a^2b + 24a^2b^2}{4a^2} = \frac{20a^2}{4a^2} - \frac{16a^2b}{4a^2} + \frac{24a^2b^2}{4a^2}$$

$$= 5 - 4b + 6b^2$$

The answer: $5 - 4b + 6b^2$

96.

$$\frac{15x^5 - 10x^2 + 20x}{5x^5} = \frac{15x^5}{5x^5} - \frac{10x^2}{5x^5} + \frac{20x}{5x^5}$$

$$= 3 - \frac{2}{x^3} + \frac{4}{x^4}$$

The answer: $3 - \frac{2}{x^3} + \frac{4}{x^4}$

97.

$$
\begin{array}{r}
x+9 \\
x+6{\overline{\smash{\big)}\,x^2+15x+54}} \\
\underline{--} \\
x^2 \not+ 6x \\
9x+54 \\
\underline{--} \\
9x \not+ 54 \\
0
\end{array}
$$

The answer: $x + 9$

98.

$$
\begin{array}{r}
x-7 \\
x+4{\overline{\smash{\big)}\,x^2-3x-28}} \\
\underline{--} \\
x^2 \not+ 4x \\
-7x-28 \\
\underline{++} \\
\not+ 7x \not+ 28 \\
0
\end{array}
$$

The answer: $x - 7$

99.

$$
\begin{array}{r}
2x+5 \\
3x-1{\overline{\smash{\big)}\,6x^2+13x-5}} \\
\underline{-+} \\
6x^2 \not- 2x \\
15x-5 \\
\underline{-+} \\
15x \not- 5 \\
0
\end{array}
$$

The answer: $2x + 5$

100.

$$
\begin{array}{r}
2x+1 \\
4x+3{\overline{\smash{\big)}\,8x^2+10x+3}} \\
\underline{--} \\
8x^2 \not+ 6x \\
4x+3 \\
\underline{--} \\
4x \not+ 3 \\
0
\end{array}
$$

The answer: $2x + 1$

101.

$$x+4 \overline{\smash{)}\,x^3+0x^2+0x+64} \quad \dfrac{x^2-4x+16}{}$$

$$\underline{x^3 \cancel{+} 4x^2}$$

$$-4x^2+0x$$

$$\underline{\cancel{-}4x^2 \cancel{-} 16x}$$

$$16x+64$$

$$\underline{16x \cancel{+} 64}$$

$$0$$

The answer: $x^2-4x+16$

102.

$$x-3 \overline{\smash{)}\,x^3+0x^2+0x-27} \quad \dfrac{x^2+3x+9}{}$$

$$\underline{x^3 \cancel{-} 3x^2}$$

$$3x^2+0x$$

$$\underline{3x^2 \cancel{-} 9x}$$

$$9x-27$$

$$\underline{9x \cancel{-} 27}$$

$$0$$

The answer: x^2+3x+9

103.

$$3x+2 \overline{\smash{)}\,3x^2-7x+10} \quad \dfrac{x-3}{}$$

$$\underline{3x^2 \cancel{+} 2x}$$

$$-9x+10$$

$$\underline{\cancel{-}9x \cancel{-} 6}$$

$$16$$

The answer: $x-3+\frac{16}{3x+2}$

104.

$$2x-3 \overline{\smash{)}\,4x^2+8x-10} \quad \dfrac{2x+7}{}$$

$$\underline{4x^2 \cancel{-} 6x}$$

$$14x-10$$

$$\underline{14x \cancel{-} 21}$$

$$11$$

The answer: $2x+7+\frac{11}{2x-3}$

105.

$$2x+1 \overline{\smash{)}\,2x^3-7x^2+6x+10} \quad \dfrac{x^2-4x+5}{}$$

$$\underline{2x^3 \cancel{+} x^2}$$

$$-8x^2+6x$$

$$\underline{\cancel{-}8x^2 \cancel{-} 4x}$$

$$10x+10$$

$$\underline{10x \cancel{+} 5}$$

$$5$$

The answer: $x^2-4x+5+\frac{5}{2x+1}$

106.

$$3x-1 \overline{\smash{)}\,3x^3-7x^2+14x-10} \quad \dfrac{x^2-2x+4}{}$$

$$\underline{3x^3 \cancel{-} x^2}$$

$$-6x^2+14x$$

$$\underline{\cancel{-}6x^2 \cancel{+} 2x}$$

$$12x-10$$

$$\underline{12x \cancel{-} 4}$$

$$-6$$

The answer: $x^2-2x+4+\frac{-6}{3x-1}$

107. $V=3x^3$ when length $=3x$, and height and width are equal to x.

108. Yes

109. Yes

110. Answers will vary

111. 3 because the solid rectangle was 3 times as long.

112. 24 with the length half as long it will fit eight times as many cubes as question 111.

CHAPTER 4 TEST

1. $(-3)^4 = (-3)(-3)(-3)(-3) = 81$

2. $\left(\dfrac{3}{4}\right)^2 = \left(\dfrac{3}{4}\right)\left(\dfrac{3}{4}\right) = \dfrac{9}{16}$

3. $\begin{aligned}(3x^3)^2(2x^4)^3 &= 3^2 \cdot x^{3 \cdot 2} \cdot 2^3 \cdot x^{4 \cdot 3}\\ &= 9x^6 \cdot 8x^{12}\\ &= 9 \cdot 8 x^{6+12}\\ &= 72x^{18}\end{aligned}$

4. $3^{-2} = \dfrac{1}{3^2} = \dfrac{1}{9}$

5. $(3a^4b^2)^0 = 1 \quad$ Definition $a^0 = 1 \quad (a \neq 0)$

6. $\begin{aligned}\dfrac{a^{-3}}{a^{-5}} &= a^{-3-(-5)}\\ &= a^{-3+5}\\ &= a^2\end{aligned}$

7. $\begin{aligned}\dfrac{(x^{-2})^3(x^{-3})^{-5}}{(x^{-4})^{-2}} &= \dfrac{x^{-2(3)}x^{-3(-5)}}{x^{-4(-2)}}\\ &= \dfrac{x^{-6}x^{15}}{x^8}\\ &= \dfrac{x^9}{x^8}\\ &= x\end{aligned}$

8. $0.0278 = 2.78 \times 10^{-2}$

9. $2.43 \times 10^5 = 2.43 \times 100,000 = 243,000$

10. $\begin{aligned}\dfrac{35x^2y^4z}{70x^6y^2z} &= \dfrac{35}{70} \cdot \dfrac{x^2}{x^6} \cdot \dfrac{y^4}{y^2} \cdot \dfrac{z}{z}\\ &= \dfrac{1}{2} \cdot \dfrac{1}{x^4} \cdot \dfrac{y^2}{1} \cdot 1\\ &= \dfrac{y^2}{2x^4}\end{aligned}$

11. $\dfrac{\left(6a^2b\right)\left(9a^3b^2\right)}{18a^4b^3} = \dfrac{54a^5b^3}{18a^4b^3}$

$\phantom{\dfrac{\left(6a^2b\right)\left(9a^3b^2\right)}{18a^4b^3}} = \dfrac{54}{18}\cdot\dfrac{a^5}{a^4}\cdot\dfrac{b^3}{b^3}$

$\phantom{\dfrac{\left(6a^2b\right)\left(9a^3b^2\right)}{18a^4b^3}} = \dfrac{3}{1}\cdot\dfrac{a}{1}\cdot 1$

$\phantom{\dfrac{\left(6a^2b\right)\left(9a^3b^2\right)}{18a^4b^3}} = 3a$

12. $\dfrac{24x^7}{3x^2} + \dfrac{14x^9}{7x^4} = 8x^5 + 2x^5 = 10x^5$

13. $\dfrac{\left(2.4\times10^5\right)\left(4.5\times10^{-2}\right)}{1.2\times10^{-6}} = \dfrac{(2.4)(4.5)}{1.2}\times\dfrac{10^5\cdot10^{-2}}{10^{-6}} = 9.0\times10^9$

14. $8x^2 - 4x + 6x + 2 = 8x^2 + 2x + 2$

15. $\left(5x^2 - 3x + 4\right) - \left(2x^2 - 7x - 2\right)$

$ = 5x^2 - 3x + 4 - 2x^2 + 7x + 2$

$ = \left(5x^2 - 2x^2\right) + (-3x + 7x) + (4 + 2)$

$ = 3x^2 + 4x + 6$

16. $(6x - 8) - (3x - 4) = 6x - 8 - 3x + 4$

$ = (6x - 3x) + (-8 + 4)$

$ = 3x - 4$

17. $2y^2 - 3y - 4$ when $y = -2$

$2(-2)^2 - 3(-2) - 4 = 2\cdot4 + 6 - 4$

$ = 8 + 6 - 4$

$ = 10$

18. $2a^2\left(3a^2 - 5a + 4\right) = 2a^2\left(3a^2\right) + 2a^2(-5a) + 2a^2(4)$

$ = 6a^4 - 10a^3 + 8a^2$

$$\textbf{F}\qquad\textbf{O}\qquad\textbf{I}\qquad\textbf{L}$$

19. $\left(x + \dfrac{1}{2}\right)\left(x + \dfrac{1}{3}\right) = x(x) + x\left(\dfrac{1}{3}\right) + x\left(\dfrac{1}{2}\right) + \left(\dfrac{1}{2}\right)\left(\dfrac{1}{3}\right)$

$ = x^2 + \dfrac{1}{3}x + \dfrac{1}{2}x + \dfrac{1}{6}$

$ = x^2 + \dfrac{5}{6}x + \dfrac{1}{6}\qquad \dfrac{1}{3} + \dfrac{1}{2} = \dfrac{2}{6} + \dfrac{3}{6} = \dfrac{5}{6}$

20. $(4x - 5)(2x + 3) = 4x(2x) + 4x(3) + (-5)(2x) + (-5)(3)$

$ = 8x^2 + 12x - 10x - 15$

$ = 8x^2 + 2x - 15$

21.
$$
\begin{array}{r}
x^2 + 3x + 9 \\
x - 3 \\
\hline
x^3 + 3x^2 + 9x \\
-3x^2 - 9x - 27 \\
\hline
x^3 \qquad\qquad\quad -27
\end{array}
$$

22. $(x + 5)^2 = (x + 5)(x + 5)$

$ = x^2 + 2\cdot5x + 5^2$

$ = x^2 + 10x + 25$

23. $(3a - 2b)^2 = (3a)^2 + 2(3a)(-2b) + (-2b)^2$

$ = 9a^2 - 12ab + 4b^2$

24. $(3x - 4y)(3x + 4y) = (3x)^2 - (4y)^2$
$$= 9x^2 - 16y^2$$

25. $(a^2 - 3)(a^2 + 3) = (a^2)^2 - 3^2$
$$= a^4 - 9$$

26. $\dfrac{10x^3 + 15x^2 - 5x}{5x} = \dfrac{10x^3}{5x} + \dfrac{15x^2}{5x} - \dfrac{5x}{5x}$
$$= 2x^2 + 3x - 1$$

27. $\dfrac{8x^2 - 6x - 5}{2x - 3}$

$$
\begin{array}{r}
4x + 3 \\
2x - 3 \overline{) 8x^2 - 6x - 5} \\
-\ + \\
\underline{8x^2 \diagup 12x} \\
6x - 5 \\
-\ + \\
\underline{6x \diagup 9} \\
4
\end{array}
$$

Our answer: $4x + 3 + \frac{4}{2x-3}$

28. $\dfrac{3x^3 - 2x + 1}{x - 3}$

$$
\begin{array}{r}
3x^2 + 9x + 25 \\
x - 3 \overline{) 3x^3 + 0x^2 - 2x + 1} \\
-\ + \\
\underline{3x^3 \diagup 9x^2} \\
9x^2 - 2x \\
-\ + \\
\underline{9x^2 \diagup 27x} \\
25x + 1 \\
-\ + \\
\underline{25x \diagup 75} \\
76
\end{array}
$$

Our answer: $3x^2 + 9x + 25 + \frac{76}{x-3}$

29. $V = s^3$

$V = (2.5)^3$

$V = (2.5)(2.5)(2.5)$

$V = 15.625 \text{ centimeters}^3$

30. $x = \text{width}$

$5x = \text{length}$

$\dfrac{x}{5} = \text{height}$

$V = W \cdot L \cdot H$

$V = x(5x)\left(\dfrac{x}{5}\right)$

$V = x^3$

CHAPTER 5

SECTION 5.1

1. $15x + 25 = 5 \cdot 3x + 5 \cdot 5$
   $= 5(3x + 5)$

5. $4x - 8y = 4(x) - 4(2y)$
   $= 4(x - 2y)$

9. $3a^2 - 3a - 60 = 3(a) - 3(20)$
   $= 3(a^2 - a - 20)$

13. $9x^2 - 8x^3 = x^2(9) - x^2(8x)$
    $= x^2(9 - 8x)$

17. $21x^2y - 28xy^2 = 7xy(3x) - 7xy(4y)$
    $= 7xy(3x - 4y)$

21. $7x^3 + 21x^2 - 28x = 7x(x^2) + 7x(3x) - 7x(4)$
    $= 7x(x^2 + 3x - 4)$

25. $100x^4 - 50x^3 + 25x^2 = 25x^2(4x^2) - 25x^2(2x) + 25x^2(1)$
    $= 25x^2(4x^2 - 2x + 1)$

29. $4a^2b - 16ab^2 + 32a^2b^2 = 4ab(a) - 4ab(4b) + 4ab(8ab)$
    $= 4ab(4a - 4b + 8ab)$

33. $12x^2y^3 - 72x^5y^3 - 36x^4y^4 = 12x^2y^3(1) - 12x^2y^3(6x^3) - 12x^2y^3(3x^2y)$
    $= 12x^2y^3(1 - 6x^3 - 3x^2y)$

37. $xy + 6x + 2y + 12 = x(y + 6) + 2(y + 6)$
    $= (y + 6)(x + 2)$

41. $ax - bx + ay - by = x(a - b) + y(a - b)$
    $= (a - b)(x + y)$

45. $3xb - 4b - 6x + 8 = b(3x - 4) - 2(3x - 4)$
    $= (b - 2)(3x - 4)$

49. $x^2 - ax - bx + ab = x(x - a) - b(x - a)$
    $= (x - b)(x - a)$

53. $6x^2 + 9x + 4x + 6$
    $= 3x(2x + 3) + 2(2x + 3)$
    $= (3x + 2)(2x + 3)$

57. $20x^2 + 4x + 25x + 5$
    $= 4x(5x + 1) + 5(5x + 1)$
    $= (4x + 5)(5x + 1)$

61. $6x^3 - 4x^2 + 15x - 10$

    $= 2x^2(3x - 2) + 5(3x - 2)$

    $= (3x - 2)(2x^2 + 5)$

65. $12x^2 + 6x + 3 = 3(4x^2 + 2x + 1)$

69. $(x + 3)(x + 4) = x \cdot x + x(4) + 3x + 3(4)$

                     F   O   I   L

               $= x^2 + 4x + 3x + 12$

               $= x^2 + 7x + 12$

73. $(x - 7)(x + 2) = x \cdot x + x(2) + (-7)x + (-7)(2)$

                     F   O   I   L

               $= x^2 + 2x - 7x - 14$

               $= x^2 - 5x - 14$

77. $\begin{array}{r} x^2 - 3x + 9 \\ x + 3 \\ \hline x^3 - 3x^2 + 9x \\ 3x^2 - 9x + 27 \\ \hline x^3 \qquad\qquad\qquad + 27 \end{array}$

SECTION 5.2

1. We need to find a pair of numbers whose sum is 7 and whose product is 12.

    $x^2 + 7x + 12 = (x + 3)(x + 4)$

5. We need to find a pair of numbers whose sum is 10 and whose product is 21.

    $a^2 + 10a + 21 = (a + 3)(a + 7)$

9. We need to find a pair of numbers whose sum is -10 and whose product is 21.

    $y^2 - 10y + 21 = (y - 3)(y - 7)$

13. We need to find a pair of numbers whose sum is 1 and whose product is -12.

    $y^2 + y - 12 = (y + 4)(y - 3)$

17. We need to find a pair of numbers whose sum is -8 and whose product is -9.

    $r^2 - 8r - 9 = (r - 9)(r + 1)$

21. We need to find a pair of numbers whose sum is 15 and whose product is 56.

    $a^2 + 15a + 56 = (a + 7)(a + 8)$

25. We need to find a pair of numbers whose sum is 13 and whose product is 42.

    $x^2 + 13x + 42 = (x + 6)(x + 7)$

29. $3a^2 - 3a - 60 = 3(a^2) - 3(a) - 3(20)$

$$= 3(a^2 - a - 20)$$

We need to find a pair of numbers whose sum is -1 and whose product is -20.

$$3(a^2 - a - 20) = 3(a+4)(a-5)$$

33. $100p^2 - 1300p + 4000 = 100(p^2) - 100(13p) + 100(40)$

$$= 100(p^2 - 13p + 40)$$

We need to find a pair of numbers whose sum is -13 and whose product is 40.

$$100(p^2 - 13p + 40) = 100(p-5)(p-8)$$

37. $2r^3 + 4r^2 - 30r = 2r(r^2) + 2r(2r) - 2r(15)$

$$= 2r(r^2 + 2r - 15)$$

We need to find a pair of numbers whose sum is 2 and whose product is -15.

$$2r(r^2 + 2r - 15) = 2r(r+5)(r-3)$$

41. $x^5 + 4x^4 + 4x^3 = x^3(x^2) + x^3(4x) + x^3(4)$

$$= x^3(x^2 + 4x + 4)$$

We need to find a pair of numbers whose sum is 4 and whose product is 4.

$$x^3(x^2 + 4x + 4) = x^3(x+2)(x+2)$$

45. $4x^4 - 52x^3 + 144x^2 = 4x^2(x^2) - 4x^2(13x) + 4x^2(36)$

$$= 4x^2(x^2 - 13x + 36)$$

We need to find a pair of numbers whose sum is -13 and whose product is 36.

$$4x^2(x^2 - 13x + 36) = 4x^2(x-4)(x-9)$$

49. We need to find a pair of numbers whose sum is -9 and whose product is 20.

$$x^2 - 9xy + 20y^2 = (x-4y)(x-5y)$$

53. We need to find a pair of numbers whose sum is -10 and whose product is 25.

$$a^2 - 10ab + 25b^2 = (a-5b)(a-5b)$$

57. We need to find a pair of numbers whose sum is 2 and whose product is -48.

$$x^2 + 2xa - 48a^2 = (x-6a)(x+8a)$$

61. $x^4 - 5x^2 + 6 = \left(x^2 - 3\right)\left(x^2 - 2\right)$

65. $x^2 - x + \dfrac{1}{4} = \left(x - \dfrac{1}{2}\right)\left(x - \dfrac{1}{2}\right)$ $\left(\dfrac{1}{2} \cdot \dfrac{1}{2} = \dfrac{1}{4}\right)$

69. $x^2 + 24x + 128 = (x + 8)(x + 16)$

The other factor is $x + 16$.

73. $h = 64 + 48t - 16t^2$

$$= 16\left(4 + 3t - t^2\right)$$

$$= 16(4 - t)(1 + t)$$

When $\quad\quad t = 4$

the equation $\quad h = 16(4 - t)(1 + t)$

becomes $\quad\quad h = 16(4 - 4)(1 + 4)$

$$= 16(0)(5)$$

$$= 0$$

When the time is 4 seconds, the height of the arrow is 0.

77. $(3a + 2)(2a + 1) = 3a(2a) + 3a(1) + 2(2a) + 2(1)$

$$\quad\quad\quad\quad\quad\quad \textbf{F}\quad\quad \textbf{O}\quad\quad \textbf{I}\quad\quad \textbf{L}$$

$$= 6a^2 + 3a + 4a + 2$$

$$= 6a^2 + 7a + 2$$

81. $\left(5x^2 + 5x - 4\right) - \left(3x^2 - 2x + 7\right) = 5x^2 + 5x - 4 - 3x^2 + 2x - 7$ Subtract

$$= 2x^2 + 7x - 11$$

85. $\left(5x^2 - 5\right) - \left(2x^2 - 4x\right) = 5x^2 - 5 - 2x^2 + 4x$ Subtract

$$= 3x^2 + 4x - 5$$

SECTION 5.3

1. Factor: $2x^2 + 7x + 3$

Possible Factors	Middle Term when Multiplied
$(2x - 1)(x - 3)$	$-7x$
$(2x + 1)(x + 3)$	$7x$
$(2x - 3)(x - 1)$	$-5x$
$(2x + 3)(x + 1)$	$5x$

$$2x^2 + 7x + 3 = (2x + 1)(x + 3)$$

5. Factor: $3x^2 + 2x - 5$

Possible Factors	Middle Term when Multiplied
$(3x+1)(x-5)$	$-14x$
$(3x+5)(x-1)$	$2x$
$(x+1)(3x-5)$	$-2x$
$(x+5)(3x-1)$	$14x$

$$3x^2 + 2x - 5 = (3x+5)(x-1)$$

9. Factor: $6x^2 + 13x + 6$

Possible Factors	Middle Term when Multiplied
$(6x+1)(x+6)$	$37x$
$(6x-1)(x-6)$	$-37x$
$(2x-3)(3x-2)$	$-13x$
$(2x+3)(3x+2)$	$13x$

$$6x^2 + 13x + 6 = (2x+3)(3x+2)$$

13. Factor: $4y^2 - 11y - 3$

Possible Factors	Middle Term when Multiplied
$(4y-1)(y+3)$	$11y$
$(4y+1)(y-3)$	$-11y$
$(4y+3)(y-1)$	$-y$
$(4y-3)(y+1)$	y
$(2y+1)(2y-3)$	$-4y$
$(2y-1)(2y+3)$	$4y$

$$4y^2 - 11y - 3 = (4y+1)(y-3)$$

17. Factor: $20a^2 + 48ab - 5b^2$

Possible Factors	Middle Term when Multiplied
$(20a-b)(a+5b)$	$99ab$
$(20a+b)(a-5b)$	$-99ab$
$(4a+5b)(5a-b)$	$21ab$
$(4a-5b)(5a+b)$	$-21ab$
$(10a+b)(2a-5b)$	$-48ab$
$(10a-b)(2a+5b)$	$48ab$

$$20a^2 + 48ab - 5b^2 = (10a-b)(2a+5b)$$

21. Factor: $12m^2 + 16m - 3$

Possible Factors	Middle Term when Multiplied
$(12m+1)(m-3)$	$-35m$
$(12m-1)(m+3)$	$35m$
$(3m+1)(4m-3)$	$-5m$
$(3m-1)(4m+3)$	$5m$
$(6m+1)(2m-3)$	$-16m$
$(6m-1)(2m+3)$	$16m$

$$12m^2 + 16m - 3 = (6m-1)(2m+3)$$

25. Factor: $12a^2 - 25ab + 12b^2$

Possible Factors	Middle Term when Multiplied
$(12a-b)(a-12b)$	$-145ab$
$(12a+b)(a+12b)$	$145ab$
$(3a+4b)(4a+3b)$	$25ab$
$(3a-4b)(4a-3b)$	$-25ab$

$$12a^2 - 25ab + 12b^2 = (3a-4b)(4a-3b)$$

29. Factor: $14x^2 + 29x - 15$

Possible Factors	Middle Term when Multiplied
$(x+1)(14x-15)$	$-x$
$(x-1)(14x+15)$	x
$(2x-5)(7x+3)$	$-29x$
$(2x+5)(7x-3)$	$29x$
$(2x-15)(7x+1)$	$-103x$
$(2x+15)(7x-1)$	$103x$
$(2x+3)(7x-5)$	$11x$
$(2x-3)(7x+5)$	$-11x$
$(2x+1)(7x-15)$	$-23x$
$(2x-1)(7x+15)$	$23x$
$(14x+3)(x-5)$	$-67x$
$(14x-3)(x+5)$	$67x$
$(14x+1)(x-15)$	$-209x$
$(14x-1)(x+15)$	$209x$
$(14x-5)(x+3)$	$37x$
$(14x+5)(x-3)$	$-37x$

$$14x^2 + 29x - 15 = (2x+5)(7x-3)$$

33. Factor: $15t^2 - 67t + 38$

Possible Factors	Middle Term when Multiplied
$(5t - 19)(3t - 2)$	$-67t$
$(5t + 19)(3t + 2)$	$67t$
$(5t - 2)(3t - 19)$	$-101t$
$(5t + 2)(3t + 19)$	$101t$
$(t + 1)(15t + 38)$	$53t$
$(t - 1)(15t - 38)$	$-53t$
$(t + 38)(15t + 1)$	$571t$
$(t - 38)(15t - 1)$	$-571t$
$(t + 2)(15t + 19)$	$49t$
$(t - 2)(15t - 19)$	$-49t$
$(t + 19)(15t + 2)$	$287t$
$(t - 19)(15t - 2)$	$-287t$
$(3t + 1)(5t + 38)$	$119t$
$(3t - 1)(5t - 38)$	$-119t$
$(3t + 38)(5t + 1)$	$193t$
$(3t - 38)(5t - 1)$	$-193t$

$$15t^2 - 67t + 38 = (5t - 19)(3t - 2)$$

37. Factor: $24a^2 - 50a + 24 = 2(12a^2 - 25a + 12)$

See possible factors in #25.

$$2(12a^2 - 25a + 12) = 2(4a - 3)(3a - 4)$$

41. Factor: $6x^4 - 11x^3 - 10x^2 = x^2(6x^2 - 11x - 10)$

Possible Factors	Middle Term when Multiplied
$(6x - 5)(x + 2)$	$7x$
$(6x + 5)(x - 2)$	$-7x$
$(3x - 2)(2x + 5)$	$11x$
$(3x + 2)(2x - 5)$	$-11x$
$(x - 10)(6x + 1)$	$-59x$
$(x + 10)(6x - 1)$	$59x$
$(2x + 1)(3x - 10)$	$-17x$
$(2x - 1)(3x + 10)$	$17x$

$$x^2(6x^2 - 11x - 10) = x^2(3x + 2)(2x - 5)$$

45. Factor: $15x^3 - 102x^2 - 21x = 3x(5x^2 - 34x - 7)$

Possible Factors	Middle Term when Multiplied
$(5x-1)(x+7)$	$34x$
$(5x+1)(x-7)$	$-34x$
$(5x+7)(x-1)$	$2x$
$(5x-7)(x+1)$	$-2x$

$3x(5x^2 - 34x - 7) = 3x(5x+1)(x-7)$

49. Factor: $15a^4 - 2a^3 - a^2 = a^2(15a^2 - 2a - 1)$

Possible Factors	Middle Term when Multiplied
$(3a+1)(5a-1)$	$2a$
$(3a-1)(5a+1)$	$-2a$
$(a-1)(15a+1)$	$-14a$
$(a+1)(15a-1)$	$14a$

$a^2(15a^2 - 2a - 1) = a^2(3a-1)(5a+1)$

53. Factor: $12x^2y - 34xy^2 + 14y^3 = 2y(6x^2 - 17xy + 7y^2)$

Possible Factors	Middle Term when Multiplied
$(2x-7y)(3x-1)$	$-23xy$
$(2x+7y)(3x+1)$	$23xy$
$(2x-y)(3x-7y)$	$-17xy$
$(2x+y)(3x+7y)$	$17xy$
$(6x-y)(x-7y)$	$-43xy$
$(6x+y)(x+7y)$	$43xy$
$(6x-7y)(x-y)$	$-13xy$
$(6x+7y)(x+y)$	$13xy$

$2y(6x^2 - 17xy + 7xy^2) = 2y(2x-y)(3x-7y)$

57. $(2x+3)(2x-3) = 4x^2 - 6x + 6x - 9$
$ = 4x^2 - 9$

61. $(x+3)(x-3) = x^2 - 3x + 3x - 9$
$ = x^2 - 9$

65. $(x+4)^2 = (x+4)(x+4)$
$ = x^2 + 4x + 4x + 16$
$ = x^2 + 8x + 16$

SECTION 5.4

1. $x^2 - 9 = (x)^2 - (3)^2 = (x+3)(x-3)$

5. $x^2 - 49 = (x)^2 - (7)^2 = (x+7)(x-7)$

9. $9x^2 + 25$ Cannot be factored

13. $9a^2 - 16b^2 = (3a+4b)(3a-4b)$
$$= (3a)^2 - (4b)^2$$

17. $25 - 4x^2 = (5)^2 - (2x)^2 = (5+2x)(5-2x)$

21. $32a^2 - 128 = 32(a^2 - 4)$
$$= 32(a^2 - 2^2)$$
$$= 32(a+2)(a-2)$$

25. $a^4 - b^4 = (a^2)^2 - (b^2)^2$
$$= (a^2 + b^2)(a^2 - b^2)$$
$$= (a^2 + b^2)(a+b)(a-b)$$

29. $3x^3y - 75xy^3 = 3xy(x^2 - 25y^2)$
$$= 3xy\left[x^2 - (5y)^2\right]$$
$$= 3xy(x+5y)(x-5y)$$

33. $x^2 + 2x + 1 = (x+1)(x+1)$
$$= (x+1)^2$$

37. $y^2 + 4y + 4 = (y+2)(y+2)$
$$= (y+2)^2$$

41. $m^2 - 12m + 36 = (m-6)(m-6)$
$$= (m-6)^2$$

45. $49x^2 - 14x + 1 = (7x-1)(7x-1)$
$$= (7x-1)^2$$

49. $x^2 + 10xy + 25y^2 = (x+5y)(x+5y)$
$$= (x+5y)^2$$

53. $3a^2 + 18a + 27 = 3(a^2 + 6a + 9)$
$$= 3(a+3)(a+3)$$
$$= 3(a+3)^2$$

57. $5x^3 + 30x^2y + 45xy^2 = 5x(x^2 + 6xy + 9y^2)$
$$= 5x(x+3y)(x+3y)$$
$$= 5x(x+3y)^2$$

61. $x^2 + 2xy + y^2 - 9$

$= (x^2 + 2xy + y^2) - 9$ Group first 3 terms together

$= (x+y)^2 - 9$ This has the form $a^2 - b^2$

$= [(x+y)+3][(x+y)-3]$ Factoring according to the formula
$$a^2 - b^2 = (a+b)(a-b)$$

$= (x+y+3)(x+y-3)$ Simplify

65. $x^2 + 10x + c = (x+5)^2$

   $x^2 + 10x + c = x^2 + 10x + 25$

   $c = 25$

69.
$$\begin{array}{r} 3x - 2 + \dfrac{9}{2x+3} \\ 2x+3 \overline{\smash{\big)}\, 6x^2 + 5x + 3} \\ \underline{--} \\ \underline{6x^2 \cancel{+} 9x} \\ -4x + 3 \\ ++ \\ \underline{\cancel{-}4x \cancel{-} 6} \\ 9 \end{array}$$

SECTION 5.5

1. $x^2 - 81 = (x+9)(x-9)$

5. $x^2 + 6x + 9 = (x+3)(x+3) = (x+3)^2$

9. $2a^3b + 6a^2b + 2ab = 2ab(a^2 + 3a + 1)$

13. $12a^2 - 75 = 3(4a^2 - 25) = 3(2a+5)(2a-5)$

17. $4x^3 + 16xy^2 = 4x(x^2 + 4y^2)$

21. $a^6 + 4a^4b^2 = a^4(a^2 + 4b^2)$

25. $x^4 - 16 = (x^2 + 4)(x^2 - 4) = (x^2 + 4)(x+2)(x-2)$

29. $5a^2 + 10ab + 5b^2 = 5(a^2 + 2ab + b^2)$

   $= 5(a+b)(a+b)$

   $= 5(a+b)^2$

33. $3x^2 + 15xy + 18y^2 = 3(x^2 + 5xy + 6y^2)$

   $= 3(x+2y)(x+3y)$

37. $100x^2 - 300x + 200 = 100(x^2 - 3x + 2)$

   $= 100(x-2)(x-1)$

41. $x^2 + 3x + ax + 3a = (x^2 + 3x) + (ax + 3a)$

   $= x(x+3) + a(x+3)$

   $= (x+a)(x+3)$

45. $49x^2 + 9y^2$ Cannot be factored

49. $xa - xb + ay - by = (xa - xb) + (ay - by)$

   $= x(a-b) + y(a-b)$

   $= (x+y)(a-b)$

53. $20x^4 - 45x^2 = 5x^2(4x^2 - 9)$

   $= 5x^2(2x+3)(2x-3)$

57. $16x^5 - 44x^4 + 30x^3 = 2x^3(8x^2 - 22x + 15)$

   $= 2x^3(2x-3)(4x-5)$

61. $y^4 - 1 = (y^2 + 1)(y^2 - 1)$

   $= (y^2 + 1)(y+1)(y-1)$

65. $3x - 6 = 9$

    $3x = 15$ Add 6 to both sides

     $x = 5$ Divide each side by 3

69. $4x + 3 = 0$

     $4x = -3$ Add -3 to both sides

     $x = -\dfrac{3}{4}$ Divide each side by 4

73. $\left(3x^3\right)^2\left(2x^4\right)^3 = 3^2 \cdot x^{3(2)} \cdot 2^3 \cdot x^{4(3)}$

$$= 9x^6 \cdot 8x^{12}$$

$$= 72x^{6+12}$$

$$= 72x^{18}$$

SECTION 5.6

1. $(x + 2)(x - 1) = 0$

    $x + 2 = 0$ or $x - 1 = 0$ Set factors to 0

       $x = -2$ $x = 1$ Solve the resulting equations

5. $x(x + 1)(x - 3) = 0$

    $x = 0$ or $x + 1 = 0$ or $x - 3 = 0$ Set factors to 0

    $x = 0$ $x = -1$ $x = 3$ Solve the resulting equations

9. $m(3m + 4)(3m - 4) = 0$

    $m = 0$ or $3m + 4 = 0$ or $3m - 4 = 0$ Set factors to 0

    $m = 0$ $3m = -4$ $3m = 4$ Solve the resulting equations

    $m = 0$ $m = -\dfrac{4}{3}$ $m = \dfrac{4}{3}$

13. $x^2 + 3x + 2 = 0$

    $(x + 2)(x + 1) = 0$ Factor the left side

    $x + 2 = 0$ or $x + 1 = 0$ Set factors to 0

       $x = -2$ $x = -1$ Solve the resulting equations

17. $a^2 - 2a - 24 = 0$

    $(x + 4)(x - 6) = 0$ Factor the left side

    $x + 4 = 0$ or $x - 6 = 0$ Set factors to 0

       $x = -4$ $x = 6$ Solve the resulting equations

21.
$$x^2 = -6x - 9$$

$x^2 + 6x + 9 = 0$ Standard form

$(x + 3)(x + 3) = 0$ Factor the left side

$x + 3 = 0$ or $x + 3 = 0$ Set factors to 0

$x = -3$ $x = -3$ Solve the resulting equations

25. $2x^2 + 5x - 12 = 0$

$(x + 4)(2x - 3) = 0$ Factor the left side

$x + 4 = 0$ or $2x - 3 = 0$ Set factors to 0

$x = -4$ $2x = 3$

$x = -4$ $x = \dfrac{3}{2}$ Solutions

29.
$$a^2 + 25 = 10a$$

$a^2 - 10a + 25 = 0$ Standard form

$(a - 5)(a - 5) = 0$ Factor left side

$a - 5 = 0$ or $a - 5 = 0$ Set factors to 0

$a = 5$ $a = 5$ Solutions

33.
$$3m^2 = 20 - 7m$$

$3m^2 + 7m - 20 = 0$ Standard form

$(m + 4)(3m - 5) = 0$ Factor left side

$m + 4 = 0$ or $3m - 5 = 0$ Set factors to 0

$m = -4$ $3m = 5$

$m = -4$ $m = \dfrac{5}{3}$ Solutions

37. $x^2 + 6x = 0$

$x(x + 6) = 0$ Factor left side

$x = 0$ or $x + 6 = 0$ Set factors to 0

$x = 0$ $x = -6$ Solutions

41.
$$2x^2 = 8x$$

$2x^2 - 8x = 0$ Standard form

$2x(x - 4) = 0$ Factor left side

$2x = 0$ or $x - 4 = 0$ Set factors to 0

$x = 0$ $x = 4$ Solutions

45.
$$1400 = 400 + 700x - 100x^2$$

$100x^2 - 700x + 1000 = 0$	Standard form
$100(x^2 - 7x + 10) = 0$	Factor 100 from each term on left side
$100(x - 2)(x - 5) = 0$	Factor left side
$x - 2 = 0$ or $x - 5 = 0$	Set factors to 0
$x = 2$ $x = 5$	Solutions

49.
$$x(2x - 3) = 20$$

$2x^2 - 3x = 20$	Distributive property
$2x^2 - 3x - 20 = 0$	Standard form
$(2x + 5)(x - 4) = 0$	Factor left side
$2x + 5 = 0$ or $x - 4 = 0$	Set factors to 0
$2x = -5$ $x = 4$	
$x = -\dfrac{5}{2}$ $x = 4$	Solutions

53.
$$4000 = (1300 - 100p)p$$
$$4000 = 1300p - 100p^2$$

$100p^2 - 1300p + 4000 = 0$	Standard form
$100(p^2 - 13p + 40) = 0$	Factor 100 from each term on left side
$100(p - 5)(p - 8) = 0$	Factor left side
$p - 5 = 0$ or $p - 8 = 0$	Set factors to 0
$p = 5$ $p = 8$	Solutions

57.
$$(x + 5)^2 = 2x + 9$$

$x^2 + 10x + 25 = 2x + 9$	Multiply left side
$x^2 + 8x + 16 = 0$	Standard form
$(x + 4)(x + 4) = 0$	Factor left side
$x + 4 = 0$ or $x + 4 = 0$	Set factors to 0
$x = -4$ $x = -4$	Solutions

61.
$$10^2 = (x + 2)^2 + x^2$$
$$100 = x^2 + 4x + 4 + x^2$$

$2x^2 + 4x - 96 = 0$	Standard form
$2(x^2 + 2x - 48) = 0$	Factor 2 from each term on left side
$2(x + 8)(x - 6) = 0$	Factor left side

$$x + 8 = 0 \quad \text{or} \quad x - 6 = 0 \qquad \text{Set factors to 0}$$
$$x = -8 \qquad\qquad x = 6 \qquad \text{Solutions}$$

65.
$$4y^3 - 2y - 30y = 0$$
$$y(4y^2 - 2y - 30) = 0 \qquad \text{Factor } y \text{ from each term on left side}$$
$$y(4y + 10)(y - 3) = 0 \qquad \text{Factor left side}$$
$$y = 0 \quad \text{or} \quad 4y + 10 = 0 \quad \text{or} \quad y - 3 = 0 \qquad \text{Set factors to 0}$$
$$y = 0 \qquad\qquad 4y = -10 \qquad\qquad y = 3$$
$$y = 0 \qquad\qquad y = -\frac{5}{2} \qquad\qquad y = 3 \qquad \text{Solutions}$$

69.
$$20a^3 = -18a^2 + 18a$$
$$20a^3 + 18a^2 - 18a = 0 \qquad \text{Standard form}$$
$$2a(10a^2 + 9a - 9) = 0 \qquad \text{Factor } 2a \text{ from each term on left side}$$
$$2a(2a + 3)(5a - 3) = 0 \qquad \text{Factor left side}$$
$$2a = 0 \quad \text{or} \quad 2a + 3 = 0 \quad \text{or} \quad 5a - 3 = 0 \qquad \text{Set factors to 0}$$
$$a = 0 \qquad\qquad 2a = -3 \qquad\qquad 5a = 3$$
$$a = 0 \qquad\qquad a = -\frac{3}{2} \qquad\qquad a = \frac{3}{5} \qquad \text{Solutions}$$

73.
$$x^3 + x^2 - 16x - 16 = 0$$
$$x^2(x + 1) - 16(x + 1) = 0 \qquad \text{Factor by grouping}$$
$$(x^2 - 16)(x + 1) = 0 \qquad \text{Factor the left side}$$
$$(x + 4)(x - 4)(x + 1) = 0 \qquad \text{Factor the left side}$$
$$x + 4 = 0 \quad \text{or} \quad x - 4 = 0 \quad \text{or} \quad x + 1 = 0 \qquad \text{Set factors to 0}$$
$$x = -4 \qquad\qquad x = 4 \qquad\qquad x = -1 \qquad \text{Solutions}$$

77. Step 1: The house and lot cost $3,000. The house cost four times the lot. We do not know the cost of the house or lot.

   Step 2 & 3: Lot $= x$. The house costs 4 times as much as the lot translates to $y = 4x$. A house and a lot cost $3000.00 translates to $x + y = 3000$.

   Step 4: Substitute
$$4x + x = 3000$$
$$x = \$600$$
$$y = 4x = \$2400$$

   Step 5: The lot cost $600 and the house cost $2400.00.

   Step 6: The lot ($600) plus the house ($2400) totals $3,000. The lot ($600) four times equals the house ($2400).

81. $\dfrac{x^5}{x^{-3}} = x^{5-(-3)} = x^8$

85. $0.0056 = 5.6 \times 10^{-3}$

SECTION 5.7

1. Let $x =$ the first even integer, then $x + 2 =$ the second even integer.

$$x(x+2) = 80$$
$$x^2 + 2x = 80$$
$$x^2 + 2x - 80 = 0$$
$$(x+10)(x-8) = 0$$

$$x + 10 = 0 \quad \text{or} \quad x - 8 = 0$$
$$x = -10 \qquad\qquad x = 8$$
$$x + 2 = -8 \qquad\quad x + 2 = 10$$

The solutions are $-10, -8,$ and $8, 10.$

5. Let $x =$ the first even integer, then $x + 2 =$ the second even integer.

$$x(x+2) = 5(x + x + 2) - 10$$
$$x^2 + 2x = 5(2x + 2) - 10$$
$$x^2 + 2x = 10x + 10 - 10$$
$$x^2 - 8x = 0$$
$$x(x-8) = 0$$

$$x = 0 \quad \text{or} \quad x - 8 = 0$$
$$x + 2 = 2 \qquad\qquad x = 8$$
$$\qquad\qquad\qquad x + 2 = 10$$

The solutions are $0, 2,$ and $8, 10.$

9. Let one number $= x$, then the other number is $5x + 2$.

$$x(5x + 2) = 24$$
$$5x^2 + 2x = 24$$
$$5x^2 + 2x - 24 = 0$$
$$(5x + 12)(x - 2) = 0$$

$$5x + 12 = 0 \quad \text{or} \quad x - 2 = 0$$
$$5x = -12 \qquad\qquad x = 2$$
$$x = \frac{-12}{5} \qquad 5x + 2 = 12$$
$$5x + 2 = -10$$

The solutions are $2, 12$ and $-\frac{12}{5}, -10.$

13. Let the width $= x$, then the length $= x + 1$

Area $=$ Length $\cdot$ Width

$$12 = (x+1) \cdot x$$
$$12 = x^2 + x$$
$$0 = x^2 + x - 12$$
$$0 = (x+4)(x-3)$$
$$0 = x+4 \quad \text{or} \quad 0 = x-3$$
$$-4 = x \qquad\qquad 3 = x$$

The solution $x = -4$ cannot be used, since the length and width are always given in positive units.

Width = 3 inches and Length = 4 inches.

17. Let $x =$ the shortest side, then $x + 2 =$ the middle side.

$$10^2 = x^2 + (x+2)^2$$
$$100 = x^2 + x^2 + 4x + 4$$
$$0 = 2x^2 + 4x - 96$$
$$0 = 2(x^2 + 2x - 48)$$
$$0 = 2(x+8)(x-6)$$
$$0 = x+8 \quad \text{or} \quad 0 = x-6$$
$$-8 = x \qquad\qquad 6 = x$$

The solution $x = -8$ cannot be used, since the length is always given in positive units. The shorter two sides of the triangle are 6 inches and 8 inches.

21. When $C = 1400$, $C = 400 + 700x - 100x^2$ becomes

$$1400 = 400 + 700x - 100x^2$$
$$100x^2 - 700x + 1000 = 0$$
$$100(x^2 - 7x + 10) = 0$$
$$100(x-2)(x-5) = 0$$
$$x - 2 = 0 \quad \text{or} \quad x - 5 = 0$$
$$x = 2 \qquad\qquad x = 5$$

The solutions are 2 hundred items and 5 hundred items.

25. When $R = 3200,$ $R = xp = (1200 - 100p)p$ becomes

$$(1200 - 100p)p = 3200$$
$$1200p - 100p^2 = 3200$$
$$0 = 100p^2 - 1200p + 3200$$
$$0 = 100(p^2 - 12p + 32)$$
$$0 = 100(p - 8)(p - 4)$$

$0 = p - 8$ or $0 = p - 4$

$8 = p$ $4 = p$

     The solutions are \$8 and \$4 for the price of the ribbons.

29. $(6a^2b)(7a^3b^2) = 42a^{2+3}b^{1+2}$

$$= 42a^5b^3$$

33. $\dfrac{(5x^4y^4)(10x^3y^3)}{2x^2y^7} = \dfrac{50x^{4+3}y^{4+3}}{2x^2y^7}$

$$= \dfrac{50x^7y^7}{2x^2y^7}$$
$$= 25x^{7-2}y^{7-7}$$
$$= 25x^5$$

37. $\dfrac{45a^6}{9a^3} - \dfrac{15a^8}{5a^5} = 5a^3 - 3a^3$

$$= 2a^3$$

CHAPTER 5 REVIEW

1. $10x - 20 = 10 \cdot x - 10 \cdot 2$

$$= 10(x - 2)$$

2. $4x^3 - 9x^2 = x^2 \cdot 4x - x^2 \cdot 9$

$$= x^2(4x - 9)$$

3. $5x - 5y = 5(x - y)$

4. $7x^3 + 2x = x \cdot 7x^2 + x \cdot 2$

$$= x(7x^2 + 2)$$

5. $8x + 4 = 4 \cdot 2x + 4 \cdot 1$

$$= 4(2x + 1)$$

6. $2x^2 + 14x + 6 = 2 \cdot x^2 + 2 \cdot 7x + 2 \cdot 3$

$$= 2(x^2 + 7x + 3)$$

7. $24y^2 - 40y + 48 = 8 \cdot 3y^2 - 8 \cdot 5y + 8 \cdot 6$

$$= 8(3y^2 - 5y + 6)$$

8. $18a^4 - 30a^3 = 6a^3 \cdot 3a - 6a^3 \cdot 5$

$$= 6a^3(3a - 5)$$

9. $30xy^3 - 45x^3y^2 = 15xy^2 \cdot 2y - 15xy^2 \cdot 3x^2$

$$= 15xy^2(2y - 3x^2)$$

10. $8y^4 - 20y^2 - 32y = 4y \cdot 2y^3 - 4y \cdot 5y - 4y \cdot 8$

$$= 4y(2y^3 - 5y - 8)$$

11. $49a^3 - 14b^3 = 7 \cdot 7a^3 - 7 \cdot 2b^3$
$\qquad = 7(7a^3 - 2b^3)$

12. $48x^5 - 36x^3 - 60x^2 = 12x^2 \cdot 4x^3 - 12x^2 \cdot 3x - 12x^2 \cdot 5$
$\qquad\qquad = 12x^2(4x^3 - 3x - 5)$

13. $6ab^2 + 18a^3b^3 - 24a^2b = 6ab \cdot b + 6ab \cdot 3a^2b^2 - 6ab \cdot 4a$
$\qquad\qquad = 6ab(b + 3a^2b^2 - 4a)$

14. $33a^4b^5 - 48a^5b^3 + 18a^4b^4 = 3a^4b^3 \cdot 11b^2 - 3a^4b^3 \cdot 16a + 3a^4b^3 \cdot 6b$
$\qquad\qquad = 3a^4b^3(11b^2 - 16a + 6b)$

15. $xy + bx + ay + ab = x(y + b) + a(y + b)$
$\qquad\qquad = (x + a)(y + b)$

16. $xy + 5x + ay + 5a = x(y + 5) + a(y + 5)$
$\qquad\qquad = (x + a)(y + 5)$

17. $xy + 4x - 5y - 20 = x(y + 4) - 5(y + 4)$
$\qquad\qquad = (x - 5)(y + 4)$

18. $xy - 4x + 5y - 20 = x(y - 4) + 5(y - 4)$
$\qquad\qquad = (x + 5)(y - 4)$

19. $2xy + 10x - 3y - 15 = 2x(y + 5) - 3(y + 5)$
$\qquad\qquad = (2x - 3)(y + 5)$

20. $3xy + 15x - 2y - 10 = 3x(y + 5) - 2(y + 5)$
$\qquad\qquad = (3x - 2)(y + 5)$

21. $5x^2 - 4ax - 10bx + 8ab = x(5x - 4a) - 2b(5x - 4a)$
$\qquad\qquad = (5x - 4a)(x - 2b)$

22. $15x^2 - 10bx - 3bx + 2b^2 = 5x(3x - 2b) - b(3x - 2b)$
$\qquad\qquad = (3x - 2b)(5x - b)$

23. $y^2 + 9y + 14$ The coefficient of y^2 is 1.
We need two numbers whose sum is 9 and whose product is 14.
The numbers are 2 and 7.
$\qquad y^2 + 9y + 14 = (y + 7)(y + 2)$

24. $w^2 + 15w + 50$ The coefficient of w^2 is 1.
We need two numbers whose sum is 15 and whose product is 50.
The numbers are 5 and 10.
$\qquad w^2 + 15w + 50 = (w + 5)(w + 10)$

25. $a^2 - 14a + 48$ The coefficient of a^2 is 1.

We need two numbers whose sum is -14 and whose product is 48.

The numbers are -6 and -8.

$$a^2 - 14a + 48 = (a - 6)(a - 8)$$

26. $r^2 + r - 20$ The coefficient of r^2 is 1.

We need two numbers whose sum is 1 and whose product is -20.

The numbers are -4 and 5

$$r^2 + r - 20 = (r - 4)(r + 5)$$

27. $r^2 - 18r + 72$ The coefficient of r^2 is 1.

We need two numbers whose sum is -18 and whose product is 72.

The numbers are -6 and -12.

$$r^2 - 18r + 72 = (r - 6)(r - 12)$$

28. $a^2 + 2a - 35$ The coefficient of a^2 is 1.

We need two numbers whose sum is 2 and whose product is -35.

The numbers are -5 and 7.

$$a^2 + 2a - 35 = (a + 7)(a - 5)$$

29. $y^2 + 20y + 99$ The coefficient of y^2 is 1.

We need two numbers whose sum is 20 and whose product is 99.

The numbers are 9 and 11.

$$y^2 + 20y + 99 = (y + 9)(y + 11)$$

30. $x^2 - 5x - 24$ The coefficient of x^2 is 1.

We need two numbers whose sum is -5 and whose product is -24.

The numbers are -8 and 3.

$$x^2 - 5x - 24 = (x - 8)(x + 3)$$

31. $y^2 + 8y + 12$ The coefficient of y^2 is 1.

We need two numbers whose sum is 8 and whose product is 12.

The numbers are 2 and 6.

$$y^2 + 8y + 12 = (y + 6)(y + 2)$$

32. $n^2 - 5n - 36$ The coefficient of n^2 is 1.

We need two numbers whose sum is -5 and whose product is -36.

The numbers are -9 and 4.

$$n^2 - 5n - 36 = (n-9)(n+4)$$

33. $2x^2 + 13x + 15$

Possible Factors	Middle Term when Multiplied
$(2x+3)(x+5)$	$13x$
$(2x-3)(x-5)$	$-13x$
$(2x+5)(x+3)$	$11x$
$(2x-5)(x-3)$	$-11x$
$(2x+1)(x+15)$	$31x$
$(2x-1)(x-15)$	$-31x$
$(2x+15)(x+1)$	$17x$
$(2x-15)(x-1)$	$-17x$

The complete problem is $2x^2 + 13x + 15 = (2x+3)(x+5)$

34. $4y^2 - 12y + 5$

Possible Factors	Middle Term when Multiplied
$(2y-5)(2y-1)$	$-12y$
$(2y+5)(2y+1)$	$12y$
$(4y-5)(y-1)$	$-9y$
$(4y+5)(y+1)$	$9y$
$(4y-1)(y-5)$	$-21y$
$(4y+1)(y+5)$	$21y$

The complete problem is $4y^2 - 12y + 15 = (2y-5)(2y-1)$

35. $5y^2 + 11y + 6$

Possible Factors	Middle Term when Multiplied
$(5y + 2)(y + 3)$	$17y$
$(5y - 2)(y - 3)$	$-17y$
$(y + 2)(5y + 3)$	$13y$
$(y - 2)(5y - 3)$	$-13y$
$(5y + 1)(y + 6)$	$31y$
$(5y - 1)(y - 6)$	$-31y$
$(5y + 6)(y + 1)$	$11y$
$(5y - 6)(y + 1)$	$-11y$

The complete problem is $5y^2 + 11y + 6 = (5y + 6)(y + 1)$

36. $8n^2 - 6mn - 5m^2$

Possible Factors	Middle Term when Multiplied
$(8n + m)(n - 5m)$	$-39mn$
$(8n - m)(n + 5m)$	$39mn$
$(n + m)(8n - 5m)$	$3mn$
$(n - m)(8n + 5m)$	$-3mn$
$(4n - 5m)(2n + m)$	$-6mn$
$(4n + 5m)(2n - m)$	$6mn$
$(4n + m)(2n - 5m)$	$-18mn$
$(4n - m)(2n + 5m)$	$18mn$

The complete problem is $8n^2 - 6mn - 5m^2 = (4n - 5m)(2n + m)$

37. $20a^2 - 27a + 9$

Possible Factors	Middle Term when Multiplied
$(a + 3)(20a + 3)$	$63a$
$(a - 3)(20a - 3)$	$-63a$
$(5a + 3)(4a + 3)$	$27a$
$(5a - 3)(4a - 3)$	$-27a$
$(2a + 3)(10a + 3)$	$36a$
$(2a - 3)(10a - 3)$	$-36a$
$(a + 1)(20a + 9)$	$29a$
$(a - 1)(20a - 9)$	$-29a$
$(20a + 1)(a + 9)$	$181a$
$(20a - 1)(a - 9)$	$-181a$
$(4a + 1)(5a + 9)$	$41a$
$(4a - 1)(5a - 9)$	$-41a$

$(5a+1)(5a-9)$	$49a$
$(5a-1)(4a-9)$	$-49a$
$(2a+1)(10a+9)$	$28a$
$(2a-1)(10a-9)$	$-28a$
$(10a+1)(2a+9)$	$92a$
$(10a-1)(2a-9)$	$-92a$

The complete problem is $20a^2 - 27a + 9 = (5a-3)(4a-3)$

38. $14x^2 + 31xy - 10y^2$

Possible Factors	Middle Term when Multiplied
$(7x+10)(2x-y)$	$13xy$
$(7x-10)(2x+y)$	$-13xy$
$(7x+2y)(2x-5y)$	$-31xy$
$(7x-2y)(2x+5y)$	$31xy$
$(x-2y)(14x+5y)$	$-23xy$
$(x+2y)(14x-5y)$	$23xy$
$(x-10y)(14x+1)$	$-139xy$
$(x+10y)(14x-1)$	$139xy$

The complete problem is $14x^2 + 31xy - 10y^2 = (7x-2y)(2x+5y)$

39. $6r^2 + 5rt - 6t^2$

Possible Factors	Middle Term when Multiplied
$(2r+3t)(3r-2t)$	$5rt$
$(2r-3t)(3r+2t)$	$-5rt$
$(r+6t)(6r-t)$	$35rt$
$(r-6t)(6r+t)$	$-35rt$

The complete problem is $6r^2 + 5rt - 6t^2 = (2r+3t)(3r-2t)$

40. $20y^2 - 36y + 9$

Possible Factors	**Middle Term when Multiplied**
$(y+1)(20y+9)$	$29y$
$(y-1)(20y-9)$	$-29y$
$(y+9)(20y+1)$	$181y$
$(y-9)(20y-1)$	$-181y$
$(4y-3)(5y-3)$	$-27y$
$(4y+3)(5y+3)$	$27y$
$(2y-3)(10y-3)$	$-36y$
$(2y+3)(10y+3)$	$36y$
$(y-3)(20y-3)$	$-63y$
$(y+3)(20y+3)$	$63y$
$(4y-1)(5y-9)$	$-41y$
$(4y+1)(5y+9)$	$41y$
$(4y-9)(5y-1)$	$-49y$
$(4y+9)(5y+1)$	$49y$
$(2y-1)(10y-9)$	$-28y$
$(2y+1)(10y+9)$	$28y$
$(2y-9)(10y-1)$	$-92y$
$(2y+9)(10y+1)$	$92y$

The complete problem is $20y^2 - 36y + 9 = (2y-3)(10y-3)$

41. $10x^2 - 29x - 21$

Possible Factors	**Middle Term when Multiplied**
$(2x+1)(5x-21)$	$-37x$
$(2x-1)(5x+21)$	$37x$
$(2x+21)(5x-1)$	$103x$
$(2x-21)(5x+1)$	$-103x$
$(2x+3)(5x-7)$	x
$(2x-3)(5x+7)$	$-x$
$(2x+7)(5x-3)$	$29x$
$(2x-7)(5x+3)$	$-29x$

$(x+7)(10x-3)$	$67x$
$(x-7)(10x+3)$	$-67x$
$(x+3)(10x-7)$	$23x$
$(x-3)(10x+7)$	$-23x$
$(x+1)(10x-21)$	$-11x$
$(x-1)(10x+21)$	$11x$
$(x+21)(10x-1)$	$209x$
$(x-21)(10x+1)$	$-209x$

The complete problem is $10x^2 - 29x - 21 = (2x-7)(5x+3)$

42. $12x^2 - 11xy - 15y^2$

Possible Factors	Middle Term when Multiplied
$(3x+5y)(4x-3y)$	$11xy$
$(3x-5y)(4x+3y)$	$-11xy$
$(2x+3y)(6x-5y)$	$8xy$
$(2x-3y)(6x+5y)$	$-8xy$
$(3x+y)(4x-15y)$	$-41xy$
$(3x-y)(4x+15y)$	$41xy$
$(x-15y)(12x+y)$	$-179xy$
$(x+15y)(12x-y)$	$179xy$
$(x+3y)(12x-5y)$	$31xy$
$(x-3y)(12x+5y)$	$-31xy$
$(2x-15y)(6x+y)$	$-88xy$
$(2x+15y)(6x-y)$	$88xy$

The complete problem is $12x^2 - 11xy - 15y^2 = (3x-5y)(4x+3y)$

43. $n^2 - 81 = (n+9)(n-9)$

44. $4y^2 - 9 = (2y+3)(2y-3)$

45. $x^2 + 49$: Prime - Cannot be factored

46. $4r^2 - 9t^2 = (2r+3t)(2r-3t)$

47. $36y^2 - 121x^2 = (6y+11x)(6y-11x)$

48. $25 - y^2 = (5+y)(5-y)$

49. $64a^2 - 121b^2 = (8a+11b)(8a-11b)$

50. $y^4 - 81 = (y^2+9)(y^2-9)$
$= (y^2+9)(y+3)(y-3)$

51. $64 - 9m^2 = (8+3m)(8-3m)$

52. $16 - a^4 = (4+a^2)(4-a^2)$
$= (4+a^2)(2+a)(2-a)$

53. $y^2 + 20y + 100 = (y + 10)(y + 10) = (y + 10)^2$

54. $m^2 - 16m + 64 = (m - 8)(m - 8) = (m - 8)^2$

55. $64t^2 + 16t + 1 = (8t + 1)(8t + 1) = (8t + 1)^2$

56. $49x^2 - 14x + 1 = (7x - 1)(7x - 1) = (7x - 1)^2$

57. $16n^2 - 24n + 9 = (4n - 3)(4n - 3) = (4n - 3)^2$

58. $49w^2 + 112w + 64 = (7w + 8)(7w + 8) = (7w + 8)^2$

59. $4r^2 - 12rt + 9t^2 = (2r - 3t)(2r - 3t) = (2r - 3t)^2$

60. $16x^2 + 72xy + 81y^2 = (4x + 9y)(4x + 9y) = (4x + 9y)^2$

61. $9m^2 + 30mn + 25n^2 = (3m + 5n)(3m + 5n) = (3m + 5n)^2$

62. $25x^2 - 30xy + 9y^2 = (5x - 3y)(5x - 3y) = (5x - 3y)^2$

63. $2x^2 + 20x + 48 = 2(x^2 + 10x + 24) = 2(x + 4)(x + 6)$

64. $a^3 - 10a^2 + 21a = a(a^2 - 10a + 21) = a(a - 3)(a - 7)$

65. $3m^3 - 18m^2 - 21m = 3m(m^2 - 6m - 7) = 3m(m + 1)(m - 7)$

66. $x^5 - x^4 - 30x^3 = x^3(x^2 - x - 30) = x^3(x - 6)(x + 5)$

67. $5y^4 + 10y^3 - 40y^2 = 5y^2(y^2 + 2y - 8) = 5y^2(y - 2)(y + 4)$

68. $6y^4 + 30y^3 - 36y^2 = 6y^2(y^2 + 5y - 6) = 6y^2(y + 6)(y - 1)$

69. $8x^2 + 16x + 6 = 2(4x^2 + 8x + 3) = 2(2x + 1)(2x + 3)$

70. $3a^3 - 14a^2 - 5a = a(3a^2 - 14a - 5) = a(3a + 1)(a - 5)$

71. $20m^3 - 34m^2 + 6m = 2m(10m^2 - 17m + 3) = 2m(2m - 3)(5m - 1)$

72. $18x^3 - 3x^2y - 3xy^2 = 3x(6x^2 - xy - y^2) = 3x(3x + y)(2x - y)$

73. $30x^2y - 55xy^2 + 15y^3 = 5y(6x^2 - 11xy + 3y^2) = 5y(2x - 3y)(3x - y)$

74. $32x^2y - 56xy^2 + 12y^3 = 4y(8x^2 - 14xy + 3y^2) = 4y(4x - y)(2x - 3y)$

75. $4x^2 + 40x + 100 = 4(x^2 + 10x + 25) = 4(x + 5)(x + 5) = 4(x + 5)^2$

76. $4x^3 + 12x^2 + 9x = x(4x^2 + 12x + 9) = x(2x + 3)(2x + 3) = x(2x + 3)^2$

77. $5x^2 - 45 = 5(x^2 - 9) = 5(x + 3)(x - 3)$

78. $45x^2y - 30xy^2 + 5y^3 = 5y(9x^2 - 6xy + y^2) = 5y(3x - y)(3x - y) = 5y(3x - y)^2$

79. $12x^3 - 27xy^2 = 3x(4x^2 - 9y^2) = 3x(2x + 3y)(2x - 3y)$

80. $80x^3 - 5xy^2 = 5x(16x^2 - y^2) = 5x(4x + y)(4x - y)$

81. $6a^3b + 33a^2b^2 + 15ab^3 = 3ab(2a^2 + 11ab + 5b^2) = 3ab(2a + b)(a + 5b)$

82. $x^5 - x^3 = x^3(x^2 - 1) = x^3(x + 1)(x - 1)$ 83. $4y^6 + 9y^4 = y^4(4y^2 + 9)$

84. $12x^5 + 20x^4y - 8x^3y^2 = 4x^3(3x^2 + 5xy - 2y^2) = 4x^3(x + 2y)(3x - y)$

85. $30a^4b + 35a^3b^2 - 15a^2b^3 = 5a^2b(6a^2 + 7ab - 3b^2) = 5a^2b(3a - b)(2a + 3b)$

86. $18a^3b^2 + 3a^2b^3 - 6ab^4 = 3ab^2(6a^2 + ab - 2b^2) = 3ab^2(2a - b)(3a + 2b)$

87. $(x - 5)(x + 2) = 0$

    $x - 5 = 0$ or $x + 2 = 0$ Set each factor to 0

       $x = 5$ $x = -2$ Solutions

88. $x(3x + 2)(x - 4) = 0$

    $x = 0$ or $3x + 2 = 0$ or $x - 4 = 0$ Set each factor to 0

    $x = 0$ $3x = -2$ $x = 4$ Solutions

$$x = -\frac{2}{3}$$

89. $3(2y+5)(2y-5)=0$

$2y+5=0 \quad$ or $\quad 2y-5=0 \qquad$ Set each factor to 0

$\quad 2y=-5 \qquad\qquad 2y=5$

$\quad y=-\dfrac{5}{2} \qquad\qquad y=\dfrac{5}{2} \qquad$ Solutions

90. $4m(m-7)(2m-7)=0$

$4m=0 \quad$ or $\quad m-7=0 \quad$ or $\quad 2m-7=0 \qquad$ Set each factor to 0

$\quad m=0 \qquad\qquad m=7 \qquad\qquad 2m=7$

$\qquad\qquad\qquad\qquad\qquad\qquad\qquad\qquad m=\dfrac{7}{2} \qquad$ Solutions

91. $\qquad m^2+3m=10$

$m^2+3m-10=0 \qquad$ Add -10 to each side

$(m+5)(m-2)=0 \qquad$ Factor completely

$m+5=0 \quad$ or $\quad m-2=0 \qquad$ Set each factor to 0

$\quad m=-5 \qquad\qquad m=2 \qquad$ Solutions

92. $\qquad 3x^2=7x+20$

$3x^2-7x-20=0 \qquad$ Add $-7x-20$ to each side

$(3x+5)(x-4)=0 \qquad$ Factor completely

$3x+5=0 \quad$ or $\quad x-4=0 \qquad$ Set each factor to 0

$\quad 3x=-5 \qquad\qquad x=4 \qquad$ Solutions

$\quad x=-\dfrac{5}{3}$

93. $\qquad a^2-49=0$

$(a+7)(a-7)=0 \qquad$ Factor completely

$a+7=0 \quad$ or $\quad a-7=0 \qquad$ Set each factor to 0

$\quad a=-7 \qquad\qquad a=7 \qquad$ Solutions

94. $\qquad 16x^2-81=0$

$(4x+9)(4x-9)=0 \qquad$ Factor completely

$4x+9=0 \quad$ or $\quad 4x-9=0 \qquad$ Set each factor to 0

$\quad 4x=-9 \qquad\qquad 4x=9$

$\quad x=-\dfrac{9}{4} \qquad\qquad x=\dfrac{9}{4} \qquad$ Solutions

95. $m^2 - 9m = 0$

    $m(m - 9) = 0$ Factor completely

   $m = 0$ or $m - 9 = 0$ Set each factor to 0

                    $m = 9$ Solutions

96. $98r^2 - 18 = 0$

       $2(49r^2 - 9) = 0$

   $2(7r + 3)(7r - 3) = 0$ Factor completely

  $7r + 3 = 0$ or $7r - 3 = 0$ Set each factor to 0

    $7r = -3$ $7r = 3$

    $r = -\dfrac{3}{7}$ $r = \dfrac{3}{7}$ Solutions

97. $6y^2 = -13y - 6$

    $6y^2 + 13y + 6 = 0$ Add $13y + 6$ to each side

  $(3y + 2)(2y + 3) = 0$ Factor completely

  $3y + 2 = 0$ or $2y + 3 = 0$ Set each factor to 0

    $3y = -2$ $2y = -3$

    $y = -\dfrac{2}{3}$ $y = -\dfrac{3}{2}$ Solutions

98. $5x^2 = -15x$

   $5x^2 + 15x = 0$ Add $15x$ to each side

   $5x(x + 3) = 0$ Factor completely

  $5x = 0$ or $x + 3 = 0$ Set each factor to 0

   $x = 0$ $x = -3$ Solutions

99. $9x^4 + 9x^3 = 10x^2$

   $9x^4 + 9x^3 - 10x^2 = 0$ Add $-10x^2$ to each side

  $x^2(9x^2 + 9x - 10) = 0$

  $x^2(3x + 5)(3x - 2) = 0$ Factor completely

  $x^2 = 0$ or $3x + 5 = 0$ or $3x - 2 = 0$ Set each factor to 0

   $x = 0$ $3x = -5$ $3x = 2$

               $x = -\dfrac{5}{3}$ $x = \dfrac{2}{3}$ Solutions

100.
$$6x^4 = 33x^3 - 42x^2$$

$$6x^4 - 33x^3 + 42x^2 = 0 \qquad \text{Add } -33x^3 + 42x^2 \text{ to each side}$$

$$3x^2(2x^2 - 11x + 14) = 0$$

$$3x^2(x-2)(2x-7) = 0 \qquad \text{Factor completely}$$

$3x^2 = 0$ or $x - 2 = 0$ or $2x - 7 = 0$ Set each factor to 0

    $x = 0$ $x = 2$ $2x = 7$

$$x = \frac{7}{2} \quad \text{Solutions}$$

101. Let x = the first even integer, then $x + 2$ = the second even integer.

$$x(x+2) = 120 \qquad \text{Their product is 120}$$

$$x^2 + 2x = 120 \qquad \text{Distributive property}$$

$$x^2 + 2x - 120 = 0 \qquad \text{Add } -120 \text{ to each side}$$

$$(x-10)(x+12) = 0 \qquad \text{Factor completely}$$

$x - 10 = 0$ or $x + 12 = 0$ Set each factor to 0

    $x = 10$ $x = -12$ First even integers

  $x + 2 = 12$ $x + 2 = -10$ Second even integers

The solutions are 10, 12 and $-12, -10$.

102. Let x = the first integer, then $x + 1$ = the second integer.

$$x(x+1) = 110 \qquad \text{Their product is 110}$$

$$x^2 + x = 110 \qquad \text{Distributive property}$$

$$x^2 + x - 110 = 0 \qquad \text{Add } -110 \text{ to each side}$$

$$(x+11)(x-10) = 0 \qquad \text{Factor completely}$$

$x + 11 = 0$ or $x - 10 = 0$ Set each factor to 0

    $x = -11$ $x = 10$ First integers

  $x + 1 = -10$ $x + 1 = 11$ Second integers

The solutions are $-11, -10$, and 10, 11.

103. Let x = the first odd integer, then $x + 2$ = the second odd integer.

$$x(x+2) = 3[x+(x+2)] - 1$$

$$x^2 + 2x = 6x + 6 - 1 \qquad \text{Distributive property}$$

$$x^2 - 4x - 5 = 0 \qquad \text{Add } -6x - 5 \text{ to each side}$$

$$(x+1)(x-5) = 0 \qquad \text{Factor completely}$$

$$x + 1 = 0 \quad \text{or} \quad x - 5 = 0 \qquad \text{Set each factor to 0}$$
$$x = -1 \qquad\qquad x = 5 \qquad\qquad \text{First odd integers}$$
$$x + 2 = 1 \qquad\qquad x + 2 = 7 \qquad\qquad \text{Second odd integers}$$

The solutions are $-1, \ 1$ and $5, \ 7.$

104. Let $x =$ the first number, then $20 - x =$ the second number.

$$x(20 - x) = 75 \qquad \text{Their product is 75}$$
$$20x - x^2 = 75 \qquad \text{Distributive property}$$
$$x^2 - 20x + 75 = 0 \qquad \text{Add } x^2 - 20x \text{ to each side}$$
$$(x - 5)(x - 15) = 0 \qquad \text{Factor completely}$$

$$x - 5 = 0 \quad \text{or} \quad x - 15 = 0 \qquad \text{Set each factor to 0}$$
$$x = 5 \qquad\qquad x = 15 \qquad \text{First number}$$
$$20 - x = 15 \qquad 20 - x = 5 \qquad \text{Second numbers}$$

The two numbers are 5 and 15.

105. Let $x =$ one number, then $2x - 1 =$ another number

$$x(2x - 1) = 66 \qquad \text{Their product is 66}$$
$$2x^2 - x = 66 \qquad \text{Distributive property}$$
$$2x^2 - x - 66 = 0 \qquad \text{Add } -66 \text{ to each side}$$
$$(2x + 11)(x - 6) = 0 \qquad \text{Factor completely}$$

$$2x + 11 = 0 \quad \text{or} \quad x - 6 = 0 \qquad \text{Set each factor to 0}$$
$$x = -\frac{11}{2} \qquad\qquad x = 6 \qquad \text{one number}$$
$$2x - 1 = -12 \qquad 2x - 1 = 11 \qquad \text{another number}$$

The two numbers are $-\frac{11}{2}, \ -12$ and $6, \ 11.$

106. Let $x =$ the width, then $5x + 1 =$ the length. Remember $A = LW$.

$$48 = (5x + 1)(x)$$
$$48 = 5x^2 + x \qquad \text{Distributive property}$$
$$5x^2 + x - 48 = 0 \qquad \text{Add } -48 \text{ to each side}$$
$$(5x + 16)(x - 3) = 0 \qquad \text{Factor completely}$$

$$5x + 16 = 0 \quad \text{or} \quad x - 3 = 0 \qquad \text{Set each factor to 0}$$
$$5x = -16 \qquad\qquad x = 3 \qquad \text{inches (width)}$$
$$x = -\frac{16}{5} \qquad 5x + 1 = 16 \qquad \text{inches(length)}$$

$\uparrow$ (Width cannot be negative)

107. Let x = the base, then $8x$ = the height. Remember $A = \frac{1}{2}bh$.

$$16 = \frac{1}{2}(x)(8x)$$

$16 = 4x^2$ Multiply

$4x^2 - 16 = 0$ Add -16 to each side

$4(x^2 - 4) = 0$

$4(x + 2)(x - 2) = 0$ Factor completely

$x + 2 = 0$ or $x - 2 = 0$ Set each factor to 0

   $x = -2$ $x = 2$ inches (base)

   ↑ a base cannot be negative

108. Let 6 = shorter leg, then x = longer leg and $x + 2$ = hypotenuse. Remember $a^2 + b^2 = c^2$.

$6^2 + x^2 = (x + 2)^2$

$36 + x^2 = x^2 + 4x + 4$

   $36 = 4x + 4$ Add $-x^2$ to each side

   $32 = 4x$ Add -4 to each side

     $x = 8$ meters (longer leg)

CHAPTER 5 TEST

1. $5x - 10 = 5(x - 2)$

2. $18x^2y - 9xy - 36xy^2 = 9xy(2x - 1 - 4y)$

3. $x^2 + 2ax - 3bx - 6ab = (x^2 + 2ax) + (-3bx - 6ab)$
$$= x(x + 2a) - 3b(x + 2a)$$
$$= (x + 2a)(x - 3b)$$

4. $xy + 4x - 7y - 28 = (xy + 4x) + (-7y - 28)$
$$= x(y + 4) - 7(y + 4)$$
$$= (x - 7)(y + 4)$$

5. $x^2 - 5x + 6 = (x - 2)(x - 3)$

6. $x^2 - x - 6 = (x - 3)(x + 2)$

7. $a^2 - 16 = (a - 4)(a + 4)$

8. $x^2 + 25$ cannot be factored

9. $x^4 - 81 = (x^2 + 9)(x^2 - 9) = (x^2 + 9)(x - 3)(x + 3)$

10. $27x^2 - 75y^2 = 3(9x^2 - 25y^2)$
$$= 3(3x - 5y)(3x + 5y)$$

11. $x^3 + 5x^2 - 9x - 45 = (x^3 + 5x^2) + (-9x - 45)$

$\qquad\qquad\qquad\quad = x^2(x+5) - 9(x+5)$

$\qquad\qquad\qquad\quad = (x+5)(x^2 - 9)$

$\qquad\qquad\qquad\quad = (x+5)(x+3)(x-3)$

12. $x^2 - bx + 5x - 5b = (x^2 - bx) + (5x - 5b)$

$\qquad\qquad\qquad\quad = x(x-b) + 5(x-b)$

$\qquad\qquad\qquad\quad = (x-b)(x+5)$

13. $4a^2 + 22a + 10 = 2(2a^2 + 11a + 5)$

$\qquad\qquad\qquad = 2(2a+1)(a+5)$

14. $3m^2 - 3m - 18 = 3(m^2 - m - 6)$

$\qquad\qquad\qquad = 3(m-3)(m+2)$

15. $6y^2 + 7y - 5 = (2y-1)(3y+5)$

16. $12x^3 - 14x^2 - 10x = 2x(6x^2 - 7x - 5)$

$\qquad\qquad\qquad\qquad = 2x(2x+1)(3x-5)$

17. $x^2 + 7x + 12 = 0$

$\quad (x+3)(x+4) = 0 \quad$ Factor the left side

$\quad x+3 = 0 \quad$ or $\quad x+4 = 0 \qquad$ Set factors to 0

$\qquad x = -3 \qquad\qquad x = -4 \qquad$ Solutions

18. $x^2 - 4x + 4 = 0$

$\quad (x-2)(x-2) = 0 \quad$ Factor the left side

$\quad x-2 = 0 \quad$ or $\quad x-2 = 0 \qquad$ Set factors to 0

$\qquad x = 2 \qquad\qquad$ same $\qquad$ Solutions

19. $x^2 - 36 = 0$

$\quad (x+6)(x-6) = 0 \qquad$ Factor the left side

$\quad x+6 = 0 \quad$ or $\quad x-6 = 0$

$\qquad x = -6 \qquad\qquad x = 6 \quad$ Solutions

20. $x^2 = x + 20$

$\quad x^2 - x - 20 = 0 \qquad$ Standard form

$\quad (x+4)(x-5) = 0 \qquad$ Factor the left side

$\quad x+4 = 0 \quad$ or $\quad x-5 = 0 \quad$ Set factors to 0

$\qquad x = -4 \qquad\qquad x = 5 \quad$ Solutions

21. $x^2 - 11x = -30$

$\quad x^2 - 11x + 30 = 0 \qquad$ Standard form

$\quad (x-5)(x-6) = 0 \qquad$ Factor the left side

$\quad x-5 = 0 \quad$ or $\quad x-6 = 0 \quad$ Set factors to 0

$\qquad x = 5 \qquad\qquad x = 6 \quad$ Solutions

22. $y^3 = 16y$

$\quad y^3 - 16y = 0 \qquad$ Standard form

$\quad y(y^2 - 16) = 0 \qquad$ Factor a "y" from each term on the left side

$\quad y(y+4)(y-4) = 0 \qquad$ Factor the left side

$\quad y = 0 \quad$ or $\quad y+4 = 0 \quad$ or $\quad y-4 = 0 \quad$ Set factors to 0

$\qquad\qquad\qquad y = -4 \qquad\qquad y = 4 \quad$ Solutions

23.
$$2a^2 = a + 15$$

$2a^2 - a - 15 = 0$ Standard form

$(2a + 5)(a - 3) = 0$ Factor left side

$2a + 5 = 0$ or $a - 3 = 0$ Set factors to 0

$2a = -5$ $a = 3$ Solutions

$a = -\dfrac{5}{2}$

24.
$$30x^3 - 20x^2 = 10x$$

$30x^3 - 20x^2 - 10x = 0$ Standard form

$10x(3x^2 - 2x - 1) = 0$ Factor $10x$ from each term on the left side

$10x(3x + 1)(x - 1) = 0$ Factor left side

$10x = 0$ or $3x + 1 = 0$ or $x - 1 = 0$ Set factors to 0

$x = 0$ $3x = -1$ $x = 1$ Solutions

$x = -\dfrac{1}{3}$

25. If we let x represent one of the numbers, then $20 - x$ must be the other number because their sum is 20. Since their product is 64, we can write -

$$x(20 - x) = 64$$

$$20x - x^2 = 64$$

$$0 = x^2 - 20x + 64$$

$x^2 - 20x + 64 = 0$ Standard form

$(x - 4)(x - 16) = 0$ Factor left side

$x - 4 = 0$ or $x - 16 = 0$ Set factors to 0

$x = 4$ $x = 16$ Solutions

$20 - x = 16$ $20 - x = 4$

The two numbers are 4 and 16.

26. Let $x =$ the first of two consecutive odd integers, then $x + 2 =$ the second of the two consecutive odd integers.

$$x(x + 2) = (x + x + 2) + 7$$

$$x^2 + 2x = 2x + 9$$

$x^2 - 9 = 0$ Standard form

$(x + 3)(x - 3) = 0$ Factor left side

$x + 3 = 0$ or $x - 3 = 0$ Set factors to 0

$x = -3$ $x = 3$ Solutions

$x + 2 = -1$ $x + 2 = 5$

We have two pairs of consecutive odd integers that are solutions. They are -3, -1 and 3, 5.

27. Let x = width and $3x + 5$ = length. Area = length · width.

$$42 = (3x + 5)x$$
$$42 = 3x^2 + 5x$$
$$0 = 3x^2 + 5x - 42$$

$$3x^2 + 5x - 42 = 0 \quad \text{Standard form}$$
$$(x - 3)(3x + 14) = 0 \quad \text{Factor left side}$$

$$x - 3 = 0 \quad \text{or} \quad 3x + 14 = 0 \qquad \text{Set factors to 0}$$
$$x = 3 \qquad\qquad 3x = -14 \qquad \text{Solutions}$$
$$x = -\frac{14}{3}$$

The solution $x = -\frac{14}{3}$ cannot be used, since the length and width are always given in positive units.

Width = 3 feet and Length = $3x + 5$ = 14 feet

28. If we let x = the length of the shortest side, then the other side must be $2x + 2$.

$$a^2 + b^2 = c^2$$
$$x^2 + (2x + 2)^2 = 13^2 \qquad \text{Pythagorean Theorem}$$
$$x^2 + 4x^2 + 8x + 4 = 169 \qquad \text{Expand } 13^2 \text{ and } (2x + 2)^2$$
$$5x^2 + 8x + 4 = 169 \qquad \text{Simplify the left side}$$
$$5x^2 + 8x - 165 = 0 \qquad \text{Add } -169 \text{ to each side}$$
$$(x - 5)(5x + 33) = 0 \qquad \text{Factor completely}$$

$$x - 5 = 0 \quad \text{or} \quad 5x + 33 = 0 \quad \text{Set factors to 0}$$
$$x = 5 \qquad\qquad 5x = -33$$
$$x = -\frac{33}{5}$$

Since a triangle cannot have a side with a negative number for a length, we cannot use $\frac{-33}{5}$. Therefore, the shortest side must be 5 meters. The next side is $2x + 2 = 2(5) + 2 = 12$ meters.

29. When $\qquad C = 800$

the equation $\qquad C = 200 + 500x - 100x^2$

becomes $\qquad 800 = 200 + 500x - 100x^2$

We can write this equation in standard form by adding -200, $-500x$, and $100x^2$ to each side. The result looks like this:

$$100x^2 - 500x + 600 = 0$$

$$100(x^2 - 5x + 6) = 0 \qquad \text{Begin factoring}$$

$$100(x - 2)(x - 3) = 0 \qquad \text{Factor completely}$$

$$x - 2 = 0 \quad \text{or} \quad x - 3 = 0 \qquad \text{Set factors to 0}$$

$$x = 2 \qquad\qquad x = 3$$

Our solutions are 2 and 3, which means that the company can manufacture 2 hundred items or 3 hundred items for a total cost of $800.

30. First, we must find the revenue equation. The equation for total revenue is $R = xp$ where x is the number of units sold and p is the price per unit. Since we want R in terms of p, we substitute $900 - 100p$ for x in the equation $R = xp$:

$$\begin{aligned} \text{If} \quad & R = xp \\ \text{and} \quad & x = 900 - 100p \\ \text{then} \quad & R = (900 - 100p)p \end{aligned}$$

We want to find p when R is $1,800. Substituting $1,800 for R in the equation gives us

$$1,800 = (900 - 100p)p$$

If we multiply out the right side we have

$$1,800 = 900p - 100p^2$$

To write this equation in standard form we add $100p^2$ and $-900p$ to each side.

$$100p^2 - 900p + 1800 = 0 \qquad \text{Add } 100p^2 \text{ and } -900p \text{ to each side}$$

$$100(p^2 - 9p + 18) = 0 \qquad \text{Begin factoring}$$

$$100(p - 3)(p - 6) = 0 \qquad \text{Factor completely}$$

$$p - 3 = 0 \quad \text{or} \quad p - 6 = 0 \qquad \text{Set variable factors to 0}$$

$$p = 3 \qquad\qquad p = 6$$

If the manufacturer sells the item for $3 each or for $6 each he will have a weekly revenue of $1,800.

CHAPTER 6

SECTION 6.1

1. $\dfrac{5}{5x-10} = \dfrac{5}{5(x-2)} = \dfrac{1}{x-2}$ $x \neq 2$

5. $\dfrac{x+5}{x^2-25} = \dfrac{x+5}{(x+5)(x-5)} = \dfrac{1}{x-5}$ $x \neq -5, 5$

9. $\dfrac{2x-10}{3x-6} = \dfrac{2(x-5)}{3(x-2)}$ $x \neq 2$

13. $\dfrac{5x^2-5}{4x+4} = \dfrac{5(x+1)(x-1)}{4(x+1)} = \dfrac{5(x-1)}{4}$

17. $\dfrac{3x+15}{3x^2+24x+45} = \dfrac{3(x+5)}{3(x+5)(x+3)} = \dfrac{1}{x+3}$

21. $\dfrac{3x-2}{9x^2-4} = \dfrac{3x-2}{(3x+2)(3x-2)} = \dfrac{1}{3x+2}$

25. $\dfrac{2m^3-2m^2-12m}{m^2-5m+6} = \dfrac{2m(m-3)(m+2)}{(m-2)(m-3)} = \dfrac{2m(m+2)}{m-2}$

29. $\dfrac{4x^3-10x^2+6x}{2x^3+x^2-3x} = \dfrac{2x(2x-3)(x-1)}{x(2x+3)(x-1)} = \dfrac{2(2x-3)}{2x+3}$

33. $\dfrac{x+3}{x^4-81} = \dfrac{x+3}{(x^2+9)(x+3)(x-3)} = \dfrac{1}{(x^2+9)(x-3)}$

37. $\dfrac{42x^3-20x^2-48x}{6x^2-5x-4} = \dfrac{2x(21x^2-10x-24)}{(2x+1)(3x-4)} = \dfrac{2x(7x+6)(3x-4)}{(2x+1)(3x-4)} = \dfrac{2x(7x+6)}{2x+1}$

41. $\dfrac{x^2-3x+ax-3a}{x^2-3x+bx-3b} = \dfrac{(x^2-3x)+(ax-3a)}{(x^2-3x)+(bx-3b)}$ Group items

$= \dfrac{x(x-3)+a(x-3)}{x(x-3)+b(x-3)}$ Factor

$= \dfrac{(x-3)(x+a)}{(x-3)(x+b)}$ Factor

$= \dfrac{x+a}{x+b}$ Simplify

45. $\dfrac{8}{6} = \dfrac{4}{3}$ Divide the numerator and denominator by 2

49. $\dfrac{32}{4} = \dfrac{8}{1}$ Divide the numerator and denominator by 4

53. 122 miles ÷ 3 hours = 40.7 miles/hour

57. 518 feet ÷ 40 seconds = 12.95 feet/second

61. 168 miles ÷ 3.5 gallons = 48 miles/gallon

65. $\dfrac{a-7}{7-a}$ when $a = 10$

$\dfrac{10-7}{7-10} = \dfrac{3}{-3} = -1$

69. $\dfrac{27x^5}{9x^2} - \dfrac{45x^8}{15x^5} = 3x^3 - 3x^3 = 0$

73. $\dfrac{38x^7 + 42x^5 - 84x^3}{2x^3} = \dfrac{38x^7}{2x^3} + \dfrac{42x^5}{2x^3} - \dfrac{84x^3}{2x^3} = 19x^4 + 21x^2 - 42$

SECTION 6.2

1. $\dfrac{x+y}{3} \cdot \dfrac{6}{x+y} = \dfrac{6(x+y)}{3(x+y)} = 2$

5. $\dfrac{9}{2a-8} \div \dfrac{3}{a-4} = \dfrac{9}{2(a-4)} \cdot \dfrac{a-4}{3} = \dfrac{3}{2}$

9. $\dfrac{a^2 + 5a}{7a} \cdot \dfrac{4a^2}{a^2 + 4a} = \dfrac{a(a+5)}{7a} \cdot \dfrac{4a^2}{a(a+4)} = \dfrac{4a(a+5)}{7(a+4)}$

13. $\dfrac{2x-8}{x^2-4} \cdot \dfrac{x^2+6x+8}{x-4} = \dfrac{2(x-4)}{(x+2)(x-2)} \cdot \dfrac{(x+2)(x+4)}{(x-4)} = \dfrac{2(x+4)}{x-2}$

17. $\dfrac{a^2 + 10a + 25}{a+5} \div \dfrac{a^2 - 25}{a-5} = \dfrac{(a+5)(a+5)}{a+5} \cdot \dfrac{a-5}{(a+5)(a-5)} = 1$

21. $\dfrac{2x^2 + 17x + 21}{x^2 + 2x - 35} \cdot \dfrac{x^2 - 25}{2x^2 - 7x - 15}$

$= \dfrac{(2x+3)(x+7)}{(x+7)(x-5)} \cdot \dfrac{(x+5)(x-5)}{(2x+3)(x-5)}$ Factor completely

$= \dfrac{(2x+3)(x+7)(x+5)(x-5)}{(x+7)(x-5)(2x+3)(x-5)}$ Divide out common factors

$= \dfrac{x+5}{x-5}$

25. $\dfrac{2a^2+7a+3}{a^2-16} \div \dfrac{4a^2+8a+3}{2a^2-5a-12} = \dfrac{(2a+1)(a+3)}{(a-4)(a+4)} \cdot \dfrac{(2a+3)(a-4)}{(2a+1)(2a+3)} = \dfrac{a+3}{a+4}$

29. $\dfrac{x^2-1}{6x^2+42x+60} \cdot \dfrac{7x^2+17x+6}{x+1} \cdot \dfrac{6x+30}{7x^2-11x-6}$

$= \dfrac{(x+1)(x-1)}{6(x+2)(x+5)} \cdot \dfrac{(7x+3)(x+2)}{x+1} \cdot \dfrac{6(x+5)}{(7x+3)(x-2)}$ Factor completely

$= \dfrac{(x+1)(x-1)(7x+3)(x+2)6(x+5)}{6(x+2)(x+5)(x+1)(7x+3)(x-2)}$ Divide out common factors

$= \dfrac{x-1}{x-2}$

33. $(x^2-9)\left(\dfrac{2}{x+3}\right) = \dfrac{x^2-9}{1} \cdot \dfrac{2}{x+3}$

$\qquad = \dfrac{(x+3)(x-3)2}{x+3}$

$\qquad = 2(x-3)$

37. $(x^2-x-6)\left(\dfrac{x+1}{x-3}\right) = \dfrac{x^2-x-6}{1} \cdot \dfrac{x+1}{x-3}$

$\qquad = \dfrac{(x-3)(x+2)(x+1)}{x-3}$

$\qquad = (x+2)(x+1)$

41. $\dfrac{x^2-9}{x^2-3x} \cdot \dfrac{2x+10}{xy+5x+3y+15} = \dfrac{(x+3)(x-3)}{x(x-3)} \cdot \dfrac{2(x+5)}{(xy+5x)+(3y+15)}$

$\qquad = \dfrac{(x+3)(x-3)}{x(x-3)} \cdot \dfrac{2(x+5)}{x(y+5)+3(y+5)}$

$\qquad = \dfrac{(x+3)(x-3)}{x(x-3)} \cdot \dfrac{2(x+5)}{(x+3)(y+5)}$

$\qquad = \dfrac{2(x+5)}{x(y+5)}$

45. $\dfrac{x^3-3x^2+4x-12}{x^4-16} \cdot \dfrac{3x^2+5x-2}{3x^2-10x+3}$

$= \dfrac{x^2(x-3)+4(x-3)}{(x^2+4)(x^2-4)} \cdot \dfrac{(x+2)(3x-1)}{(3x-1)(x-3)}$ Factor

$= \dfrac{(x-3)(x^2+4)}{(x^2+4)(x+2)(x-2)} \cdot \dfrac{(x+2)(3x-1)}{(3x-1)(x-3)}$ Factor

$= \dfrac{(x-3)(x^2+4)(x+2)(3x-1)}{(x^2+4)(x+2)(x-2)(3x-1)(x-3)}$ Divide out common factors

$= \dfrac{1}{x-2}$

49. $\left(1-\dfrac{1}{2}\right)\left(1-\dfrac{1}{3}\right)\left(1-\dfrac{1}{4}\right)\cdots\left(1-\dfrac{1}{99}\right)\left(1-\dfrac{1}{100}\right)$

$=\left(\dfrac{1}{2}\right)\left(\dfrac{2}{3}\right)\left(\dfrac{3}{4}\right)\cdots\left(\dfrac{98}{99}\right)\left(\dfrac{99}{100}\right)$ Subtract

$=\left(\dfrac{1}{2}\right)\left(\dfrac{2}{3}\right)\left(\dfrac{3}{4}\right)\cdots\left(\dfrac{98}{99}\right)\left(\dfrac{99}{100}\right)$ Divide out common factors

$=\dfrac{1}{100}$

53. Speed $= 1{,}088$ ft/sec $= \dfrac{1{,}088 \text{ ft}}{1 \text{ sec}} \cdot \dfrac{1 \text{ mi}}{5{,}280 \text{ ft}} \cdot \dfrac{3600 \text{ sec}}{1 \text{ hr}} = 741.8$ 742 miles/hour (approx.)

57. Average speed $= \dfrac{518 \text{ ft}}{40 \text{ sec}} \cdot \dfrac{1 \text{ mi}}{5{,}280 \text{ ft}} \cdot \dfrac{3{,}600 \text{ sec}}{1 \text{ hr}} = 8.8$ miles/hour

61. $\dfrac{1}{2}+\dfrac{5}{2}=\dfrac{6}{2}=3$

65. $\dfrac{1}{10}+\dfrac{3}{14}=\dfrac{1\cdot 7}{10\cdot 7}+\dfrac{3\cdot 5}{14\cdot 5}$ LCD $=2\cdot 5\cdot 7$

$=\dfrac{7}{70}+\dfrac{15}{70}$

$=\dfrac{22}{70}$

$=\dfrac{11}{35}$

69. $\dfrac{12a^2b^5}{3ab^3}+\dfrac{14a^4b^7}{7a^3b^5}=4ab^2+2ab^2=6ab^2$

SECTION 6.3

1. $\dfrac{3}{x}+\dfrac{4}{x}=\dfrac{7}{x}$

5. $\dfrac{1}{x+1}+\dfrac{x}{x+1}=\dfrac{x+1}{x+1}=1$

9. $\dfrac{x^2}{x+2}+\dfrac{4x+4}{x+2}=\dfrac{x^2+4x+4}{x+2}$

$=\dfrac{(x+2)(x+2)}{x+2}$

$=x+2$

13. $\dfrac{x+2}{x+6}-\dfrac{x-4}{x+6}=\dfrac{x+2-x+4}{x+6}=\dfrac{6}{x+6}$

17. $\dfrac{1}{2}+\dfrac{a}{3}=\dfrac{3}{6}+\dfrac{2a}{6}=\dfrac{3+2a}{6}$

21. $\dfrac{x+1}{x-2}-\dfrac{4x+7}{5x-10}=\dfrac{x+1}{x-2}-\dfrac{4x+7}{5(x-2)}$

$\qquad\qquad =\dfrac{\mathbf{5}}{\mathbf{5}}\left(\dfrac{x+1}{x-2}\right)-\dfrac{4x+7}{5(x-2)}$

$\qquad\qquad =\dfrac{5x+5}{5(x-2)}-\dfrac{4x+7}{5(x-2)}$

$\qquad\qquad =\dfrac{x-2}{5(x-2)}$

$\qquad\qquad =\dfrac{1}{5}$

25. $\dfrac{6}{x(x-2)}+\dfrac{3}{x}=\dfrac{6}{x(x-2)}+\dfrac{3(x-2)}{x(x-2)}$

$\qquad\qquad =\dfrac{6+3x-6}{x(x-2)}$

$\qquad\qquad =\dfrac{3x}{x(x-2)}$

$\qquad\qquad =\dfrac{3}{x-2}$

29. $\dfrac{2}{x+5}-\dfrac{10}{x^{2}-25}=\dfrac{2}{x+5}-\dfrac{10}{(x+5)(x-5)}$

$\qquad\qquad =\dfrac{2(x-5)}{(x+5)(x-5)}-\dfrac{10}{(x+5)(x-5)}$

$\qquad\qquad =\dfrac{2x-10-10}{(x+5)(x-5)}$

$\qquad\qquad =\dfrac{2x-20}{(x+5)(x-5)}$

$\qquad\qquad =\dfrac{2(x-10)}{(x+5)(x-5)}$

33. $\dfrac{a-4}{a-3}+\dfrac{5}{a^{2}-a-6}=\dfrac{a-4}{a-3}+\dfrac{5}{(a+2)(a-3)}$

$\qquad\qquad =\dfrac{\mathbf{a+2}}{\mathbf{a+2}}\left(\dfrac{a-4}{a-3}\right)+\dfrac{5}{(a+2)(a-3)}$

$\qquad\qquad =\dfrac{a^{2}-2a-8}{(a+2)(a-3)}+\dfrac{5}{(a+2)(a-3)}$

$\qquad\qquad =\dfrac{a^{2}-2a-3}{(a+2)(a-3)}$

$\qquad\qquad =\dfrac{(a+1)(a-3)}{(a+2)(a-3)}$

$\qquad\qquad =\dfrac{a+1}{a+2}$

37. $\dfrac{4y}{y^2+6y+5}-\dfrac{3y}{y^2+5y+4}=\dfrac{4y}{(y+5)(y+1)}-\dfrac{3y}{(y+4)(y+1)}$

$\qquad\qquad\qquad=\dfrac{4y(\mathbf{y+4})}{(y+5)(y+1)(\mathbf{y+4})}-\dfrac{3y(\mathbf{y+5})}{(y+1)(y+4)(\mathbf{y+5})}$

$\qquad\qquad\qquad=\dfrac{4y^2+16y-3y^2-15y}{(y+5)(y+1)(y+4)}$

$\qquad\qquad\qquad=\dfrac{y^2+y}{(y+5)(y+1)(y+4)}$

$\qquad\qquad\qquad=\dfrac{y(y+1)}{(y+5)(y+1)(y+4)}$

$\qquad\qquad\qquad=\dfrac{y}{(y+5)(y+4)}$

41. $\dfrac{1}{x}+\dfrac{x}{3x+9}-\dfrac{3}{x^2+3x}=\dfrac{1}{x}+\dfrac{x}{3(x+3)}-\dfrac{3}{x(x+3)}$

$\qquad\qquad\qquad=\dfrac{\mathbf{3(x+3)}}{\mathbf{3(x+3)}}\left(\dfrac{1}{x}\right)+\dfrac{\mathbf{x}}{\mathbf{x}}\left(\dfrac{x}{3(x+3)}\right)-\dfrac{\mathbf{3}}{\mathbf{3}}\left(\dfrac{3}{x(x+3)}\right)$

$\qquad\qquad\qquad=\dfrac{3x+9}{3x(x+3)}+\dfrac{x^2}{3x(x+3)}-\dfrac{9}{3x(x+3)}$

$\qquad\qquad\qquad=\dfrac{x^2+3x}{3x(x+3)}$

$\qquad\qquad\qquad=\dfrac{x(x+3)}{3x(x+3)}$

$\qquad\qquad\qquad=\dfrac{1}{3}$

45. $1-\dfrac{1}{x+1}=\dfrac{\mathbf{x+1}}{\mathbf{x+1}}(1)-\dfrac{1}{x+1}$

$\qquad\qquad=\dfrac{x+1}{x+1}-\dfrac{1}{x+1}$

$\qquad\qquad=\dfrac{x+1-1}{x+1}$

$\qquad\qquad=\dfrac{x}{x+1}$

49. $1-\dfrac{1}{x+3}=\dfrac{\mathbf{x+3}}{\mathbf{x+3}}(1)-\dfrac{1}{x+3}$

$\qquad\qquad=\dfrac{x+3}{x+3}-\dfrac{1}{x+3}$

$\qquad\qquad=\dfrac{x+3-1}{x+3}$

$\qquad\qquad=\dfrac{x+2}{x+3}$

53. $\dfrac{1}{x}+\dfrac{1}{2x}=\dfrac{2\cdot 1}{2x}+\dfrac{1}{2x}=\dfrac{3}{2x}$

57. $x-3(x+3)=x-3$

$\qquad x-3x-9=x-3$

$\qquad\quad -2x-9=x-3 \qquad$ Add $2x$ to both sides

$\qquad\qquad\quad -9=3x-3$

$\qquad\qquad\quad -6=3x \qquad\qquad$ Add 3 to both sides

$\qquad\qquad\quad -2=x \qquad\qquad$ Multiply both sides by $\dfrac{1}{3}$

61. $x^2+5x+6=0$

$(x+2)(x+3)=0 \qquad$ Factor the left side

$x+2=0 \quad$ or $\quad x+3=0 \qquad$ Set factors to 0

$x=-2 \qquad\qquad x=-3 \qquad$ Solutions

65. $x^2-5x=0$

$x(x-5)=0 \qquad$ Factor left side

$x=0 \quad$ or $\quad x-5=0 \quad$ Set factors to 0

$x=0 \qquad\qquad x=5 \quad$ Solutions

SECTION 6.4

1. $\qquad\dfrac{x}{3}+\dfrac{1}{2}=-\dfrac{1}{2}$

$6\left(\dfrac{x}{3}\right)+6\left(\dfrac{1}{2}\right)=6\left(-\dfrac{1}{2}\right) \qquad$ Multiply both sides by 6

$\qquad\quad 2x+3=-3$

$\qquad\qquad 2x=-6$

$\qquad\qquad\; x=-3$

The solution set is $\{-3\}$.

5. $\qquad\dfrac{3}{x}+1=\dfrac{2}{x}$

$x\left(\dfrac{3}{x}\right)+x(1)=x\left(\dfrac{2}{x}\right) \qquad$ Multiply both sides by x

$\qquad\quad 3+x=2$

$\qquad\qquad\; x=-1$

The solution set is $\{-1\}$.

9. $\qquad\dfrac{3}{x}+2=\dfrac{1}{2}$

$2x\left(\dfrac{3}{x}\right)+2x(2)=2x\left(\dfrac{1}{2}\right) \qquad$ Multiply both sides by $2x$

$\qquad\quad 6+4x=x$

$\qquad\qquad\; 6=-3x$

$\qquad\qquad -2=x$

13.
$$1 - \frac{8}{x} = \frac{-15}{x^2}$$

$x^2(1) - x^2\left(\frac{8}{x}\right) = x^2\left(\frac{-15}{x^2}\right)$ Multiply both sides by x^2

$x^2 - 8x = -15$

$x^2 - 8x + 15 = 0$ Standard form

$(x - 3)(x - 5) = 0$ Factor the left side

$x - 3 = 0$ or $x - 5 = 0$ Set factors to 0

$x = 3$ $x = 5$ Solutions

The solutions are 3 and 5.

17.
$$\frac{x-3}{2} + \frac{2x}{3} = \frac{5}{6}$$

$6\left(\frac{x-3}{2}\right) + 6\left(\frac{2x}{3}\right) = 6\left(\frac{5}{6}\right)$ Multiply both sides by 6

$3(x - 3) + 2(2x) = 5$

$3x - 9 + 4x = 5$ Distributive property

$7x - 9 = 5$ Combine like terms

$7x = 14$

$x = \frac{14}{7}$ Divide both sides by 7

$x = 2$

The solution set is {2}.

21.
$$\frac{6}{x+2} = \frac{3}{5}$$

$5(x + 2)\left(\frac{6}{x+2}\right) = 5(x + 2)\left(\frac{3}{5}\right)$ Multiply both sides by $5(x + 2)$

$30 = 3(x + 2)$

$30 = 3x + 6$ Distributive property

$24 = 3x$ Add -6 to each side

$8 = x$ Divide each side by 3

The solution set is {8}.

25.
$$\frac{x}{x-2}+\frac{2}{3}=\frac{2}{x-2}$$

$$3(x-2)\left(\frac{x}{x-2}\right)+3(x-2)\left(\frac{2}{3}\right)=3(x-2)\left(\frac{2}{x-2}\right) \qquad \text{Multiply both sides by } 3(x-2)$$

$$3x+2(x-2)=6$$

$$3x+2x-4=6 \qquad\qquad\qquad \text{Distributive property}$$

$$5x=10 \qquad\qquad\qquad\qquad \text{Combine like terms and add 4 to both sides}$$

$$x=2 \qquad\qquad\qquad\qquad\quad \text{Divide both sides by 5}$$

Possible solution 2, which does not check, $\varnothing$.

29.
$$\frac{5}{x+2}+\frac{1}{x+3}=\frac{-1}{x^2+5x+6}$$

$$(x+2)(x+3)\frac{5}{x+2}+(x+2)(x+3)\frac{1}{x+3}=(x+2)(x+3)\frac{-1}{(x+2)(x+3)} \qquad \text{Multiply both sides by } (x+2)(x+3)$$

$$5(x+3)+x+2=-1$$

$$5x+15+x+2=-1 \qquad\qquad \text{Distributive property}$$

$$6x+17=-1 \qquad\qquad\qquad \text{Combine like terms}$$

$$6x=-18 \qquad\qquad\qquad\quad \text{Add } -17 \text{ to each side}$$

$$x=-3 \qquad\qquad\qquad\qquad \text{Divide each side by 6}$$

Possible solution -3, which does not check, $\varnothing$.

33.
$$\frac{a}{2}+\frac{3}{a-3}=\frac{a}{a-3}$$

$$2(a-3)\frac{a}{2}+2(a-3)\left(\frac{3}{a-3}\right)=2(a-3)\frac{a}{a-3} \qquad \text{Multiply both sides by } 2(a-3)$$

$$a(a-3)+6=2a$$

$$a^2-3a+6=2a \qquad\qquad \text{Distributive property}$$

$$a^2-5a+6=0 \qquad\qquad \text{Standard form}$$

$$(a-2)(a-3)=0 \qquad\qquad \text{Factor the left side}$$

$$a-2=0 \quad \text{or} \quad a-3=0 \qquad \text{Set factors to 0}$$

$$a=2 \qquad\qquad a=3$$

Possible solutions are 2 and 3, but only 2 checks, 2.

37.
$$\frac{2}{a^2-9}=\frac{3}{a^2+a-12}$$

$$\frac{2}{(a+3)(a-3)}=\frac{3}{(a+4)(a-3)}$$

$$(a+3)(a-3)(a+4)\cdot\frac{2}{(a+3)(a-3)}=(a+3)(a-3)(a+4)\cdot\frac{3}{(a+4)(a-3)}$$

$$2(a+4)=3(a+3)$$

$2a+8=3a+9$ Distributive property

$8=a+9$ Add $-2a$ to each side

$-1=a$ Add -9 to each side

The solution set is $\{-1\}$.

41.
$$\frac{2x}{x+2}=\frac{x}{x+3}-\frac{3}{x^2+5x+6}$$

$$(x+2)(x+3)\frac{2x}{x+2}=(x+2)(x+3)\frac{x}{x+3}-(x+2)(x+3)\frac{3}{(x+2)(x+3)}$$

$2x(x+3)=x(x+2)-3$

$2x^2+6x=x^2+2x-3$ Distributive property

$x^2+4x+3=0$ Standard form

$(x+3)(x+1)=0$ Factor the left side

$x+3=0$ or $x+1=0$ Set factors to 0

$x=-3$ $x=-1$

Possible solutions are -3 and -1, but only -1 checks, -1.

45. Step 1: We know the length is 5 more than twice the width. The perimeter is 34 inches. The perimeter formula is
$P=2L+2W$. We do not know the length or width.

Step 2: Let $x=$ the width.

"Length is 5 more than twice the width" translates to
$$L=2x+5$$
Replacing W with x and L with $2x+5$ in $P=2L+2W$, we have
$$34=2(2x+5)+2x$$

Step 3: The equation is
$$34=2(2x+5)+2x$$

Step 4: $34=2(2x+5)+2x$

$34=4x+10+2x$ Distributive property

$34=6x+10$ Combine like terms

$24=6x$ Add -10 to each side

$4=x$ Divide each side by 6

$13=2x+5$

Step 5: The width is 4 inches and the length is 13 inches.

Step 6: $34 \overset{?}{=} 2(13) + 2(4)$

$\qquad 34 \overset{?}{=} 26 + 8$

$\qquad 34 = 34$

49. $10^2 = x^2 + (x+2)^2$ Pythagorean theorem

$100 = x^2 + x^2 + 4x + 4$ Expand $(x+2)^2$

$100 = 2x^2 + 4x + 4$ Simplify the right side

$0 = 2x^2 + 4x - 96$ Add -100 to both sides

$0 = 2(x^2 + 2x - 48)$ Begin factoring

$0 = 2(x+8)(x-6)$ Factor completely

$0 = x+8$ or $0 = x - 6$ Set variable factors to 0

$-8 = x$ $6 = x$

$x = -8$ is not a possible solution.

The two legs are 6 inches and 8 inches.

SECTION 6.5

1. Let $x =$ a number, then $3x$ is the other number. Their reciprocals are $\frac{1}{x}$ and $\frac{1}{3x}$.

$$\frac{1}{x} + \frac{1}{3x} = \frac{16}{3}$$

$$3x\left(\frac{1}{x}\right) + 3x\left(\frac{1}{3x}\right) = 3x\left(\frac{16}{3}\right)$$

$$3 + 1 = 16x$$

$$4 = 16x \qquad \text{Combine like terms}$$

$$\frac{4}{16} = x \qquad \text{Divide each side by 16}$$

$$\frac{1}{4} = x \qquad \text{Reduce}$$

$$\frac{3}{4} = 3x$$

The numbers are $\frac{1}{4}$, $\frac{3}{4}$.

5. Let $x =$ a certain number.

$$\frac{7+x}{9+x} = \frac{5}{7}$$

$$7(9+x)\frac{7+x}{9+x} = 7(9+x)\frac{5}{7}$$

$$7(7+x) = 5(9+x)$$

$$49 + 7x = 45 + 5x \qquad \text{Distributive property}$$

$$49 + 2x = 45 \qquad\qquad \text{Add } -5x \text{ to each side}$$

$$2x = -4 \qquad\qquad \text{Add } -49 \text{ to each side}$$

$$x = -2$$

The number is -2.

9.

	d	r	t
upstream	26	$x-3$	
downstream	38	$x+3$	

To fill two time blocks, remember $d = rt$. We use the formula $t = \frac{d}{r}$.

	d	r	t
upstream	26	$x-3$	$\frac{26}{x-3}$
downstream	38	$x+3$	$\frac{38}{x+3}$

The time upstream and the time downstream are equal.

$$\frac{26}{x-3} = \frac{38}{x+3}$$

$$(x+3)(x-3)\frac{26}{x-3} = (x+3)(x-3)\frac{38}{x+3} \qquad \text{LCD} = (x+3)(x-3)$$

$$(x+3)26 = (x-3)38$$

$$26x + 78 = 38x - 114 \qquad \text{Distributive property}$$

$$78 = 12x - 114 \qquad \text{Add } -26x \text{ to each side}$$

$$192 = 12x \qquad\qquad \text{Add } 114 \text{ to each side}$$

$$16 = x \qquad\qquad\quad \text{Divide each side by 12}$$

The speed of the boat is 16 mph.

13. Let $x =$ the slower plane

	d	r	t
faster plane	285	$x+20$	
slower plane	255	x	

To fill the two time blocks, remember $d = rt$. We use the formula $t = \frac{d}{r}$.

	d	r	t
faster plane	285	$x + 20$	$\frac{285}{x+20}$
slower plane	255	x	$\frac{255}{x}$

$$\frac{285}{x+20} = \frac{255}{x}$$

$$x(x+20)\frac{285}{x+20} = x(x+20)\frac{255}{x}$$

$$285x = 255(x+20)$$

$$285x = 255x + 5100 \qquad \text{Distributive property}$$

$$30x = 5100 \qquad \text{Add } -255x \text{ to each side}$$

$$x = 170 \qquad \text{Divide each side by 30}$$

The slower plane has a speed of 170 mph and the faster plane has a speed of 190 mph.

17. Let r = rate on level ground.

	d	r	t
downhill	5	$r + 2$	
level ground	4	r	

To fill the two time blocks, remember $d = rt$. We use the formula $t = \frac{d}{r}$.

	d	r	t
downhill	5	$r + 2$	$\frac{5}{r+2}$
level ground	4	r	$\frac{4}{r}$

Remember that Jerri jogs 1 hour a day.

$$\frac{5}{r+2} + \frac{4}{r} = 1$$

$$r(r+2)\frac{5}{r+2} + r(r+2)\frac{4}{r} = r(r+2)1$$

$$5r + 4(r+2) = r(r+2)$$

$$5r + 4r + 8 = r^2 + 2r \qquad \text{Distributive property}$$

$$0 = r^2 - 7r - 8 \qquad \text{Standard form}$$

$$0 = (r+1)(r-8) \qquad \text{Factor right side}$$

$$r + 1 = 0 \quad \text{or} \quad r - 8 = 0 \qquad \text{Set factors equal to 0}$$

$$r = -1 \qquad\qquad r = 8$$

Since a rate of jogging cannot be negative the only answer is 8 miles per hour.

21. Let $x =$ amount of time to fill the tub with both faucets open in one hour.

$$\underset{\substack{\text{Amount of time} \\ \text{for cold water}}}{\frac{1}{10}} \quad + \quad \underset{\substack{\text{Amount of time} \\ \text{for hot water}}}{\frac{1}{12}} \quad = \quad \underset{\substack{\text{Amount of time} \\ \text{for both waters}}}{\frac{1}{x}}$$

$$60x\left(\frac{1}{10}\right) + 60x\left(\frac{1}{12}\right) = 60x\left(\frac{1}{x}\right) \quad \text{LCD} = 60x$$

$$6x + 5x = 60$$

$$11x = 60$$

$$x = \frac{60}{11}$$

It will take $\frac{60}{11}$ minutes to fill the tub if both faucets are open.

25. $y = \frac{2}{x}$

x	y
2	1
1	2
-1	-2
-2	-1

See the graph in the back of the textbook.

29. $y = \frac{8}{x}$

x	y
4	2
2	4
1	8
-1	-8
-2	-4
-4	-2

See the graph in the back of the textbook.

33. $2x + y = 3$ $2x + y = 3$

$\underline{3x - y = 7}$ $2(2) + y = 3$

$5x \quad = 10$ $\quad y = -1$

$x \quad = 2$

The solution to the system is $(2, -1)$. It satisfies both equations.

37. Substituting $y = 3x - 2$ in the first equation, we have

$$5x + 2(3x - 2) = 7 \qquad y = 3x - 2$$
$$5x + 6x - 4 = 7 \qquad y = 3(1) - 2$$
$$11x - 4 = 7 \qquad y = 3 - 2$$
$$11x = 11 \qquad y = 1$$
$$x = 1$$

The solution to the system is $(1, 1)$. It satisfies both equations.

SECTION 6.6

1. Method 1 $\dfrac{\frac{3}{4}}{\frac{1}{8}} = \dfrac{8 \cdot \frac{3}{4}}{8 \cdot \frac{1}{8}}$ $LCD = 8$ Method 2 $\dfrac{\frac{3}{4}}{\frac{1}{8}} = \dfrac{3}{4} \cdot \dfrac{8}{1}$

$\qquad\qquad\qquad = \dfrac{6}{1}$ $\qquad\qquad\qquad\qquad\qquad = \dfrac{6}{1}$

$\qquad\qquad\qquad = 6$ $\qquad\qquad\qquad\qquad\qquad = 6$

5. Method 1 $\dfrac{\frac{x^2}{y}}{\frac{x}{y^3}} = \dfrac{y^3 \cdot \frac{x^2}{y}}{y^3 \cdot \frac{x}{y^3}}$ $LCD = y^3$ Method 2 $\dfrac{\frac{x^2}{y}}{\frac{x}{y^3}} = \dfrac{x^2}{y} \cdot \dfrac{y^3}{x}$

$\qquad\qquad\quad = \dfrac{x^2 y^2}{x}$ $\qquad\qquad\qquad\qquad\qquad = xy^2$

$\qquad\qquad\quad = xy^2$

9. $\dfrac{y + \frac{1}{x}}{x + \frac{1}{y}} = \dfrac{xy\left(y + \frac{1}{x}\right)}{xy\left(x + \frac{1}{y}\right)}$ $LCD = xy$

$\qquad = \dfrac{xy \cdot y + xy \cdot \frac{1}{x}}{xy \cdot x + xy \cdot \frac{1}{y}}$

$\qquad = \dfrac{xy^2 + y}{x^2 y + x}$

$\qquad = \dfrac{y(xy + 1)}{x(xy + 1)}$

$\qquad = \dfrac{y}{x}$

13. <u>Method 1</u> $\dfrac{\frac{x+1}{x^2-9}}{\frac{2}{x+3}} = \dfrac{(x+3)(x-3)\left(\frac{x+1}{x^2-9}\right)}{(x+3)(x-3)\left(\frac{2}{x+3}\right)}$ $LCD = (x+3)(x-3)$

$$= \dfrac{x+1}{2(x-3)}$$

<u>Method 2</u> $\dfrac{\frac{x+1}{x^2-9}}{\frac{2}{x+3}} = \dfrac{x+1}{x^2-9} \cdot \dfrac{x+3}{2}$

$$= \dfrac{x+1}{(x+3)(x-3)} \cdot \dfrac{x+3}{2}$$

$$= \dfrac{x+1}{2(x-3)}$$

17. $\dfrac{1-\frac{9}{y^2}}{1-\frac{1}{y}-\frac{6}{y^2}} = \dfrac{y^2\left(1-\frac{9}{y^2}\right)}{y^2\left(1-\frac{1}{y}-\frac{6}{y^2}\right)}$ $LCD = y^2$

$$= \dfrac{y^2-9}{y^2-y-6}$$

$$= \dfrac{(y+3)(y-3)}{(y-3)(y+2)}$$

$$= \dfrac{y+3}{y+2}$$

21. $\dfrac{1-\frac{1}{a^2}}{1-\frac{1}{a}} = \dfrac{a^2\left(1-\frac{1}{a^2}\right)}{a^2\left(1-\frac{1}{a}\right)}$ $LCD = a^2$

$$= \dfrac{a^2-1}{a^2-a}$$

$$= \dfrac{(a+1)(a-1)}{a(a-1)}$$

$$= \dfrac{a+1}{a}$$

25. $\dfrac{\frac{1}{a+1}+2}{\frac{1}{a+1}+3} = \dfrac{(a+1)\left(\frac{1}{a+1}+2\right)}{(a+1)\left(\frac{1}{a+1}+3\right)}$ $LCD = a+1$

$$= \dfrac{1+2(a+1)}{1+3(a+1)}$$

$$= \dfrac{1+2a+2}{1+3a+3}$$

$$= \dfrac{2a+3}{3a+4}$$

29. $\left(1+\dfrac{1}{x+3}\right)\left(1+\dfrac{1}{x+2}\right)\left(1+\dfrac{1}{x+1}\right)$

$=\left(\dfrac{x+3+1}{x+3}\right)\left(\dfrac{x+2+1}{x+2}\right)\left(\dfrac{x+1+1}{x+1}\right)$ Common denominators

$=\left(\dfrac{x+4}{x+3}\right)\left(\dfrac{x+3}{x+2}\right)\left(\dfrac{x+2}{x+1}\right)$ Add

$=\left(\dfrac{x+4}{x+3}\right)\left(\dfrac{x+3}{x+2}\right)\left(\dfrac{x+2}{x+1}\right)$ Divide out common factors

$=\dfrac{x+4}{x+1}$

33. $2x+3<5$

$\quad 2x<2$

$\qquad x<1$

37. $-2x+8>-4$ Remember to reverse the inequality

$\quad -2x>-12$ symbol when multiplying by a negative

$\qquad x<6$ number.

SECTION 6.7

1. $\dfrac{x}{2}=\dfrac{6}{12}$ Extremes are x and 12; means are 2 and 6

$\quad 12x=12$ Product of extremes = product of means

$\qquad x=1$ Divide both sides by 12

5. $\dfrac{10}{20}=\dfrac{20}{x}$ Extremes are 10 and x; means are 20 and 20

$\quad 10x=400$ Product of extremes = product of means

$\qquad x=40$ Divide both sides by 10

9. $\dfrac{2}{x}=\dfrac{6}{7}$ Extremes are 2 and 7; means are x and 6

$\quad 14=6x$ Product of extremes = product of means

$\quad \dfrac{14}{6}=x$ Divide both sides by 6

$\quad \dfrac{7}{3}=x$ Reduce to lowest terms

13. $\dfrac{x}{2}=\dfrac{8}{x}$ Extremes are x and x; means are 2 and 8

$\qquad x^2=16$ Product of extremes = product of means

$\quad x^2-16=0$ Standard form for a quadratic equation

$(x+4)(x-4)=0$ Factor

$x + 4 = 0$ or $x - 4 = 0$ Set factors equal to 0

$\qquad x = -4 \qquad\qquad x = 4$

This time we have two solutions, -4 and 4.

17. $\qquad\qquad \dfrac{1}{x} = \dfrac{x-5}{6}$ Extremes are 1 and 6; means are x and $x - 5$

$\qquad\qquad\qquad 6 = x^2 - 5x$ Product of extremes $=$ product of means

$\qquad x^2 - 5x - 6 = 0$ $\qquad$ Standard form for a quadratic equation

$(x + 1)(x - 6) = 0$ $\qquad$ Factor

$\qquad x + 1 = 0$ or $x - 6 = 0$ Set factors equal to 0

$\qquad\quad x = -1 \qquad\qquad x = 6$

The solutions are -1 and 6.

21. $\dfrac{\text{alcohol}}{\text{water}}$ $\qquad \dfrac{12}{16} = \dfrac{x}{28}$ Extremes are 12 and 28; means are 16 and x

$\qquad\qquad\quad 336 = 16x$ Product of extremes $=$ product of means

$\qquad\qquad\quad\ 21 = x$

It will take 21 milliliters of alcohol.

25. $\dfrac{\text{inches}}{\text{miles}}$ $\qquad \dfrac{3.5}{100} = \dfrac{x}{420}$

$\qquad\qquad$ Using the means - extremes property, we have

$\qquad\qquad 100x = 1470$

$\qquad\qquad\quad\ x = 14.7$

The distance between the two cities is 14.7 inches.

29. $\dfrac{x^2 - x - 6}{x^2 - 9} = \dfrac{(x-3)(x+2)}{(x-3)(x+3)} = \dfrac{x+2}{x+3}$

33. $\dfrac{x}{x^2 - 16} + \dfrac{4}{x^2 - 16} = \dfrac{x+4}{(x+4)(x-4)} = \dfrac{1}{x-4}$

CHAPTER 6 REVIEW

1. $\dfrac{7}{14x - 28} = \dfrac{7}{14(x-2)} = \dfrac{1}{2(x-2)}$ $x \neq 2$

2. $\dfrac{-6}{6x + 18} = \dfrac{-6}{6(x+3)} = \dfrac{-1}{x+3}$ $x \neq -3$

3. $\dfrac{a+6}{a^2 - 36} = \dfrac{a+6}{(a+6)(a-6)} = \dfrac{1}{a-6}$ $a \neq -6, 6$

4. $\dfrac{a-9}{a^2 - 81} = \dfrac{a-9}{(a-9)(a+9)} = \dfrac{1}{a+9}$ $a \neq -9, 9$

5. $\dfrac{8x-4}{4x+12} = \dfrac{4(2x-1)}{4(x+3)} = \dfrac{2x-1}{x+3}$ $x \neq -3$

6. $\dfrac{5x+15}{x+3} = \dfrac{5(x+3)}{x+3} = 5$ $x \neq -3$

7. $\dfrac{x+4}{x^2+8x+16} = \dfrac{x+4}{(x+4)(x+4)} = \dfrac{1}{x+4}$

8. $\dfrac{x^2-14x+49}{x-7} = \dfrac{(x-7)(x-7)}{x-7} = x-7$

9. $\dfrac{3x^3+16x^2-12x}{2x^3+9x^2-18x} = \dfrac{x(3x^2+16x-12)}{x(2x^2+9x-18)}$ Factor

       $= \dfrac{x(x+6)(3x-2)}{x(x+6)(2x-3)}$ Factor

       $= \dfrac{x(x+6)(3x-2)}{x(x+6)(2x-3)}$

       $= \dfrac{3x-2}{2x-3}$

10. $\dfrac{8x^3+10x^2-3x}{20x^3+3x^2-2x} = \dfrac{x(8x^2+10x-3)}{x(20x^2+3x-2)}$ Factor

       $= \dfrac{x(2x+3)(4x-1)}{x(5x+2)(4x-1)}$ Factor

       $= \dfrac{2x+3}{5x+2}$

11. $\dfrac{x+2}{x^4-16} = \dfrac{x+2}{(x^2+4)(x+2)(x-2)} = \dfrac{1}{(x^2+4)(x-2)}$

12. $\dfrac{x^2-9}{x^4-81} = \dfrac{(x+3)(x-3)}{(x^2+9)(x+3)(x-3)} = \dfrac{1}{x^2+9}$

13. $\dfrac{x^2+5x-14}{x+7} = \dfrac{(x+7)(x-2)}{x+7} = \dfrac{(x+7)(x-2)}{x+7} = x-2$

14. $\dfrac{x^2+5x-24}{x+8} = \dfrac{(x+8)(x-3)}{x+8} = x-3$

15. $\dfrac{a^2+16a+64}{a+8} = \dfrac{(a+8)(a+8)}{a+8} = a+8$

16. $\dfrac{a^2-12a+36}{a-6} = \dfrac{(a-6)(a-6)}{a-6} = a-6$

17. $\dfrac{xy+bx+ay+ab}{xy+5x+ay+5a} = \dfrac{(xy+bx)+(ay+ab)}{(xy+5x)+(ay+5a)}$

       $= \dfrac{x(y+b)+a(y+b)}{x(y+5)+a(y+5)}$

       $= \dfrac{(y+b)(x+a)}{(y+5)(x+a)}$

       $= \dfrac{(y+b)(x+a)}{(y+5)(x+a)}$

       $= \dfrac{y+b}{y+5}$

18. $\dfrac{2xy+10x+3y+15}{3xy+15x+2y+10} = \dfrac{2x(y+5)+3(y+5)}{3x(y+5)+2(y+5)}$

       $= \dfrac{(y+5)(2x+3)}{(y+5)(3x+2)}$

       $= \dfrac{2x+3}{3x+2}$

19. $\dfrac{3x+9}{x^2} \cdot \dfrac{x^3}{6x+18} = \dfrac{3(x+3)}{x^2} \cdot \dfrac{x^3}{6(x+3)} = \dfrac{x}{2}$

20. $\dfrac{6x-12}{6x+12} \cdot \dfrac{3x+3}{12x-24} = \dfrac{6(x-2)}{6(x+2)} \cdot \dfrac{3(x+1)}{12(x-2)} = \dfrac{x+1}{4(x+2)}$

21. $\dfrac{x^2+8x+16}{x^2+x-12} \div \dfrac{x^2-16}{x^2-x-6}$

$= \dfrac{x^2+8x+16}{x^2+x-12} \cdot \dfrac{x^2-x-6}{x^2-16}$ Division is multiplication by the reciprocal

$= \dfrac{(x+4)(x+4)}{(x+4)(x-3)} \cdot \dfrac{(x-3)(x+2)}{(x+4)(x-4)}$

$= \dfrac{(x+4)(x+4)(x-3)(x+2)}{(x+4)(x-3)(x+4)(x-4)}$ Factor and multiply

$= \dfrac{x+2}{x-4}$ Divide out common factors

22. $\dfrac{x^2+3x}{x^2+4x+4} \cdot \dfrac{x^2-5x-14}{x^2+6x+9} = \dfrac{x(x+3)}{(x+2)(x+2)} \cdot \dfrac{(x+2)(x-7)}{(x+3)(x+3)} = \dfrac{x(x-7)}{(x+2)(x+3)}$

23. $(a^2-4a-12)\left(\dfrac{a-6}{a+2}\right) = \dfrac{(a+2)(a-6)}{1}\left(\dfrac{a-6}{a+2}\right) = (a-6)^2$

24. $(a^2+5a-24)\left(\dfrac{a+8}{a-3}\right) = \dfrac{(a+8)(a-3)}{1}\left(\dfrac{a+8}{a-3}\right)$ Factor and multiply

$= (a+8)^2$ Divide out common factors

25. $\dfrac{3x^2-2x-1}{x^2+6x+8} \div \dfrac{3x^2+13x+4}{x^2+8x+16}$

$= \dfrac{3x^2-2x-1}{x^2+6x+8} \cdot \dfrac{x^2+8x+16}{3x^2+13x+4}$ Division is multiplication by the reciprocal

$= \dfrac{(3x+1)(x-1)}{(x+2)(x+4)} \cdot \dfrac{(x+4)(x+4)}{(3x+1)(x+4)}$

$= \dfrac{(3x+1)(x-1)(x+4)(x+4)}{(x+2)(x+4)(3x+1)(x+4)}$ Factor and multiply

$= \dfrac{x-1}{x+2}$

26. $\dfrac{x^2+5x+6}{x^2-x-6} \div \dfrac{x^2+6x+9}{2x^2-5x-3}$

$\quad = \dfrac{x^2+5x+6}{x^2-x-6} \cdot \dfrac{2x^2-5x-3}{x^2+6x+9}$ Division is multiplication by the reciprocal

$\quad = \dfrac{(x+2)(x+3)}{(x+2)(x-3)} \cdot \dfrac{(2x+1)(x-3)}{(x+3)(x+3)}$

$\quad = \dfrac{(x+2)(x+3)(2x+1)(x-3)}{(x+2)(x-3)(x+3)(x+3)}$ Factor and multiply

$\quad = \dfrac{2x+1}{x+3}$

27. $\dfrac{2x}{2x+3} + \dfrac{3}{2x+3} = \dfrac{2x+3}{2x+3} = 1$

28. $\dfrac{3x}{3x-4} - \dfrac{4}{3x-4} = \dfrac{3x-4}{3x-4} = 1$

29. $\dfrac{x^2}{x-9} - \dfrac{18x-81}{x-9} = \dfrac{x^2-18x+81}{x-9}$ Subtract

$\qquad\qquad = \dfrac{(x-9)(x-9)}{x-9}$

$\qquad\qquad = x-9$

30. $\dfrac{x^2}{x-7} - \dfrac{14x-49}{x-7} = \dfrac{x^2-14x+49}{x-7}$ Subtract

$\qquad\qquad = \dfrac{(x-7)(x-7)}{x-7}$

$\qquad\qquad = x-7$

31. $\dfrac{a+4}{a+8} - \dfrac{a-9}{a+8} = \dfrac{a+4-a+9}{a+8}$ Subtract

$\qquad\qquad = \dfrac{13}{a+8}$

32. $\dfrac{a-3}{a-7} - \dfrac{a+10}{a-7} = \dfrac{a-3-a-10}{a-7}$ Subtract

$\qquad\qquad = \dfrac{-13}{a-7}$

33. $\dfrac{x}{x+9} + \dfrac{5}{x} = \dfrac{x \cdot \mathbf{x}}{(x+9)\mathbf{x}} + \dfrac{5(\mathbf{x+9})}{x(x+9)}$ LCD $= x(x+9)$

$\qquad\qquad = \dfrac{x^2+5x+45}{x(x+9)}$

34. $\dfrac{x}{x-3} - \dfrac{9}{x} = \dfrac{x \cdot \mathbf{x}}{(x-3)\mathbf{x}} - \dfrac{9(\mathbf{x-3})}{x(\mathbf{x-3})}$ LCD $= x(x-3)$

$\qquad\qquad = \dfrac{x^2-9x+27}{x(x-3)}$

35. $\dfrac{5}{4x+20} + \dfrac{x}{x+5} = \dfrac{5}{4(x+5)} + \dfrac{x}{x+5}$

$\qquad\qquad = \dfrac{5}{4(x+5)} + \dfrac{x \cdot 4}{(x+5)4}$ LCD $= 4(x+5)$

$\qquad\qquad = \dfrac{4x+5}{4(x+5)}$

36. $\dfrac{6}{10x+30} + \dfrac{x}{x+3}$

$\quad = \dfrac{6}{10(x+3)} + \dfrac{x}{x+3}$

$\quad = \dfrac{6}{10(x+3)} + \dfrac{\mathbf{10}x}{\mathbf{10}(x+3)}$ LCD $= 10(x+3)$

$\quad = \dfrac{10x+6}{10(x+3)}$

$\quad = \dfrac{2(5x+3)}{10(x+3)}$

$\quad = \dfrac{5x+3}{5(x+3)}$

37. $\dfrac{3}{x^2 - 36} - \dfrac{2}{x^2 - 4x - 12}$

$= \dfrac{3}{(x+6)(x-6)} - \dfrac{2}{(x+2)(x-6)}$ $\text{LCD} = (x+6)(x-6)(x+2)$

$= \dfrac{3(\mathbf{x+2})}{(x+6)(x-6)(\mathbf{x+2})} - \dfrac{2(\mathbf{x+6})}{(x+2)(x-6)(\mathbf{x+6})}$

$= \dfrac{3x + 6 - 2x - 12}{(x+6)(x-6)(x+2)}$

$= \dfrac{x - 6}{(x+6)(x-6)(x+2)}$

$= \dfrac{1}{(x+6)(x+2)}$

38. $\dfrac{-1}{x^2 - 4} - \dfrac{-2}{x^2 - 4x - 12}$

$= \dfrac{-1}{(x+2)(x-2)} - \dfrac{-2}{(x+2)(x-6)}$ $\text{LCD} = (x+2)(x-2)(x-6)$

$= \dfrac{-1(\mathbf{x-6})}{(x+2)(x-2)(\mathbf{x-6})} - \dfrac{-2(\mathbf{x-2})}{(x+2)(x-6)(\mathbf{x-2})}$

$= \dfrac{-x + 6 + 2x - 4}{(x+2)(x-2)(x-6)}$

$= \dfrac{x + 2}{(x+2)(x-2)(x-6)}$

$= \dfrac{1}{(x-2)(x-6)}$

39. $\dfrac{3a}{a^2 + 8a + 15} - \dfrac{2}{a+5}$

$= \dfrac{3a}{(a+3)(a+5)} - \dfrac{2}{a+5}$ $\text{LCD} = (a+3)(a+5)$

$= \dfrac{3a}{(a+3)(a+5)} - \dfrac{2(\mathbf{a+3})}{(a+5)(\mathbf{a+3})}$

$= \dfrac{3a - 2a - 6}{(a+3)(a+5)}$

$= \dfrac{a - 6}{(a+5)(a+3)}$

40. $\dfrac{7a}{a^2 - 3a - 54} + \dfrac{5}{a-9}$

$= \dfrac{7a}{(a-9)(a+6)} + \dfrac{5}{a-9}$ $\text{LCD} = (a-9)(a+6)$

$= \dfrac{7a}{(a-9)(a+6)} + \dfrac{5(\mathbf{a+6})}{(a-9)(\mathbf{a+6})}$

$= \dfrac{7a + 5a + 30}{(a-9)(a+6)}$

$= \dfrac{12a + 30}{(a-9)(a+6)}$

$= \dfrac{6(2a + 5)}{(a-9)(a+6)}$

41.
$$\frac{3}{x} + \frac{1}{2} = \frac{5}{x} \qquad \text{LCD} = 2x$$

$$2x\left(\frac{3}{x}\right) + 2x\left(\frac{1}{2}\right) = 2x\left(\frac{5}{x}\right) \qquad \text{Multiply both sides by } 2x$$

$$6 + x = 10$$

$$x = 4 \qquad \text{Add } -6 \text{ to both sides}$$

42.
$$\frac{5}{x} - \frac{1}{3} = \frac{3}{x} \qquad \text{LCD} = 3x$$

$$3x\left(\frac{5}{x}\right) - 3x\left(\frac{1}{3}\right) = 3x\left(\frac{3}{x}\right) \quad \text{Multiply both sides by } 3x$$

$$15 - x = 9$$

$$-x = -6 \qquad \text{Add } -15 \text{ to both sides}$$

$$x = 6 \qquad \text{Multiply both sides by } -1$$

43.
$$\frac{a}{a-3} = \frac{3}{2} \qquad \text{LCD} = 2(a-3)$$

$$2(a-3)\left(\frac{a}{a-3}\right) = 2(a-3)\left(\frac{3}{2}\right) \quad \text{Multiply both sides by } 2(a-3)$$

$$2a = 3a - 9$$

$$9 = a \qquad \text{Add } -2a \text{ and 9 to both sides}$$

44.
$$\frac{a}{a+4} = \frac{7}{3} \qquad \text{LCD} = 3(a+4)$$

$$3(a+4)\left(\frac{a}{a+4}\right) = 3(a+4)\left(\frac{7}{3}\right) \quad \text{Multiply both sides by } 3(a+4)$$

$$3a = 7a + 28$$

$$-28 = 4a \qquad \text{Add } -3a \text{ and } -28 \text{ to both sides}$$

$$-7 = a$$

45.
$$1 - \frac{7}{x} = \frac{-6}{x^2} \qquad \text{LCD} = x^2$$

$$x^2(1) + x^2\left(-\frac{7}{x}\right) = x^2\left(\frac{-6}{x^2}\right) \qquad \text{Multiply both sides by } x^2$$

$$x^2 - 7x = -6$$

$$x^2 - 7x + 6 = 0 \qquad \text{Add 6 to each side}$$

$$(x-1)(x-6) = 0$$

$$x - 1 = 0 \quad \text{or} \quad x - 6 = 0$$

$$x = 1 \qquad\qquad x = 6$$

The solutions are 1 and 6.

46.
$$1 + \frac{12}{x} = \frac{-35}{x^2} \qquad \text{LCD} = x^2$$

$$x^2(1) + x^2\left(\frac{12}{x}\right) = x^2\left(\frac{-35}{x^2}\right) \qquad \text{Multiply both sides by } x^2$$

$$x^2 + 12x = -35$$

$$x^2 + 12x + 35 = 0 \qquad \text{Add 35 to each side}$$

$$(x+7)(x+5) = 0$$

$$x + 7 = 0 \quad \text{or} \quad x + 5 = 0$$

$$x = -7 \qquad\qquad x = -5$$

The solutions are -5 and -7.

47. $\dfrac{3}{x+6} - \dfrac{1}{x-2} = \dfrac{-8}{x^2 + 4x - 12}$

$$\frac{3}{x+6} - \frac{1}{x-2} = \frac{-8}{(x+6)(x-2)} \qquad \text{LCD} = (x+6)(x-2)$$

$$(x+6)(x-2)\left(\frac{3}{x+6}\right) - (x+6)(x-2)\left(\frac{1}{x-2}\right)$$

$$= (x+6)(x-2)\left(\frac{-8}{(x+6)(x-2)}\right) \qquad \text{Multiply both sides by } (x+6)(x-2)$$

$$3x - 6 - x - 6 = -8$$

$$2x - 12 = -8 \qquad\qquad \text{Combine terms}$$

$$2x = 4 \qquad\qquad \text{Add 12 to both sides}$$

$$x = 2 \qquad\qquad \text{Divide both sides by 2}$$

Possible solution 2, which does not check; $\varnothing$

48. $\dfrac{4}{x-5} - \dfrac{3}{x+2} = \dfrac{28}{x^2 - 3x - 10}$

$$\frac{4}{x-5} - \frac{3}{x+2} = \frac{28}{(x+2)(x-5)} \qquad \text{LCD} = (x+2)(x-5)$$

$$(x+2)(x-5)\left(\frac{4}{x-5}\right) - (x+2)(x-5)\left(\frac{3}{x+2}\right)$$

$$= (x+2)(x-5)\left(\frac{28}{(x+2)(x-5)}\right) \qquad \text{Multiply both sides by } (x+2)(x-5)$$

$$4x + 8 - 3x + 15 = 28$$

$$x + 23 = 28 \qquad\qquad \text{Combine terms}$$

$$x = 5 \qquad\qquad \text{Add } -23 \text{ to both sides}$$

Possible solution 5, which does not check; $\varnothing$

49. $\dfrac{2}{y^2-16}=\dfrac{10}{y^2+4y}$

$\dfrac{2}{(y+4)(y-4)}=\dfrac{10}{y(y+4)}$ LCD $=y(y+4)(y-4)$

$y(y+4)(y-4)\cdot\dfrac{2}{(y+4)(y-4)}$

$=y(y+4)(y-4)\cdot\dfrac{10}{y(y+4)}$ Multiply both sides by $y(y+4)(y-4)$

$2y=10y-40$

$40=8y$ Add $-2y$ and 40 to both sides

$5=y$ Divide both sides by 8

50. $\dfrac{4}{y^2-25}=\dfrac{24}{y^2-5y}$

$\dfrac{4}{(y+5)(y-5)}=\dfrac{24}{y(y-5)}$ LCD $=y(y+5)(y-5)$

$y(y+5)(y-5)\cdot\left(\dfrac{4}{(y+5)(y-5)}\right)$

$=y(y+5)(y-5)\cdot\left(\dfrac{24}{y(y-5)}\right)$ Multiply both sides by $y(y+5)(y-5)$

$4y=24y+120$

$-120=20y$ Add $-4y$ and -120 to both sides

$-6=y$ Divide both sides by 20

51. Let $x=$ the number and $\frac{1}{x}=$ the reciprocal.

$x+7\left(\dfrac{1}{x}\right)=\dfrac{16}{3}$

$x+\dfrac{7}{x}=\dfrac{16}{3}$ LCD $=3x$

$3x(x)+3x\left(\dfrac{7}{x}\right)=3x\left(\dfrac{16}{3}\right)$ Multiply both sides by $3x$

$3x^2+21=16x$

$3x^2-16x+21=0$ Add $-16x$ to both sides

$(x-3)(3x-7)=0$

$x-3=0$ or $3x-7=0$

$x=3$ $x=\dfrac{7}{3}$

52. Let x = the number and $\frac{1}{x}$ = the reciprocal.

$$x + 5\left(\frac{1}{x}\right) = 6$$

$$x + \frac{5}{x} = 6 \qquad \text{LCD} = x$$

$$x(x) + x\left(\frac{5}{x}\right) = x(6) \qquad \text{Multiply both sides by } x$$

$$x^2 + 5 = 6x$$

$$x^2 - 6x + 5 = 0 \qquad \text{Add } -6x \text{ to both sides}$$

$$(x-1)(x-5) = 0$$

$$x - 1 = 0 \quad \text{or} \quad x - 5 = 0$$

$$x = 1 \qquad\qquad x = 5$$

53. Let x = speed of the boat in still water

	d	r	t
upstream	48	$x - 3$	$\frac{48}{x-3}$
downstream	72	$x + 3$	$\frac{72}{x+3}$

time (upstream) = time (downstream)

$$\frac{48}{x-3} = \frac{72}{x+3} \qquad\qquad \text{LCD} = (x+3)(x-3)$$

$$(x+3)(x-3) \cdot \frac{48}{x-3} = (x+3)(x-3) \cdot \frac{72}{x+3}$$

$$(x+3)48 = (x-3)72$$

$$48x + 144 = 72x - 216$$

$$360 = 24x$$

$$15 = x$$

The speed of the boat in still water is 15 mph.

54. Let x = speed of the boat in still water

	d	r	t
upstream	54	$x - 2$	$\frac{54}{x-2}$
downstream	66	$x + 2$	$\frac{66}{x+2}$

time (upstream) = time (downstream)

$$\frac{54}{x-2} = \frac{66}{x+2} \qquad\qquad LCD = (x-2)(x+2)$$

$$(x-2)(x+2)\cdot\left(\frac{54}{x-2}\right) = (x-2)(x+2)\cdot\left(\frac{66}{x+2}\right)$$

$$54x + 108 = 66x - 132$$

$$240 = 12x$$

$$20 = x$$

The speed of the boat in still water is 20 mph.

55. Let x = amount of time to fill the pool with both pipes open. In one hour,

Amount of water let in		Amount of water let out		Total amount of water in pool
$\dfrac{1}{21}$	$-$	$\dfrac{1}{28}$	$=$	$\dfrac{1}{x}$

$$84x\left(\frac{1}{21}\right) - 84x\left(\frac{1}{28}\right) = 84x\left(\frac{1}{x}\right) \qquad LCD = 84x$$

$$4x - 3x = 84$$

$$x = 84$$

It will take 84 hours to fill the pool with both pipes left open.

56. Let x = amount of time to fill the tub with both faucets open. In one hour,

Cold water in tub		Hot water in tub		Total amount of water in tub
$\dfrac{1}{6}$	$+$	$\dfrac{1}{12}$	$=$	$\dfrac{1}{x}$

$$12x\left(\frac{1}{6}\right) + 12x\left(\frac{1}{12}\right) = 12x\left(\frac{1}{x}\right)$$

$$2x + x = 12$$

$$3x = 12$$

$$x = 4$$

It takes 4 hours to fill the tub with both faucets left open.

57. $y = \frac{4}{x}$

x	y
4	1
2	2
-2	-2
-4	-1

See the graph in the back of the textbook.

58. $y = \frac{-1}{x}$

x	y
1	-1
2	-1/2
-1	1
-2	1/2

See the graph in the back of the textbook.

59. $\dfrac{\frac{x-2}{x^2+6x+8}}{\frac{4}{x+4}} = \dfrac{\frac{x-2}{(x+2)(x+4)}}{\frac{4}{x+4}}$ Factor

$= \dfrac{(x+2)(x+4)\frac{x-2}{(x+2)(x+4)}}{(x+2)(x+4)\frac{4}{x+4}}$ Multiply top and bottom by LCD $= (x+2)(x+4)$

$= \dfrac{x-2}{4(x+2)}$ Simplify

60. $\dfrac{1-\frac{9}{y^2}}{1+\frac{4}{y}-\frac{21}{y^2}} = \dfrac{y^2\left(1-\frac{9}{y^2}\right)}{y^2\left(1+\frac{4}{y}-\frac{21}{y^2}\right)}$ Multiply top and bottom by LCD $= y^2$

$= \dfrac{y^2 \cdot 1 - y^2 \cdot \frac{9}{y^2}}{y^2 \cdot 1 + y^2 \cdot \frac{4}{y} - y^2 \cdot \frac{21}{y^2}}$ Distributive property

$= \dfrac{y^2-9}{y^2+4y-21}$ Simplify

$= \dfrac{(y+3)(y-3)}{(y+7)(y-3)}$ Factor

$= \dfrac{y+3}{y+7}$ Reduce

61. $\dfrac{1-\frac{16}{y^2}}{1-\frac{4}{y}-\frac{32}{y^2}} = \dfrac{y^2\left(1-\frac{16}{y^2}\right)}{y^2\left(1-\frac{4}{y}-\frac{32}{y^2}\right)}$ Multiply top and bottom by LCD $= y^2$

$= \dfrac{y^2(1) - y^2\left(\frac{16}{y^2}\right)}{y^2(1) - y^2\left(\frac{4}{y}\right) - y^2\left(\frac{32}{y^2}\right)}$ Distributive property

$= \dfrac{y^2-16}{y^2-4y-32}$ Simplify

$= \dfrac{(y+4)(y-4)}{(y+4)(y-8)}$ Factor

$= \dfrac{y-4}{y-8}$ Reduce

62. $\dfrac{\frac{1}{a-2}+4}{\frac{1}{a-2}+1}=\dfrac{(a-2)\left(\frac{1}{a-2}+4\right)}{(a-2)\left(\dfrac{1}{a-2}+1\right)}$ Multiply top and bottom by $a-2$

$=\dfrac{(a-2)\left(\frac{1}{a-2}\right)+(a-2)(4)}{(a-2)\left(\frac{1}{a-2}\right)+(a-2)(1)}$ Distributive property

$=\dfrac{1+4a-8}{1+a-2}$ Simplify

$=\dfrac{4a-7}{a-1}$

63. 40 to 100 $\dfrac{40}{100}=\dfrac{2}{5}$

64. $\dfrac{10}{3}$ to $\dfrac{5}{2}$ $\dfrac{\frac{10}{3}}{\frac{5}{2}}=\dfrac{6\left(\frac{10}{3}\right)}{6\left(\frac{5}{2}\right)}$ Multiply top and bottom by 6

$=\dfrac{20}{15}$ Simplify

$=\dfrac{4}{3}$ Reduce

65. $\dfrac{40\text{ sec}}{3\text{ min}}=\dfrac{40}{3(60)}=\dfrac{40}{180}=\dfrac{2}{9}$

66. $\dfrac{10\text{ in}}{5\text{ ft}}=\dfrac{10}{5(12)}=\dfrac{10}{60}=\dfrac{1}{6}$

67. $\dfrac{x}{9}=\dfrac{4}{3}$

$3x=36$ Product of extremes $=$ product of means

$x=12$

68. $\dfrac{4}{x}=\dfrac{7}{3}$

$12=7x$ Product of extremes $=$ product of means

$x=\dfrac{12}{7}$

69. $\dfrac{a}{3}=\dfrac{12}{a}$

$a^2=36$ Product of extremes $=$ Product of means

$a^2-36=0$ Standard form for a quadratic equation

$(a+6)(a-6)=0$ Factor

$a+6=0$ or $a-6=0$ Set factors equal to 0

$a=-6$ $a=6$

The solutions are -6 and 6.

70. $$\frac{4}{a} = \frac{a}{16}$$

$64 = a^2$ Product of extremes = product of means

$a^2 - 64 = 0$ Standard form for a quadratic equation

$(a+8)(a-8) = 0$ Factor

$a+8 = 0$ or $a-8 = 0$ Set factors equal to 0

$a = -8$ $a = 8$

The solutions are −8 and 8.

71. $\frac{8}{x-2} = \frac{x}{6}$

$48 = x(x-2)$ Product of extremes = product of means

$48 = x^2 - 2x$ Distributive property

$0 = x^2 - 2x - 48$ Standard form for a quadratic equation

$0 = (x+6)(x-8)$ Factor

$x+6 = 0$ or $x-8 = 0$ Set factors equal to 0

$x = -6$ $x = 8$

The solutions are −6 and 8.

72. $$\frac{x}{3} = \frac{6}{x-3}$$

$x(x-3) = 18$ Product of extremes = product of means

$x^2 - 3x - 18 = 0$ Standard form for a quadratic equation

$(x+3)(x-6) = 0$ Factor

$x+3 = 0$ or $x-6 = 0$ Set factors equal to 0

$x = -3$ $x = 6$

The solutions are −3 and 6.

73. $\frac{\text{Miles}}{\text{Hours}}$ $\frac{124}{4} = \frac{x}{9}$

$1116 = 4x$ Product of extremes = product of means

$x = 279$ miles

74. $\frac{\text{Miles}}{\text{Hours}}$ $\frac{1860}{5} = \frac{x}{7}$

$13{,}020 = 5x$ Product of extremes = product of means

$x = 2{,}604$ miles

CHAPTER 6 TEST

1. $\dfrac{x^2-16}{x^2-8x+16} = \dfrac{(x+4)(x-4)}{(x-4)(x-4)} = \dfrac{x+4}{x-4}$

2. $\dfrac{10a+20}{5a^2+20a+20} = \dfrac{10(a+2)}{5(a+2)(a+2)} = \dfrac{2}{a+2}$

3. $\dfrac{xy+7x+5y+35}{x^2+ax+5x+5a} = \dfrac{(xy+7x)+(5y+35)}{(x^2+ax)+(5x+5a)}$

$\qquad = \dfrac{x(y+7)+5(y+7)}{x(x+a)+5(x+a)}$

$\qquad = \dfrac{(x+5)(y+7)}{(x+5)(x+a)}$

$\qquad = \dfrac{y+7}{x+a}$

4. $\dfrac{3x-12}{4} \cdot \dfrac{8}{2x-8} = \dfrac{3(x-4)}{4} \cdot \dfrac{8}{2(x-4)}$

$\qquad = \dfrac{24(x-4)}{8(x-4)}$

$\qquad = 3$

5. $\dfrac{x^2-49}{x+1} \div \dfrac{x+7}{x^2-1} = \dfrac{(x+7)(x-7)}{x+1} \cdot \dfrac{(x+1)(x-1)}{x+7}$

$\qquad = (x-7)(x-1)$

6. $\dfrac{x^2-3x-10}{x^2-8x+15} \div \dfrac{3x^2+2x-8}{x^2+x-12} = \dfrac{x^2-3x-10}{x^2-8x+15} \cdot \dfrac{x^2+x-12}{3x^2+2x-8}$

$\qquad = \dfrac{(x-5)(x+2)}{(x-3)(x-5)} \cdot \dfrac{(x+4)(x-3)}{(x+2)(3x-4)}$

$\qquad = \dfrac{x+4}{3x-4}$

7. $(x^2-9)\left(\dfrac{x+2}{x+3}\right) = \dfrac{(x+3)(x-3)}{1}\left(\dfrac{x+2}{x+3}\right) = (x-3)(x+2)$

8. $\dfrac{3}{x-2} - \dfrac{6}{x-2} = \dfrac{3-6}{x-2} = \dfrac{-3}{x-2}$

9. $\dfrac{x}{x^2-9} + \dfrac{4}{4x-12} = \dfrac{x}{(x+3)(x-3)} + \dfrac{4}{4(x-3)}$

$\qquad = \dfrac{x}{(x+3)(x-3)} + \dfrac{1}{x-3}$

$\qquad = \dfrac{x}{(x+3)(x-3)} + \dfrac{1(x+3)}{(x-3)(x+3)}$

$\qquad = \dfrac{2x+3}{(x+3)(x-3)}$

10. $\dfrac{2x}{x^2-1}+\dfrac{x}{x^2-3x+2}$

$=\dfrac{2x}{(x+1)(x-1)}+\dfrac{x}{(x-1)(x-2)}$

$\dfrac{2x(\mathbf{x-2})}{(x+1)(x-1)(\mathbf{x-2})}+\dfrac{x(\mathbf{x+1})}{(x-1)(x-2)(\mathbf{x+1})}$

$=\dfrac{2x^2-4x+x^2+x}{(x+1)(x-1)(x-2)}$

$=\dfrac{3x^2-3x}{(x+1)(x-1)(x-2)}$

$=\dfrac{3x(x-1)}{(x+1)(x-1)(x-2)}$

$=\dfrac{3x}{(x+1)(x-2)}$

11. $\dfrac{7}{5}=\dfrac{x+2}{3}$

$15\left(\dfrac{7}{5}\right)=\left(\dfrac{x+2}{3}\right)15\quad \text{LCD}=15$

$21=(x+2)5$

$21=5x+10$

$11=5x$

$\dfrac{11}{5}=x$

12. $\dfrac{10}{x+4}=\dfrac{6}{x}-\dfrac{4}{x}$

$x(x+4)\left(\dfrac{10}{x+4}\right)=x(x+4)\left(\dfrac{6}{x}\right)-x(x+4)\left(\dfrac{4}{x}\right)\quad \text{LCD}=x(x+4)$

$10x=6(x+4)-4(x+4)$

$10x=6x+24-4x-16$

$10x=2x+8$

$8x=8$

$x=1$

13. $\dfrac{3}{x-2}-\dfrac{4}{x+1}$

$=\dfrac{5}{x^2-x-2}$

$(x+1)(x-2)\left(\dfrac{3}{x-2}\right)-(x+1)(x-2)\left(\dfrac{4}{x+1}\right)$

$=(x+1)(x-2)\left(\dfrac{5}{(x+1)(x-2)}\right)\quad \text{LCD}=(x+1)(x-2)$

$3(x+1)-4(x-2)=5$

$3x+3-4x+8=5$

$-x+11=5$

$-x=-6$

$x=6$

14. Let x = speed of the boat in still water.

	d	r	t
upstream	26	$x-2$	$\frac{26}{x-2}$
downstream	34	$x+2$	$\frac{34}{x+2}$

Remember $d = rt$ so $\frac{d}{r} = t$

time (upstream) = time (downstream)

$$\frac{26}{x-2} = \frac{34}{x+2}$$

$$(x+2)(x-2)\left(\frac{26}{x-2}\right) = (x+2)(x-2)\left(\frac{34}{x+2}\right)$$

$$26(x+2) = 34(x-2)$$

$$26x + 52 = 34x - 68$$

$$52 = 8x - 68$$

$$120 = 8x$$

$$15 = x$$

The speed of the boat in still water is 15 mph.

15. Let x = amount of time to empty the pool with both pipes open. In one hour,.

Amount of water let out by outlet pipe	Amount of water let in by inlet pipe	Total amount of water in pool
$\frac{1}{12}$	$-$ $\frac{1}{15}$	$=$ $\frac{1}{x}$

$$60x\left(\frac{1}{12}\right) - 60x\left(\frac{1}{15}\right) = \left(\frac{1}{x}\right)60x \quad \text{LCD} = 60x$$

$$5x - 4x = 60$$

$$x = 60 \text{ hours}$$

Remember to subtract inlet pipe from the outlet pipe because the pool is full and you are trying to empty the pool.

16. $y = \frac{8}{x}$

x	y
8	1
4	2
2	4
1	8
-1	-8
-2	-4
-4	-2
-8	-1

See the graph in the back of the textbook.

17. $\dfrac{27}{54} = \dfrac{1}{2}$ $\dfrac{\text{solution of alcohol}}{\text{solution of water}}$

$27 + 54 = 81$ total volume

$\dfrac{27}{81} = \dfrac{1}{3}$ $\dfrac{\text{alcohol}}{\text{total volume}}$

18. $\dfrac{8}{100} = \dfrac{x}{1650}$ $\dfrac{\text{defective parts}}{\text{parts produced}}$

$8(1650) = 100x$

$13{,}200 = 100x$

$132 = x$

19. $\dfrac{1 + \dfrac{1}{x}}{1 - \dfrac{1}{x}} = \dfrac{x\left(1 + \dfrac{1}{x}\right)}{x\left(1 - \dfrac{1}{x}\right)}$

$= \dfrac{x(1) + x\left(\dfrac{1}{x}\right)}{x(1) - x\left(\dfrac{1}{x}\right)}$

$= \dfrac{x + 1}{x - 1}$

20. $\dfrac{1 - \dfrac{16}{x^2}}{1 - \dfrac{2}{x} - \dfrac{8}{x^2}} = \dfrac{x^2\left(1 - \dfrac{16}{x^2}\right)}{x^2\left(1 - \dfrac{2}{x} - \dfrac{8}{x^2}\right)}$

$= \dfrac{x^2(1) - x^2\left(\dfrac{16}{x^2}\right)}{x^2(1) - x^2\left(\dfrac{2}{x}\right) - x^2\left(\dfrac{8}{x^2}\right)}$

$= \dfrac{x^2 - 16}{x^2 - 2x - 8}$

$= \dfrac{(x + 4)(x - 4)}{(x - 4)(x + 2)}$

$= \dfrac{x + 4}{x + 2}$

CHAPTER 7

SECTION 7.1

1. $\sqrt{9} = 3$ Because $3^2 = 9$.

5. $\sqrt{-25}$ Not a real number since there is no real number whose square is -25.

9. $\sqrt{625} = 25$ Because $(25)^2 = 625$.

13. $-\sqrt{64} = -8$ Because -8 is the negative number we square to get 64.

17. $\sqrt{1225} = 35$ Because $(35)^2 = 1225$.

21. $\sqrt[3]{-8} = -2$ Because $(-2)^3 = -8$.

25. $\sqrt[3]{-1} = -1$ Because $(-1)^3 = -1$.

29. $-\sqrt[4]{16} = -2$ Because -2 is the negative number we raise to the fourth power to get 16.

33. $\sqrt{9x^2} = 3x$ Because $(3x)^2 = 9x^2$

37. $\sqrt{(a+b)^2} = a + b$

41. $\sqrt[3]{x^3} = x$

45. $\sqrt{x^4} = x^2$ Because $(x^2)^2 = x^4$.

49. $\sqrt{25a^8b^4} = 5a^4b^2$ Because $(5a^4b^2)^2 = 25a^8b^4$.

53. $\sqrt[3]{27a^{12}} = 3a^4$ Because $(3a^4)^3 = 27a^{12}$.

57. $\sqrt{9} + \sqrt{16} = 3 + 4 = 7$

61. $\sqrt{144} + \sqrt{25} = 12 + 5 = 17$

65. $\dfrac{5 + \sqrt{49}}{2}$ and $\dfrac{5 - \sqrt{49}}{2}$

$= \dfrac{5 + 7}{2}$ $= \dfrac{5 - 7}{2}$

$= \dfrac{12}{2}$ $= \dfrac{-2}{2}$

$= 6$ $= -1$

69. $\sqrt{x^2 + 6x + 9} = \sqrt{(x+3)^2} = x + 3$

73. $A = \$65 \quad P = \$50$

$$r = \frac{\sqrt{A} - \sqrt{P}}{\sqrt{P}}$$

$$r = \frac{\sqrt{65} - \sqrt{50}}{\sqrt{50}}$$

$$r \approx \frac{8.062 - 7.071}{7.071}$$

$$r = \frac{0.991}{7.071}$$

$$r \approx 0.14 = 14\%$$

77. $c = \sqrt{a^2 + b^2}; \quad a = 4, \quad b = 3, \quad c = x$

$x = \sqrt{4^2 + 3^2}$

$x = \sqrt{16 + 9}$

$x = \sqrt{25}$

$x = 5$

81. $c = \sqrt{a^2 + b^2}; \quad a = 24 \text{ feet}, \quad b = 18 \text{ feet}, \quad c = x$

$x = \sqrt{(24)^2 + (18)^2}$

$x = \sqrt{576 + 324}$

$x = \sqrt{900}$

$x = 30 \text{ feet}$

85. $\dfrac{10a + 20}{5a^2 - 20} = \dfrac{10(a + 2)}{5(a + 2)(a - 2)} = \dfrac{2}{a - 2}$

89. $\dfrac{xy + 3x + 2y + 6}{xy + 3x + ay + 3a} = \dfrac{x(y + 3) + 2(y + 3)}{x(y + 3) + a(y + 3)}$

$$= \dfrac{(y + 3)(x + 2)}{(y + 3)(x + a)}$$

$$= \dfrac{x + 2}{x + a}$$

SECTION 7.2

1. $\sqrt{8} = \sqrt{4 \cdot 2}$ Factor 8 into $4 \cdot 2$

   $= \sqrt{4} \cdot \sqrt{2}$ Property 1 for radicals

   $= 2\sqrt{2}$ Since $\sqrt{4} = 2$

5. $\sqrt[3]{24} = \sqrt[3]{8 \cdot 3}$ Factor 24 into $8 \cdot 3$

   $= \sqrt[3]{8} \cdot \sqrt[3]{3}$ Property 1 for radicals

   $= 2\sqrt[3]{3}$ Since $\sqrt[3]{8} = 2$

9. $\sqrt{45a^2b^2} = \sqrt{9a^2b^2 \cdot 5}$ Factor $45a^2b^2$ into $9a^2b^2 \cdot 5$

   $= \sqrt{9a^2b^2} \cdot \sqrt{5}$ Property 1 for radicals

   $= 3ab\sqrt{5}$ Since $\sqrt{9a^2b^2} = 3ab$

13. $\sqrt{32x^4} = \sqrt{16x^4 \cdot 2}$ Factor $32x^4$ into $16x^4 \cdot 2$

   $= \sqrt{16x^4} \cdot \sqrt{2}$ Property 1 for radicals

   $= 4x^2\sqrt{2}$ Since $\sqrt{16x^4} = 4x^2$

17. $\dfrac{1}{2}\sqrt{28x^3} = \dfrac{1}{2}\sqrt{4x^2 \cdot 7x}$ Factor $28x^3$ into $4x^2 \cdot 7x$

$\qquad\qquad = \dfrac{1}{2}\sqrt{4x^2} \cdot \sqrt{7x}$ Property 1 for radicals

$\qquad\qquad = \dfrac{1}{2} \cdot 2x\sqrt{7x}$ Since $\sqrt{4x^2} = 2x$

$\qquad\qquad = x\sqrt{7x}$

21. $2a\sqrt[3]{27a^5} = 2a\sqrt[3]{27a^3 \cdot a^2}$ Factor $27a^5$ into $27a^3 \cdot a^2$

$\qquad\qquad = 2a\sqrt[3]{27a^3} \cdot \sqrt[3]{a^2}$ Property 1 for radicals

$\qquad\qquad = 2a \cdot 3a\sqrt[3]{a^2}$ Since $\sqrt[3]{27a^3} = 3a$

$\qquad\qquad = 6a^2\sqrt[3]{a^2}$

25. $3\sqrt{50xy^2} = 3\sqrt{25y^2 \cdot 2x}$ Factor $50xy^2$ into $25y^2 \cdot 2x$

$\qquad\qquad = 3\sqrt{25y^2} \cdot \sqrt{2x}$ Property 1 for radicals

$\qquad\qquad = 3(5y)\sqrt{2x}$ Since $\sqrt{25y^2} = 5y$

$\qquad\qquad = 15y\sqrt{2x}$

29. $\sqrt{\dfrac{16}{25}} = \dfrac{\sqrt{16}}{\sqrt{25}}$ Property 2

$\qquad = \dfrac{4}{5}$ $\sqrt{16} = 4$ and $\sqrt{25} = 5$

33. $\sqrt[3]{\dfrac{8}{27}} = \dfrac{\sqrt[3]{8}}{\sqrt[3]{27}}$ Property 2

$\qquad = \dfrac{2}{3}$ $\sqrt[3]{8} = 2$ and $\sqrt[3]{27} = 3$

37. $\sqrt{\dfrac{100x^2}{25}} = \dfrac{\sqrt{100x^2}}{\sqrt{25}}$ Property 2

$\qquad = \dfrac{10x}{5}$ $\sqrt{100x^2} = 10x$ and $\sqrt{25} = 5$

$\qquad = 2x$

41. $\sqrt[3]{\dfrac{27x^3}{8y^3}} = \dfrac{\sqrt[3]{27x^3}}{\sqrt[3]{8y^3}}$ Property 2

$\qquad = \dfrac{3x}{2y}$ $\sqrt[3]{27x^3} = 3x$ and $\sqrt[3]{8y^3} = 2y$

45. $\sqrt{\dfrac{75}{25}} = \sqrt{3}$ Since $\dfrac{75}{25} = 3$

$\sqrt{\dfrac{75}{25}} = \dfrac{\sqrt{75}}{\sqrt{25}}$ Property 2

$\qquad = \dfrac{\sqrt{25 \cdot 3}}{\sqrt{25}}$

$\qquad = \dfrac{\sqrt{25} \cdot \sqrt{3}}{\sqrt{25}}$ Property 1

$\qquad = \sqrt{3}$

49. $\sqrt{\dfrac{288x}{25}} = \dfrac{\sqrt{288x}}{\sqrt{25}}$ Property 2

$\qquad = \dfrac{\sqrt{144 \cdot 2x}}{\sqrt{25}}$

$\qquad = \dfrac{\sqrt{144} \cdot \sqrt{2x}}{\sqrt{25}}$ Property 1

$\qquad = \dfrac{12\sqrt{2x}}{5}$ $\sqrt{144} = 12$

53. $\dfrac{3\sqrt{50}}{2} = \dfrac{3\sqrt{25}\sqrt{2}}{2}$ Property 1

$\qquad = \dfrac{3(5)\sqrt{2}}{2}$ $\sqrt{25} = 5$

$\qquad = \dfrac{15\sqrt{2}}{2}$

57. $\dfrac{5\sqrt{72a^2b^2}}{\sqrt{36}} = \dfrac{5\sqrt{36a^2b^2 \cdot 2}}{\sqrt{36}}$

$\qquad = \dfrac{5\sqrt{36a^2b^2} \cdot \sqrt{2}}{\sqrt{36}}$ Property 1

$\qquad = \dfrac{5(6)ab\sqrt{2}}{6}$ $\sqrt{36a^2b^2} = 6ab$

$\qquad = 5ab\sqrt{2}$

61. $\dfrac{8\sqrt{12x^2y^3}}{\sqrt{100}} = \dfrac{8\sqrt{4x^2y^2 \cdot 3y}}{\sqrt{100}}$

$\qquad = \dfrac{8\sqrt{4x^2y^2}\sqrt{3y}}{\sqrt{100}}$ Property 1

$\qquad = \dfrac{8(2)xy\sqrt{3y}}{10}$ $\sqrt{4x^2y^2} = 2xy$

$\qquad = \dfrac{8xy\sqrt{3y}}{5}$

65. Let $h = 25$ feet, then

$t = \sqrt{\dfrac{h}{16}}$ becomes

$t = \sqrt{\dfrac{25}{16}}$

$t = \dfrac{5}{4}$ seconds

69. $\dfrac{x^2 + 3x - 4}{3x^2 + 7x - 20} \div \dfrac{x^2 - 2x + 1}{3x^2 - 2x - 5} = \dfrac{x^2 + 3x - 4}{3x^2 + 7x - 20} \cdot \dfrac{3x^2 - 2x - 5}{x^2 - 2x + 1}$

$\qquad\qquad = \dfrac{(x+4)(x-1)(x+1)(3x-5)}{(x+4)(3x-5)(x-1)(x-1)}$

$\qquad\qquad = \dfrac{x+1}{x-1}$

SECTION 7.3

1. $\sqrt{\dfrac{1}{2}} = \dfrac{\sqrt{1}}{\sqrt{2}}$ Property 2

$\qquad = \dfrac{1}{\sqrt{2}} \cdot \dfrac{\sqrt{2}}{\sqrt{2}}$

$\qquad = \dfrac{\sqrt{2}}{2}$ $\sqrt{2} \cdot \sqrt{2} = \sqrt{4} = 2$

5. $\sqrt{\dfrac{2}{5}} = \dfrac{\sqrt{2}}{\sqrt{5}}$ Property 2

$\qquad = \dfrac{\sqrt{2}}{\sqrt{5}} \cdot \dfrac{\sqrt{5}}{\sqrt{5}}$

$\qquad = \dfrac{\sqrt{10}}{5}$ $\sqrt{5} \cdot \sqrt{5} = \sqrt{25} = 5$

9. $\sqrt{\dfrac{20}{3}} = \dfrac{\sqrt{4}\sqrt{5}}{\sqrt{3}}$ Property 2 & 1

$\qquad = \dfrac{2\sqrt{5}}{\sqrt{3}} \cdot \dfrac{\sqrt{3}}{\sqrt{3}}$

$\qquad = \dfrac{2\sqrt{15}}{3}$ $\sqrt{3} \cdot \sqrt{3} = \sqrt{9} = 3$

13. $\sqrt{\dfrac{20}{5}} = \sqrt{4}$ $\dfrac{20}{5} = 4$

$\qquad = 2$

17. $\dfrac{\sqrt{35}}{\sqrt{7}} = \sqrt{\dfrac{35}{7}}$ Property 2

    $= \sqrt{5}$ $\dfrac{35}{7} = 5$

21. $\dfrac{6\sqrt{21}}{3\sqrt{7}} = \dfrac{6}{3}\sqrt{\dfrac{21}{7}}$

    $= 2\sqrt{3}$ $\dfrac{21}{7} = 3$

25. $\sqrt{\dfrac{4x^2 y^2}{2}} = \sqrt{2x^2 y^2}$

    $= xy\sqrt{2}$ Since $\sqrt{x^2 y^2} = xy$

29. $\sqrt{\dfrac{16a^4}{5}} = \dfrac{\sqrt{16a^4}}{\sqrt{5}}$

    $= \dfrac{4a^2}{\sqrt{5}} \cdot \dfrac{\sqrt{5}}{\sqrt{5}}$ Since $\sqrt{16a^4} = 4a^2$

    $= \dfrac{4a^2\sqrt{5}}{5}$

33. $\sqrt{\dfrac{20x^2 y^3}{3}} = \dfrac{\sqrt{4x^2 y^2 \cdot 5y}}{\sqrt{3}}$

    $= \dfrac{2xy\sqrt{5y}}{\sqrt{3}} \cdot \dfrac{\sqrt{3}}{\sqrt{3}}$ Since $\sqrt{4x^2 y^2} = 2xy$

    $= \dfrac{2xy\sqrt{15y}}{3}$

37. $\dfrac{6\sqrt{54a^2 b^3}}{5} = \dfrac{6\sqrt{9a^2 b^2 \cdot 6b}}{5}$ Property 1

    $= \dfrac{6(3)ab\sqrt{6b}}{5}$ Since $\sqrt{9a^2 b^2} = 3ab$

    $= \dfrac{18ab\sqrt{6b}}{5}$

41. $\sqrt[3]{\dfrac{1}{2}} = \dfrac{\sqrt[3]{1}}{\sqrt[3]{2}}$

    $= \dfrac{1}{\sqrt[3]{2}} \cdot \dfrac{\sqrt[3]{4}}{\sqrt[3]{4}}$

    $= \dfrac{\sqrt[3]{4}}{\sqrt[3]{8}}$

    $= \dfrac{\sqrt[3]{4}}{2}$ Since $\sqrt[3]{8} = 2$

45. $\sqrt[3]{\dfrac{3}{2}} = \dfrac{\sqrt[3]{3}}{\sqrt[3]{2}} \cdot \dfrac{\sqrt[3]{4}}{\sqrt[3]{4}}$

    $= \dfrac{\sqrt[3]{12}}{\sqrt[3]{8}}$

    $= \dfrac{\sqrt[3]{12}}{2}$ Since $\sqrt[3]{8} = 2$

49. $\dfrac{1}{\sqrt{2}} = \dfrac{1}{1.414}$

    $= .707$

    $\dfrac{\sqrt{2}}{2} = \dfrac{1.414}{2}$ Both $\dfrac{1}{\sqrt{2}}$ and $\dfrac{\sqrt{2}}{2}$ equal 0.707.

    $= 0.707$

53. See the graph in the back of the textbook.

57. $3x + 7x = (3 + 7)x = 10x$

61. $7a - 3a + 6a = (7 - 3 + 6)a = 10a$

65. $\dfrac{a}{3} + \dfrac{2}{5} = \dfrac{a}{3} \cdot \dfrac{\mathbf{5}}{\mathbf{5}} + \dfrac{2}{5} \cdot \dfrac{\mathbf{3}}{\mathbf{3}}$

$\qquad = \dfrac{5a}{15} + \dfrac{6}{15}$

$\qquad = \dfrac{5a + 6}{15}$

SECTION 7.4

1. $3\sqrt{2} + 4\sqrt{2} = (3 + 4)\sqrt{2}$ Distributive property

$\qquad\qquad\quad = 7\sqrt{2}$

5. $\sqrt{3} + 6\sqrt{3} = 1\sqrt{3} + 6\sqrt{3}$

$\qquad\qquad = (1 + 6)\sqrt{3}$ Distributive property

$\qquad\qquad = 7\sqrt{3}$

9. $14\sqrt{13} - \sqrt{13} = 14\sqrt{13} - 1\sqrt{13}$

$\qquad\qquad\quad = (14 - 1)\sqrt{13}$ Distributive property

$\qquad\qquad\quad = 13\sqrt{13}$

13. $5\sqrt{5} + \sqrt{5} = 5\sqrt{5} + 1\sqrt{5}$

$\qquad\qquad = (5 + 1)\sqrt{5}$ Distributive property

$\qquad\qquad = 6\sqrt{5}$

17. $3\sqrt{3} - \sqrt{27} = 3\sqrt{3} - \sqrt{9}\sqrt{3}$ Property 1 for radicals

$\qquad\qquad\quad = 3\sqrt{3} - 3\sqrt{3}$

$\qquad\qquad\quad = (3 - 3)\sqrt{3}$ Distributive property

$\qquad\qquad\quad = 0$

21. $-\sqrt{75} - \sqrt{3} = -\sqrt{25}\sqrt{3} - \sqrt{3}$ Property 1 for radicals

$\qquad\qquad\quad = -5\sqrt{3} - 1\sqrt{3}$ $\sqrt{25} = 5$

$\qquad\qquad\quad = (-5 - 1)\sqrt{3}$ Distributive property

$\qquad\qquad\quad = -6\sqrt{3}$

25. $\dfrac{3}{4}\sqrt{8} + \dfrac{3}{10}\sqrt{75}$

$\qquad = \dfrac{3}{4} \cdot \sqrt{4}\sqrt{2} + \dfrac{3}{10} \cdot \sqrt{25}\sqrt{3}$ Property 1 for radicals

$\qquad = \dfrac{3}{2} \cdot \sqrt{2} + \dfrac{3}{2} \cdot \sqrt{3}$ $\sqrt{4} = 2$ and $\sqrt{25} = 5$

$\qquad = \dfrac{3}{2}\sqrt{2} + \dfrac{3}{2}\sqrt{3}$

29. $\dfrac{5}{6}\sqrt{72} - \dfrac{3}{8}\sqrt{8} + \dfrac{3}{10}\sqrt{50}$

$\quad = \dfrac{5}{6}\sqrt{36}\sqrt{2} - \dfrac{3}{8}\sqrt{4}\sqrt{2} + \dfrac{3}{10}\sqrt{25}\sqrt{2}$ Property 1 for radicals

$\quad = \dfrac{5}{6}\cdot 6\sqrt{2} - \dfrac{3}{8}\cdot 2\sqrt{2} + \dfrac{3}{10}\cdot 5\sqrt{2}$ $\sqrt{36} = 6,\ \sqrt{4} = 2$ and $\sqrt{25} = 5$

$\quad = 5\sqrt{2} - \dfrac{3}{4}\sqrt{2} + \dfrac{3}{2}\sqrt{2}$

$\quad = \left(\dfrac{20}{4} - \dfrac{3}{4} + \dfrac{6}{4}\right)\sqrt{2} = \dfrac{23}{4}\sqrt{2}$ Remember $5 = \dfrac{5}{1} = \dfrac{20}{4}$

33. $6\sqrt{48} - 2\sqrt{12} + 5\sqrt{27}$

$\quad = 6\sqrt{16}\sqrt{3} - 2\sqrt{4}\sqrt{3} + 5\sqrt{9}\sqrt{3}$ Property 1 for radicals

$\quad = 6\cdot 4\sqrt{3} - 2\cdot 2\sqrt{3} + 5\cdot 3\sqrt{3}$ $\sqrt{16} = 4,\ \sqrt{4} = 2,\ \sqrt{9} = 3$

$\quad = 24\sqrt{3} - 4\sqrt{3} + 15\sqrt{3}$

$\quad = (24 - 4 + 15)\sqrt{3}$ Distributive property

$\quad = 35\sqrt{3}$

37. $\sqrt{x^3} + x\sqrt{x}$

$\quad = \sqrt{x^2}\sqrt{x} + x\sqrt{x}$ Property 1 for radicals

$\quad = x\sqrt{x} + x\sqrt{x}$ $\sqrt{x^2} = x$

$\quad = (x + x)\sqrt{x}$ Distributive property

$\quad = 2x\sqrt{x}$

41. $5\sqrt{8x^3} + x\sqrt{50x}$

$\quad = 5\sqrt{4x^2}\sqrt{2x} + x\sqrt{25}\sqrt{2x}$ Property 1 for radicals

$\quad = 5\cdot 2x\sqrt{2x} + x\cdot 5\sqrt{2x}$ $\sqrt{4x^2} = 2x,\ $ and $\sqrt{25} = 5$

$\quad = 10x\sqrt{2x} + 5x\sqrt{2x}$

$\quad = (10x + 5x)\sqrt{2x}$ Distributive property

$\quad = 15x\sqrt{2x}$

45. $\sqrt{20ab^2} - b\sqrt{45a}$

$\quad = \sqrt{4b^2}\sqrt{5a} - b\sqrt{9}\sqrt{5a}$ Property 1 for radicals

$\quad = 2b\sqrt{5a} - b\cdot 3\sqrt{5a}$ $\sqrt{4b^2} = 2b$ and $\sqrt{9} = 3$

$\quad = (2b - 3b)\sqrt{5a}$ Distributive property

$\quad = -b\sqrt{5a}$

49. $7\sqrt{50x^2y} + 8x\sqrt{8y} - 7\sqrt{32x^2y}$

$= 7\sqrt{25x^2}\sqrt{2y} + 8x\sqrt{4}\sqrt{2y} - 7\sqrt{16x^2}\sqrt{2y}$ Property 1 for radicals

$= 7 \cdot 5x\sqrt{2y} + 8x \cdot 2\sqrt{2y} - 7 \cdot 4x\sqrt{2y}$ $\sqrt{25x^2} = 5x, \quad \sqrt{4} = 2$ and $\sqrt{16x^2} = 4x$

$= 35x\sqrt{2y} + 16x\sqrt{2y} - 28x\sqrt{2y}$

$= (35x + 16x - 28x)\sqrt{2y}$ Distributive property

$= 23x\sqrt{2y}$

53. $\dfrac{6+\sqrt{8}}{2} = \dfrac{6+\sqrt{4}\sqrt{2}}{2}$ Property 1 for radicals

$= \dfrac{6+2\sqrt{2}}{2}$ $\sqrt{4} = 2$

$= \dfrac{2(3+\sqrt{2})}{2}$

$= 3 + \sqrt{2}$

57. $\sqrt{5} + \sqrt{3} \overset{?}{=} \sqrt{8}$

$2.236 + 1.732 \overset{?}{=} 2.828$

$3.968 \neq 2.828$

$\sqrt{5} + \sqrt{3}$ does not equal $\sqrt{8}$

61. $(3x+y)^2 = (3x)^2 + 2(3x)(y) + y^2$

$= 9x^2 + 6xy + y^2$

65. $\dfrac{x}{3} - \dfrac{1}{2} = \dfrac{5}{2}$

$6\left(\dfrac{x}{3}\right) - 6\left(\dfrac{1}{2}\right) = 6\left(\dfrac{5}{2}\right)$ Multiply by LCD of 6

$2x - 3 = 15$

$2x = 18$

$x = 9$

69. $\dfrac{a}{a-4} - \dfrac{a}{2} = \dfrac{4}{a-4}$

$2(a-4)\left(\dfrac{a}{a-4}\right) - 2(a-4)\left(\dfrac{a}{2}\right) = 2(a-4)\left(\dfrac{4}{a-4}\right)$

$2a - a(a-4) = 8$

$2a - a^2 + 4a = 8$

$0 = a^2 - 6a + 8$

$0 = (a-2)(a-4)$

$a - 2 = 0 \quad$ or $\quad a - 4 = 0$

$a = 2 \qquad\qquad a = 4$

Possible solutions 2 and 4, only 2 checks.

SECTION 7.5

1. $\sqrt{3}\,\sqrt{2} = \sqrt{3\cdot 2} = \sqrt{6}$ Property 1 of radicals and multiply

5. $\left(2\sqrt{3}\right)\left(5\sqrt{7}\right)$

$= \left(2\cdot 5\right)\left(\sqrt{3}\sqrt{7}\right)$ Commutative and Associative properties

$= \left(2\cdot 5\right)\left(\sqrt{3\cdot 7}\right)$ Property 1 of radicals

$= 10\sqrt{21}$ Multiply

9. $\sqrt{2}\left(\sqrt{3}-1\right)$

$= \sqrt{2}\cdot\sqrt{3} - \sqrt{2}(1)$ Distributive property

$= \sqrt{6}-\sqrt{2}$ Multiply

13. $\sqrt{3}\left(2\sqrt{2}+\sqrt{3}\right)$

$= \sqrt{3}\cdot 2\sqrt{2}+\sqrt{3}\cdot\sqrt{3}$ Distributive property

$= 2\sqrt{6}+3$ Multiply

17. $2\sqrt{3}\left(\sqrt{2}+\sqrt{5}\right)$

$= 2\sqrt{3}\cdot\sqrt{2}+2\sqrt{3}\cdot\sqrt{5}$ Distributive property

$= 2\sqrt{6}+2\sqrt{15}$

21. $\left(\sqrt{x}+3\right)^{2} = \left(\sqrt{x}+3\right)\left(\sqrt{x}+3\right)$

$= \underset{\text{F}}{\sqrt{x}\cdot\sqrt{x}}+\underset{\text{O}}{3\sqrt{x}}+\underset{\text{I}}{3\sqrt{x}}+\underset{\text{L}}{3\cdot 3}$

$= x+6\sqrt{x}+9$

25. $\left(\sqrt{a}-\dfrac{1}{2}\right)^{2} = \left(\sqrt{a}-\dfrac{1}{2}\right)\left(\sqrt{a}-\dfrac{1}{2}\right)$

$= \underset{\text{F}}{\sqrt{a}\sqrt{a}}-\underset{\text{O}}{\dfrac{1}{2}\sqrt{a}}-\underset{\text{I}}{\dfrac{1}{2}\sqrt{a}}+\underset{\text{L}}{\dfrac{1}{4}}$

$= a-\sqrt{a}+\dfrac{1}{4}$

29. $\left(\sqrt{5}+3\right)\left(\sqrt{5}+2\right) = \underset{\text{F}}{\sqrt{5}\sqrt{5}}+\underset{\text{O}}{2\sqrt{5}}+\underset{\text{I}}{3\sqrt{5}}+\underset{\text{L}}{6}$

$= 5+5\sqrt{5}+6$

$= 11+5\sqrt{5}$

33. $\left(\sqrt{3}+\dfrac{1}{2}\right)\left(\sqrt{2}+\dfrac{1}{3}\right) = \underset{\text{F}}{\sqrt{3}\sqrt{2}}+\underset{\text{O}}{\dfrac{1}{3}\sqrt{3}}+\underset{\text{I}}{\dfrac{1}{2}\sqrt{2}}+\underset{\text{L}}{\dfrac{1}{6}}$

$= \sqrt{6}+\dfrac{1}{3}\sqrt{3}+\dfrac{1}{2}\sqrt{2}+\dfrac{1}{6}$

37. $\left(\sqrt{a}+\dfrac{1}{3}\right)\left(\sqrt{a}+\dfrac{2}{3}\right) = \underset{\text{F}}{\sqrt{a}\sqrt{a}}+\underset{\text{O}}{\dfrac{2}{3}\sqrt{a}}+\underset{\text{I}}{\dfrac{1}{3}\sqrt{a}}+\underset{\text{L}}{\dfrac{2}{9}}$

$= a+\sqrt{a}+\dfrac{2}{9}$

41. $\left(2\sqrt{7}+3\right)\left(3\sqrt{7}-4\right) = \underset{\text{F}}{2\sqrt{7}\cdot 3\sqrt{7}}-\underset{\text{O}}{4\cdot 2\sqrt{7}}+\underset{\text{I}}{3\cdot 3\sqrt{7}}+\underset{\text{L}}{3(-4)}$

$= 6\cdot 7-8\sqrt{7}+9\sqrt{7}-12$

$= 42+\sqrt{7}-12$

$= 30+\sqrt{7}$

45. $\left(7\sqrt{a}+2\sqrt{b}\right)\left(7\sqrt{a}-2\sqrt{b}\right)=(7\sqrt{a})^2-\left(2\sqrt{b}\right)^2=49a-4b$

49.
$$\frac{\sqrt{5}}{\sqrt{5}+\sqrt{2}}=\frac{\sqrt{5}}{\sqrt{5}+\sqrt{2}}\cdot\frac{\left(\sqrt{5}-\sqrt{2}\right)}{\left(\sqrt{5}-\sqrt{2}\right)}$$
$$=\frac{\sqrt{5}\sqrt{5}-\sqrt{5}\sqrt{2}}{\left(\sqrt{5}\right)^2-\left(\sqrt{2}\right)^2}$$
$$=\frac{5-\sqrt{10}}{5-2}$$
$$=\frac{5-\sqrt{10}}{3}$$

53.
$$\frac{\sqrt{3}+\sqrt{2}}{\sqrt{3}-\sqrt{2}}=\frac{\sqrt{3}+\sqrt{2}}{\sqrt{3}-\sqrt{2}}\cdot\frac{\left(\sqrt{3}+\sqrt{2}\right)}{\left(\sqrt{3}+\sqrt{2}\right)}$$
$$=\frac{\sqrt{3}\sqrt{3}+\sqrt{3}\sqrt{2}+\sqrt{2}\sqrt{3}+\sqrt{2}\sqrt{2}}{\left(\sqrt{3}\right)^2-\left(\sqrt{2}\right)^2}$$
$$=\frac{3+\sqrt{6}+\sqrt{6}+2}{3-2}$$
$$=\frac{5+2\sqrt{6}}{1}$$
$$=5+2\sqrt{6}$$

57.
$$\frac{\sqrt{x}+2}{\sqrt{x}-2}=\frac{\sqrt{x}+2}{\sqrt{x}-2}\cdot\frac{\left(\sqrt{x}+2\right)}{\left(\sqrt{x}+2\right)}$$
$$=\frac{\sqrt{x}\sqrt{x}+2\sqrt{x}+2\sqrt{x}+2\cdot2}{\left(\sqrt{x}\right)^2-2^2}$$
$$=\frac{x+4\sqrt{x}+4}{x-4}$$

61. $2\left(3\sqrt{5}\right)=6\sqrt{15}$ Given statement is false

      $2\left(3\sqrt{5}\right)=6\sqrt{5}$ Corrected statement

65. $x^2+5x-6=0$

    $(x+6)(x-1)=0$ Factor the left side

    $x+6=0$ or $x-1=0$ Set factors equal to 0

       $x=-6$ $x=1$

The solutions are −6 and 1.

69. $\dfrac{x}{3}=\dfrac{27}{x}$

    $x\cdot x=3\cdot27$ Product of extremes $=$

     $x^2=81$ product of means

      $x=9,-9$

The solutions are 9 and −9.

73. $\dfrac{\text{miles}}{\text{hours}}=\dfrac{375}{15}=\dfrac{m}{20}$

   $375\cdot20=15m$ Product of extremes $=$

    $7500=15m$ product of means

     $500=m$

You would drive 500 miles in 20 hours.

SECTION 7.6

1. $\sqrt{x+1} = 2$

   $\left(\sqrt{x+1}\right)^2 = 2^2$ Square both sides

   $x+1 = 4$

   $x = 3$

   To check our solution when $x = 3$,

   $\sqrt{3+1} = 2$

   $\sqrt{4} = 2$

   $2 = 2$ A true statement

   The solution is 3.

9. $\sqrt{x-8} = 0$

   $\left(\sqrt{x-8}\right)^2 = 0^2$ Square both sides

   $x-8 = 0$

   $x = 8$

   To check our solution when $x = 8$,

   $\sqrt{8-8} = 0$

   $\sqrt{0} = 0$

   $0 = 0$ A true statement

   The solution is 8.

17. $2\sqrt{x} = 10$

    $\sqrt{x} = 5$ Divide by 2

    $\left(\sqrt{x}\right)^2 = 5^2$ Square both sides

    $x = 25$

    To check our solution when $x = 25$,

    $2\sqrt{25} = 10$

    $2 \cdot 5 = 10$

    $10 = 10$ A true statement

    The solution is 25.

5. $\sqrt{x-9} = -6$

   $\left(\sqrt{x-9}\right)^2 = (-6)^2$ Square both sides

   $x-9 = 36$

   $x = 45$

   To check our solution when $x = 45$,

   $\sqrt{45-9} = -6$

   $\sqrt{36} = -6$

   $6 = -6$ A false statement

   The solution is $\varnothing$.

13. $\sqrt{2x-3} = -5$

    $\left(\sqrt{2x-3}\right)^2 = (-5)^2$ Square both sides

    $2x-3 = 25$

    $2x = 28$

    $x = 14$

    To check our solution when $x = 14$,

    $\sqrt{2(14)-3} = -5$

    $\sqrt{28-3} = -5$

    $\sqrt{25} = -5$

    $5 = -5$ A false statement

    The solution is $\varnothing$.

21. $\sqrt{3x+4} - 3 = 2$

    $\sqrt{3x+4} = 5$ Add 3 to both sides

    $\left(\sqrt{3x+4}\right)^2 = 5^2$ Square both sides

    $3x+4 = 25$

    $3x = 21$

    $x = 7$

    To check our solution when $x = 7$,

    $\sqrt{3(7)+4} - 3 = 2$

    $\sqrt{21+4} - 3 = 2$

    $\sqrt{25} - 3 = 2$

    $5 - 3 = 2$

    $2 = 2$ A true statement

    The solution is 7.

25. $\sqrt{2x+1}+5=2$

$\quad\quad \sqrt{2x+1}=-3$ Add -5 to both sides

$\quad \left(\sqrt{2x+1}\right)^2=(-3)^2$ Square both sides

$\quad\quad\quad 2x+1=9$

$\quad\quad\quad\quad 2x=8$

$\quad\quad\quad\quad\quad x=4$

To check our solution when $x=4$,

$\quad\quad \sqrt{2(4)+1}+5=2$

$\quad\quad\quad \sqrt{8+1}+5=2$

$\quad\quad\quad\quad \sqrt{9}+5=2$

$\quad\quad\quad\quad\quad 3+5=2$

$\quad\quad\quad\quad\quad\quad 8=2$ A false statement

$\quad\quad\quad\quad$ The solution is $\varnothing$.

33. $\quad \sqrt{y-4}=y-6$

$\quad \left(\sqrt{y-4}\right)^2=(y-6)^2$ Square both sides

$\quad\quad y-4=y^2-12y+36$

$\quad\quad\quad 0=y^2-13y+40$ Add $-(y-4)$ to both sides

$\quad\quad\quad 0=(y-8)(y-5)$ Factor

$0=y-8$ or $0=y-5$

$8=y$ $5=y$

Check $y=8$

$\sqrt{8-4}=8-6$

$\quad \sqrt{4}=2$

$\quad\quad 2=2$ A true statement

Check $y=5$

$\sqrt{5-4}=5-6$

$\quad \sqrt{1}=-1$

$\quad\quad 1=-1$ A false statement

Possible solutions are 5 and 8, but only 8 checks.

29. $\quad \sqrt{a+2}=a+2$

$\quad \left(\sqrt{a+2}\right)^2=(a+2)^2$ Square both sides

$\quad\quad a+2=a^2+4a+4$

$\quad\quad\quad 0=a^2+3a+2$ Add $-(a+2)$ to both sides

$\quad\quad\quad 0=(a+2)(a+1)$ Factor

$a+2=0$ or $a+1=0$

$\quad a=-2$ $a=-1$

Check $a=-2$

$\sqrt{-2+2}=-2+2$

$\quad\quad 0=0$ A true statement

Check $a=-1$

$\sqrt{-1+2}=-1+2$

$\quad\quad \sqrt{1}=1$

$\quad\quad\quad 1=1$ A true statement

The solutions are -2 and -1.

37. $\sqrt{x+3} = \sqrt{x} + 3$

$\left(\sqrt{x+3}\right)^2 = \left(\sqrt{x}+3\right)^2$ Square both sides

$x+3 = x + 6\sqrt{x} + 9$ Remember $\left(\sqrt{x}+3\right)^2 =$

$\left(\sqrt{x}\right)^2 + 2\left(\sqrt{x}\right)(3) + 3^2$

$-6 = 6\sqrt{x}$ Adding $-x$ and -9 to both sides

$-1 = \sqrt{x}$ Divide both sides by 6

$1 = x$ Square both sides again

Check $x = 1$

$\sqrt{1+3} = \sqrt{1} + 3$

$\sqrt{4} = 1 + 3$

$2 = 4$ A false statement

The solution set is $\varnothing$.

41. $x + 2 = \sqrt{8x}$ Translation

$(x+2)^2 = \left(\sqrt{8x}\right)^2$ Square both sides

$x^2 + 4x + 4 = 8x$ Remember $(x+2)^2 = x^2 + 2(x)(2) + 2^2$

$x^2 - 4x + 4 = 0$ Add $-8x$ to both sides

$(x-2)(x-2) = 0$ Factor

$x - 2 = 0$ or $x - 2 = 0$ Set factors to 0

$x = 2$ $x = 2$

The solution is 2.

45. Let $T = 2$

then $T = \dfrac{11}{7}\sqrt{\dfrac{L}{2}}$

becomes $2 = \dfrac{11}{7}\sqrt{\dfrac{L}{2}}$

$2\left(\dfrac{7}{11}\right) = \dfrac{11}{7}\left(\dfrac{7}{11}\right)\sqrt{\dfrac{L}{2}}$ Multiply by $\dfrac{7}{11}$

$\dfrac{14}{11} = \sqrt{\dfrac{L}{2}}$

$\left(\dfrac{14}{11}\right)^2 = \left(\sqrt{\dfrac{L}{2}}\right)^2$ Square both sides

$\dfrac{196}{121} = \dfrac{L}{2}$

$2\left(\dfrac{196}{121}\right) = 2\left(\dfrac{L}{2}\right)$ Multiply by 2

$\dfrac{392}{121} = L$

$\dfrac{392}{121} \approx 3.2$ feet

49. $y = 2\sqrt{x}$

x	y
0	$2\sqrt{0} = 0$
1	$2\sqrt{1} = 2$
4	$2\sqrt{4} = 4$
9	$2\sqrt{9} = 6$

See the graph in the back of the textbook.

53. $\dfrac{\frac{2}{5}}{\frac{4}{15}} = \dfrac{2}{5} \cdot \dfrac{15}{4} = \dfrac{30}{20} = \dfrac{3}{2}$

57. Let x = a number and $\frac{1}{x}$ = its reciprocal, then

$$x + \frac{1}{x} = \frac{10}{3}$$

$$3x(x) + 3x\left(\frac{1}{x}\right) = 3x\left(\frac{10}{3}\right) \qquad \text{Multiply both sides by LCD} = 3x$$

$$3x^2 + 3 = 10x$$

$$3x^2 - 10x + 3 = 0 \qquad \text{Add } -10x \text{ to each side}$$

$$(3x - 1)(x - 3) = 0 \qquad \text{Factor}$$

$$3x - 1 = 0 \quad \text{or} \quad x - 3 = 0$$

$$x = 3$$

$$\frac{1}{x} = \frac{1}{3}$$

$$\frac{1}{x} = 3$$

The solution is 3 and $\frac{1}{3}$.

61. Let x = amount of time to fill the sink with both faucets open. In one hour,

Hot water in sink   +   Cold water in sink   =   Total amount of water in sink

$$\frac{1}{4} \qquad\qquad + \qquad\qquad \frac{1}{3} \qquad\qquad = \qquad\qquad \frac{1}{x}\text{--}$$

$$\frac{1}{4} + \frac{1}{3} = \frac{1}{x}$$

$$12x\left(\frac{1}{4}\right) + 12x\left(\frac{1}{3}\right) = 12x\left(\frac{1}{x}\right) \qquad \text{Multiply both sides by LCD} = 12x$$

$$3x + 4x = 12$$

$$7x = 12$$

$$x = \frac{12}{7}$$

It will take $\frac{12}{7}$ minutes to fill the sink.

CHAPTER 7 REVIEW

1. $\sqrt{25} = 5 \qquad 5^2 = 25$

2. $\sqrt{81} = 9 \qquad 9^2 = 81$

3. $\sqrt{169} = 13 \qquad 13^2 = 169$

4. $\sqrt{400} = 20 \qquad 20^2 = 400$

5. $\sqrt[3]{-1} = -1 \qquad (-1)^3 = -1$

6. $-\sqrt[3]{-64} = -(-4) = 4 \qquad (-4)^3 = -64$

7. $\sqrt[4]{625} = 5 \qquad 5^4 = 625$

8. $-\sqrt[4]{81} = -3 \qquad (3)^4 = 81$

9. $\sqrt{100x^2y^4} = \sqrt{100} \cdot \sqrt{x^2} \cdot \sqrt{y^4} = 10xy^2$

10. $\sqrt{121x^4y^2} = 11x^2y \qquad 11^2 = 121$

11. $\sqrt[3]{8a^3} = 2a \qquad 2^3 = 8$

12. $\sqrt[3]{-27a^3} = -3a \qquad (-3)^3 = -27$

13. $\sqrt{24} = \sqrt{4 \cdot 6}$
$= \sqrt{4} \cdot \sqrt{6} \qquad$ Property 1
$= 2\sqrt{6} \qquad \sqrt{4} = 2$

14. $\sqrt{44} = \sqrt{4 \cdot 11}$
$= \sqrt{4} \cdot \sqrt{11} \qquad$ Property 1
$= 2\sqrt{11} \qquad \sqrt{4} = 2$

15. $\sqrt{60x^2} = \sqrt{4x^2 \cdot 15}$
$= \sqrt{4x^2} \cdot \sqrt{15} \qquad$ Property 1
$= 2x\sqrt{15} \qquad \sqrt{4x^2} = 2x$

16. $\sqrt{80x^2} = \sqrt{16x^2 \cdot 5}$
$= \sqrt{16x^2} \cdot \sqrt{5} \qquad$ Property 1
$= 4x\sqrt{5} \qquad \sqrt{16x^2} = 4x$

17. $\sqrt{90x^3y^4} = \sqrt{9x^2y^4 \cdot 10x}$
$= \sqrt{9x^2y^4} \cdot \sqrt{10x} \qquad$ Property 1
$= 3xy^2\sqrt{10x} \qquad \sqrt{9x^2y^4} = 3xy^2$

18. $\sqrt{120x^4y^3} = \sqrt{4x^4y^2 \cdot 30y}$
$= \sqrt{4x^4y^2} \cdot \sqrt{30y} \qquad$ Property 1
$= 2x^2y\sqrt{30y} \qquad \sqrt{4x^4y^2} = 2x^2y$

19. $-\sqrt{32} = -\sqrt{16 \cdot 2}$
$= -\sqrt{16} \cdot \sqrt{2} \qquad$ Property 1
$= -4\sqrt{2} \qquad -\sqrt{16} = -4$

20. $-\sqrt{72} = -\sqrt{36 \cdot 2} = -\sqrt{36} \cdot \sqrt{2} = -6\sqrt{2}$

21. $3\sqrt{20x^3y} = 3\sqrt{4x^2 \cdot 5xy}$
$= 3\sqrt{4x^2} \cdot \sqrt{5xy} \qquad$ Property 1
$= 3 \cdot 2x\sqrt{5xy} \qquad \sqrt{4x^2} = 2x$
$= 6x\sqrt{5xy}$

22. $4\sqrt{28x^2y^3} = 4\sqrt{4x^2y^2 \cdot 7y}$
$= 4\sqrt{4x^2y^2} \cdot \sqrt{7y} \qquad$ Property 1
$= 4 \cdot 2xy\sqrt{7y} \qquad \sqrt{4x^2y^2} = 2xy$
$= 8xy\sqrt{7y}$

23. $\sqrt{\dfrac{3}{49}} = \dfrac{\sqrt{3}}{\sqrt{49}}$ Property 2

     $= \dfrac{\sqrt{3}}{7}$ $\sqrt{49} = 7$

24. $\sqrt{\dfrac{5}{64}} = \dfrac{\sqrt{5}}{\sqrt{64}}$ Property 2

     $= \dfrac{\sqrt{5}}{8}$ $\sqrt{64} = 8$

25. $\sqrt{\dfrac{8}{81}} = \dfrac{\sqrt{8}}{\sqrt{81}}$ Property 2

     $= \dfrac{\sqrt{4 \cdot 2}}{9}$ $\sqrt{81} = 9$

     $= \dfrac{2\sqrt{2}}{9}$ $\sqrt{4} = 2$

26. $\sqrt{\dfrac{18}{121}} = \dfrac{\sqrt{18}}{\sqrt{121}}$ Property 2

     $= \dfrac{\sqrt{9 \cdot 2}}{\sqrt{121}}$

     $= \dfrac{3\sqrt{2}}{11}$ $\sqrt{9} = 3$ and $\sqrt{121} = 11$

27. $\sqrt{\dfrac{49}{64}} = \dfrac{\sqrt{49}}{\sqrt{64}}$ Property 2

     $= \dfrac{7}{8}$ $\sqrt{49} = 7$ and $\sqrt{64} = 8$

28. $\sqrt{\dfrac{121}{144}} = \dfrac{\sqrt{121}}{\sqrt{144}}$ Property 2

     $= \dfrac{11}{12}$ $\sqrt{121} = 11$ and $\sqrt{144} = 12$

29. $\sqrt{\dfrac{49a^2b^2}{16}} = \dfrac{\sqrt{49a^2b^2}}{\sqrt{16}}$ Property 2

     $= \dfrac{7ab}{4}$ $\sqrt{49a^2b^2} = 7ab$

30. $\sqrt{\dfrac{100a^2b^2}{81}} = \dfrac{\sqrt{100a^2b^2}}{\sqrt{81}}$ Property 2

     $= \dfrac{10ab}{9}$ $\sqrt{100} = 10$ and $\sqrt{81} = 9$

31. $\sqrt{\dfrac{80}{49}} = \dfrac{\sqrt{80}}{\sqrt{49}}$ Property 2

     $= \dfrac{\sqrt{16 \cdot 5}}{7}$ $\sqrt{49} = 7$

     $= \dfrac{\sqrt{16} \cdot \sqrt{5}}{7}$ Property 1

     $= \dfrac{4\sqrt{5}}{7}$ $\sqrt{16} = 4$

32. $\sqrt{\dfrac{200}{81}} = \dfrac{\sqrt{100 \cdot 2}}{\sqrt{81}} = \dfrac{10\sqrt{2}}{9}$

33. $\sqrt{\dfrac{40a^2}{121}} = \dfrac{\sqrt{40a^2}}{\sqrt{121}}$ Property 2

     $= \dfrac{\sqrt{4a^2 \cdot 10}}{11}$

     $= \dfrac{2a\sqrt{10}}{11}$ $\sqrt{4a^2} = 2a$

34. $\sqrt{\dfrac{90a^2}{169}} = \dfrac{\sqrt{9a^2 \cdot 10}}{\sqrt{169}} = \dfrac{3a\sqrt{10}}{13}$

35. $\dfrac{5\sqrt{84}}{7} = \dfrac{5\sqrt{4\cdot 21}}{7}$

$= \dfrac{5\sqrt{4}\cdot\sqrt{21}}{7}$ Property 1

$= \dfrac{5\cdot 2\sqrt{21}}{7}$ $\sqrt{4} = 2$

$= \dfrac{10\sqrt{21}}{7}$

36. $\dfrac{3\sqrt{96}}{5} = \dfrac{3\sqrt{16\cdot 6}}{5} = \dfrac{12\sqrt{6}}{5}$ $\sqrt{16} = 4$

37. $\dfrac{3\sqrt{120a^2b^2}}{\sqrt{25}} = \dfrac{3\sqrt{4a^2b^2\cdot 30}}{\sqrt{25}}$

$= \dfrac{3\cdot 2ab\sqrt{30}}{5}$ Property 1, $\sqrt{4a^2b^2} = 2ab$, $\sqrt{25} = 5$

$= \dfrac{6ab\sqrt{30}}{5}$

38. $\dfrac{8\sqrt{150a^2b^2}}{\sqrt{81}} = \dfrac{8\sqrt{25a^2b^2\cdot 6}}{\sqrt{81}}$

$= \dfrac{8\cdot 5ab\sqrt{6}}{9}$ Property 1, $\sqrt{25a^2b^2} = 5ab$, $\sqrt{81} = 9$

$= \dfrac{40ab\sqrt{6}}{9}$

39. $\dfrac{-5\sqrt{20x^3y^2}}{\sqrt{144}} = \dfrac{-5\sqrt{4x^2y^2\cdot 5x}}{\sqrt{144}}$

$= \dfrac{-5\cdot 2xy\sqrt{5x}}{12}$ Property 1, $\sqrt{4x^2y^2} = 2xy$, $\sqrt{144} = 12$

$= \dfrac{-5xy\sqrt{5x}}{6}$

40. $\dfrac{-9\sqrt{45x^2y^3}}{\sqrt{36}} = \dfrac{-9\sqrt{9x^2y^2\cdot 5y}}{\sqrt{36}}$

$= \dfrac{-9\cdot 3xy\sqrt{5y}}{6}$ Property 1, $\sqrt{9x^2y^2} = 3xy$

$= \dfrac{-9xy\sqrt{5y}}{2}$

41. $\dfrac{2}{\sqrt{7}} = \dfrac{2}{\sqrt{7}} \cdot \dfrac{\sqrt{7}}{\sqrt{7}} = \dfrac{2\sqrt{7}}{7}$ $\left(\sqrt{7}\right)^2 = 7$

42. $\dfrac{5}{\sqrt{3}} = \dfrac{5}{\sqrt{3}} \cdot \dfrac{\sqrt{3}}{\sqrt{3}}$ Rationalize the denominator

    $= \dfrac{5\sqrt{3}}{3}$ $\left(\sqrt{3}\right)^2 = 3$

43. $\sqrt{\dfrac{32}{5}} = \dfrac{\sqrt{16 \cdot 2}}{\sqrt{5}}$ Property 2

    $= \dfrac{4\sqrt{2}}{\sqrt{5}}$ $\sqrt{16} = 4$

    $= \dfrac{4\sqrt{2}}{\sqrt{5}} \cdot \dfrac{\sqrt{5}}{\sqrt{5}}$ Rationalize the denominator

    $= \dfrac{4\sqrt{10}}{5}$ $\left(\sqrt{5}\right)^2 = 5$

44. $\sqrt{\dfrac{40}{3}} = \dfrac{\sqrt{4 \cdot 10}}{\sqrt{3}}$ Property 2

    $= \dfrac{2\sqrt{10}}{\sqrt{3}}$ $\sqrt{4} = 2$

    $= \dfrac{2\sqrt{10}}{\sqrt{3}} \cdot \dfrac{\sqrt{3}}{\sqrt{3}}$ Rationalize the denominator

    $= \dfrac{2\sqrt{30}}{3}$ $\left(\sqrt{3}\right)^2 = 3$

45. $\sqrt{\dfrac{5}{48}} = \dfrac{\sqrt{5}}{\sqrt{48}}$ Property 2

    $= \dfrac{\sqrt{5}}{\sqrt{16 \cdot 3}}$

    $= \dfrac{\sqrt{5}}{4\sqrt{3}}$ $\sqrt{16} = 4$

    $= \dfrac{\sqrt{5}}{4\sqrt{3}} \cdot \dfrac{\sqrt{3}}{\sqrt{3}}$ Rationalize the denominator

    $= \dfrac{\sqrt{15}}{4 \cdot 3}$

    $= \dfrac{\sqrt{15}}{12}$

46. $\sqrt{\dfrac{7}{50}} = \dfrac{\sqrt{7}}{\sqrt{25 \cdot 2}}$ Property 2

     $= \dfrac{\sqrt{7}}{5\sqrt{2}} \cdot \dfrac{\sqrt{2}}{\sqrt{2}}$ Rationalize the denominator

     $= \dfrac{\sqrt{14}}{5 \cdot 2}$ $\left(\sqrt{2}\right)^2 = 2$

     $= \dfrac{\sqrt{14}}{10}$

47. $\dfrac{-3\sqrt{60}}{\sqrt{5}} = -3\sqrt{\dfrac{60}{5}}$ Property 2

     $= -3\sqrt{12}$ $\dfrac{60}{5} = 12$

     $= -3\sqrt{4 \cdot 3}$

     $= -3\sqrt{4}\sqrt{3}$ Property 1

     $= -3 \cdot 2\sqrt{3}$ $\sqrt{4} = 2$

     $= -6\sqrt{3}$

48. $\dfrac{-5\sqrt{80}}{\sqrt{3}} = \dfrac{-5\sqrt{16 \cdot 5}}{\sqrt{3}}$

     $= \dfrac{-5 \cdot 4\sqrt{5}}{\sqrt{3}} \cdot \dfrac{\sqrt{3}}{\sqrt{3}}$ Rationalize the denominator

     $= \dfrac{-20\sqrt{15}}{3}$ $\left(\sqrt{3}\right)^2 = 3$

49. $\sqrt{\dfrac{32ab^2}{3}} = \dfrac{\sqrt{32ab^2}}{\sqrt{3}}$ Property 2

     $= \dfrac{\sqrt{16b^2 \cdot 2a}}{\sqrt{3}}$

     $= \dfrac{4b\sqrt{2a}}{\sqrt{3}}$

     $= \dfrac{4b\sqrt{2a}}{\sqrt{3}} \cdot \dfrac{\sqrt{3}}{\sqrt{3}}$ Rationalize the denominator

     $= \dfrac{4b\sqrt{6a}}{3}$ $\left(\sqrt{3}\right)^2 = 3$

50. $\sqrt{\dfrac{40a^2b}{7}} = \dfrac{\sqrt{4a^2 \cdot 10b}}{\sqrt{7}}$ Property 2

     $= \dfrac{2a \cdot \sqrt{10b}}{\sqrt{7}} \cdot \dfrac{\sqrt{7}}{\sqrt{7}}$ Rationalize the denominator

     $= \dfrac{2a\sqrt{70b}}{7}$ $\left(\sqrt{7}\right)^2 = 7$

51. $\sqrt[3]{\dfrac{3}{4}} = \dfrac{\sqrt[3]{3}}{\sqrt[3]{4}}$ Property 2

$\phantom{\sqrt[3]{\dfrac{3}{4}}} = \dfrac{\sqrt[3]{3}}{\sqrt[3]{4}} \cdot \dfrac{\sqrt[3]{2}}{\sqrt[3]{2}}$ Rationalize the denominator

$\phantom{\sqrt[3]{\dfrac{3}{4}}} = \dfrac{\sqrt[3]{6}}{2}$ $\sqrt[3]{8} = 2$

52. $\sqrt[3]{\dfrac{2}{3}} = \dfrac{\sqrt[3]{2}}{\sqrt[3]{3}}$ Property 2

$\phantom{\sqrt[3]{\dfrac{2}{3}}} = \dfrac{\sqrt[3]{2}}{\sqrt[3]{3}} \cdot \dfrac{\sqrt[3]{9}}{\sqrt[3]{9}}$ Rationalize the denominator

$\phantom{\sqrt[3]{\dfrac{2}{3}}} = \dfrac{\sqrt[3]{18}}{3}$ $\sqrt[3]{27} = 3$

53. $\dfrac{3}{\sqrt{3}-4} = \dfrac{3}{\sqrt{3}-4} \cdot \dfrac{\sqrt{3}+4}{\sqrt{3}+4}$ Multiply by the conjugate $\sqrt{3}+4$

$\phantom{\dfrac{3}{\sqrt{3}-4}} = \dfrac{3\sqrt{3}+12}{\left(\sqrt{3}\right)^2 - 4^2}$

$\phantom{\dfrac{3}{\sqrt{3}-4}} = \dfrac{3\sqrt{3}+12}{3-16}$

$\phantom{\dfrac{3}{\sqrt{3}-4}} = \dfrac{3\sqrt{3}+12}{-13}$

$\phantom{\dfrac{3}{\sqrt{3}-4}} = \dfrac{-3\sqrt{3}-12}{13}$ Do not leave a negative

                                           in the denominator

54. $\dfrac{-2}{\sqrt{5}+1} = \dfrac{-2}{\sqrt{5}+1} \cdot \dfrac{\sqrt{5}-1}{\sqrt{5}-1}$ Multiply by the conjugate $\sqrt{5}-1$

$\phantom{\dfrac{-2}{\sqrt{5}+1}} = \dfrac{-2\sqrt{5}+2}{\left(\sqrt{5}\right)^2 - 1^2}$

$\phantom{\dfrac{-2}{\sqrt{5}+1}} = \dfrac{-2\sqrt{5}+2}{5-1}$

$\phantom{\dfrac{-2}{\sqrt{5}+1}} = \dfrac{-2\sqrt{5}+2}{4}$

$\phantom{\dfrac{-2}{\sqrt{5}+1}} = \dfrac{2\left(-\sqrt{5}+1\right)}{4}$ Distributive property

$\phantom{\dfrac{-2}{\sqrt{5}+1}} = \dfrac{-\sqrt{5}+1}{2}$

55. $\dfrac{2}{3+\sqrt{7}} = \dfrac{2}{3+\sqrt{7}} \cdot \dfrac{\mathbf{3-\sqrt{7}}}{\mathbf{3-\sqrt{7}}}$ Multiply by the conjugate $3 - \sqrt{7}$

$= \dfrac{6-2\sqrt{7}}{3^2 - \left(\sqrt{7}\right)^2}$

$= \dfrac{6-2\sqrt{7}}{9-7}$

$= \dfrac{6-2\sqrt{7}}{2}$

$= \dfrac{2\left(3-\sqrt{7}\right)}{2}$ Distributive property

$= 3-\sqrt{7}$

56. $\dfrac{6}{2-\sqrt{5}} = \dfrac{6}{2-\sqrt{5}} \cdot \dfrac{\mathbf{2+\sqrt{5}}}{\mathbf{2+\sqrt{5}}}$ Multiply by the conjugate $2 + \sqrt{5}$

$= \dfrac{12+6\sqrt{5}}{2^2 - \left(\sqrt{5}\right)^2}$

$= \dfrac{12+6\sqrt{5}}{4-5}$

$= \dfrac{12+6\sqrt{5}}{-1}$

$= -12-6\sqrt{5}$ Do not leave a negative
in the denominator

57. $\dfrac{3}{\sqrt{5}-\sqrt{2}} = \dfrac{3}{\sqrt{5}-\sqrt{2}} \cdot \dfrac{\mathbf{\sqrt{5}+\sqrt{2}}}{\mathbf{\sqrt{5}+\sqrt{2}}}$ Multiply by the conjugate $\sqrt{5} + \sqrt{2}$

$= \dfrac{3\sqrt{5}+3\sqrt{2}}{\left(\sqrt{5}\right)^2 - \left(\sqrt{2}\right)^2}$

$= \dfrac{3\sqrt{5}+3\sqrt{2}}{5-2}$

$= \dfrac{3\sqrt{5}+3\sqrt{2}}{3}$

$= \dfrac{3\left(\sqrt{5}+\sqrt{2}\right)}{3}$ Distributive property

$= \sqrt{5}+\sqrt{2}$

58. $\dfrac{5}{\sqrt{7}+\sqrt{3}} = \dfrac{5}{\sqrt{7}+\sqrt{3}} \cdot \dfrac{\sqrt{7}-\sqrt{3}}{\sqrt{7}-\sqrt{3}}$ Multiply by the conjugate $\sqrt{7}-\sqrt{3}$

$= \dfrac{5\sqrt{7}-5\sqrt{3}}{\left(\sqrt{7}\right)^2 - \left(\sqrt{3}\right)^2}$

$= \dfrac{5\sqrt{7}-5\sqrt{3}}{7-3}$

$= \dfrac{5\sqrt{7}-5\sqrt{3}}{4}$

59. $\dfrac{\sqrt{5}}{\sqrt{3}-\sqrt{5}} = \dfrac{\sqrt{5}}{\sqrt{3}-\sqrt{5}} \cdot \dfrac{\sqrt{3}+\sqrt{5}}{\sqrt{3}+\sqrt{5}}$ Multiply by the conjugate $\sqrt{3}+\sqrt{5}$

$= \dfrac{\sqrt{15}+\left(\sqrt{5}\right)^2}{\left(\sqrt{3}\right)^2 - \left(\sqrt{5}\right)^2}$

$= \dfrac{\sqrt{15}+5}{3-5}$

$= \dfrac{\sqrt{15}+5}{-2}$

$= \dfrac{-\sqrt{15}-5}{2}$ Do not leave a negative

in the denominator

60. $\dfrac{\sqrt{2}}{\sqrt{5}+\sqrt{2}} = \dfrac{\sqrt{2}}{\sqrt{5}+\sqrt{2}} \cdot \dfrac{\sqrt{5}-\sqrt{2}}{\sqrt{5}-\sqrt{2}}$ Multiply by the conjugate $\sqrt{5}-\sqrt{2}$

$= \dfrac{\sqrt{10}-\left(\sqrt{2}\right)^2}{\left(\sqrt{5}\right)^2 - \left(\sqrt{2}\right)^2}$

$= \dfrac{\sqrt{10}-2}{5-2}$

$= \dfrac{\sqrt{10}-2}{3}$

61. $\dfrac{\sqrt{5}-\sqrt{2}}{\sqrt{5}+\sqrt{2}} = \dfrac{\sqrt{5}-\sqrt{2}}{\sqrt{5}+\sqrt{2}} \cdot \dfrac{\sqrt{5}-\sqrt{2}}{\sqrt{5}-\sqrt{2}}$ Multiply by the conjugate $\sqrt{5}-\sqrt{2}$

$= \dfrac{5-2\sqrt{10}+2}{\left(\sqrt{5}\right)^2 - \left(\sqrt{2}\right)^2}$ FOIL

$= \dfrac{7-2\sqrt{10}}{5-2}$

$= \dfrac{7-2\sqrt{10}}{3}$

62. $\dfrac{\sqrt{7}+\sqrt{3}}{\sqrt{7}+\sqrt{3}} = \dfrac{\sqrt{7}+\sqrt{3}}{\sqrt{7}-\sqrt{3}} \cdot \dfrac{\sqrt{7}+\sqrt{3}}{\sqrt{7}+\sqrt{3}}$ Multiply by the conjugate $\sqrt{7}+\sqrt{3}$

$\qquad = \dfrac{\left(\sqrt{7}\right)^2 + 2\left(\sqrt{3}\right)\left(\sqrt{7}\right) + \left(\sqrt{3}\right)^2}{\left(\sqrt{7}\right)^2 - \left(\sqrt{3}\right)^2}$ FOIL

$\qquad = \dfrac{10 + 2\sqrt{21}}{7-3}$

$\qquad = \dfrac{2\left(5 + \sqrt{21}\right)}{4}$ Distributive property

$\qquad = \dfrac{5 + \sqrt{21}}{2}$

63. $\dfrac{\sqrt{x}+3}{\sqrt{x}-3} = \dfrac{\sqrt{x}+3}{\sqrt{x}-3} \cdot \dfrac{\sqrt{\mathbf{x}}+\mathbf{3}}{\sqrt{\mathbf{x}}+\mathbf{3}}$ Multiply by the conjugate $\sqrt{x}+3$

$\qquad = \dfrac{\left(\sqrt{x}\right)^2 + 2\left(3\sqrt{x}\right) + 3^2}{\left(\sqrt{x}\right)^2 - 3^2}$ FOIL

$\qquad = \dfrac{x + 6\sqrt{x} + 9}{x - 9}$

64. $\dfrac{\sqrt{x}-2}{\sqrt{x}+2} = \dfrac{\sqrt{x}-2}{\sqrt{x}+2} \cdot \dfrac{\sqrt{\mathbf{x}}-\mathbf{2}}{\sqrt{\mathbf{x}}-\mathbf{2}}$ Multiply by the conjugate $\sqrt{x}-2$

$\qquad = \dfrac{\left(\sqrt{x}\right)^2 - 2(2)\left(\sqrt{x}\right) + 2^2}{\left(\sqrt{x}\right)^2 - 2^2}$ FOIL

$\qquad = \dfrac{x - 4\sqrt{x} + 4}{x - 4}$

65. $3\sqrt{5} - 7\sqrt{5} = (3-7)\sqrt{5} = -4\sqrt{5}$ 66. $-5\sqrt{11} + 9\sqrt{11} = (-5+9)\sqrt{11} = 4\sqrt{11}$

67. $3\sqrt{27} - 5\sqrt{48} = 3\sqrt{9 \cdot 3} - 5\sqrt{16 \cdot 3}$

$\qquad\qquad\qquad = 3 \cdot 3\sqrt{3} - 5 \cdot 4\sqrt{3}$ $\sqrt{9} = 3$ and $\sqrt{16} = 4$

$\qquad\qquad\qquad = 9\sqrt{3} - 20\sqrt{3}$

$\qquad\qquad\qquad = (9 - 20)\sqrt{3}$ Distributive property

$\qquad\qquad\qquad = -11\sqrt{3}$

68. $5\sqrt{200} + 9\sqrt{50} = 5\sqrt{100 \cdot 2} + 9\sqrt{25 \cdot 2}$

$\qquad = 5 \cdot 10\sqrt{2} + 9 \cdot 5\sqrt{2}$ $\qquad \sqrt{100} = 10$ and $\sqrt{25} = 5$

$\qquad = 50\sqrt{2} + 45\sqrt{2}$

$\qquad = (50 + 45)\sqrt{2}$ $\qquad$ Distributive property

$\qquad = 95\sqrt{2}$

69. $-2\sqrt{45} - 5\sqrt{80} + 2\sqrt{20} = -2\sqrt{9 \cdot 5} - 5\sqrt{16 \cdot 5} + 2\sqrt{4 \cdot 5}$

$\qquad = -2 \cdot 3\sqrt{5} - 5 \cdot 4\sqrt{5} + 2 \cdot 2\sqrt{5}$ $\qquad \sqrt{9} = 3,\ \sqrt{16} = 4,\ \sqrt{4} = 2$

$\qquad = -6\sqrt{5} - 20\sqrt{5} + 4\sqrt{5}$

$\qquad = (-6 - 20 + 4)\sqrt{5}$ $\qquad$ Distributive property

$\qquad = -22\sqrt{5}$

70. $5\sqrt{12} + 3\sqrt{48} - 2\sqrt{300} = 5\sqrt{4 \cdot 3} + 3\sqrt{16 \cdot 3} - 2\sqrt{100 \cdot 3}$

$\qquad = 5 \cdot 2\sqrt{3} + 3 \cdot 4\sqrt{3} - 2 \cdot 10\sqrt{3}$

$\qquad = 10\sqrt{3} + 12\sqrt{3} - 20\sqrt{3}$

$\qquad = (10 + 12 - 20)\sqrt{3}$ $\qquad$ Distributive property

$\qquad = 2\sqrt{3}$

71. $3\sqrt{50x^2} - x\sqrt{200} = 3\sqrt{25x^2 \cdot 2} - x\sqrt{100 \cdot 2}$

$\qquad = 3 \cdot 5x\sqrt{2} - x \cdot 10\sqrt{2}$ $\qquad \sqrt{25x^2} = 5x$ and $\sqrt{100} = 10$

$\qquad = 15x\sqrt{2} - 10x\sqrt{2}$

$\qquad = (15x - 10x)\sqrt{2}$ $\qquad$ Distributive property

$\qquad = 5x\sqrt{2}$

72. $5\sqrt{63x^2} - x\sqrt{28} = 5\sqrt{9x^2 \cdot 7} - x\sqrt{4 \cdot 7}$

$\qquad = 5 \cdot 3x\sqrt{7} - x \cdot 2\sqrt{7}$ $\qquad \sqrt{9x^2} = 3x$ and $\sqrt{4} = 2$

$\qquad = 15x\sqrt{7} - 2x\sqrt{7}$

$\qquad = (15x - 2x)\sqrt{7}$ $\qquad$ Distributive property

$\qquad = 13x\sqrt{7}$

73. $\sqrt{40a^3b^2} - a\sqrt{90ab^2} = \sqrt{4a^2b^2 \cdot 10a} - a\sqrt{9b^2 \cdot 10a}$

$\qquad = 2ab\sqrt{10a} - a \cdot 3b\sqrt{10a}$ $\qquad \sqrt{4a^2b^2} = 2ab$ and $\sqrt{9b^2} = 3b$

$\qquad = 2ab\sqrt{10a} - 3ab\sqrt{10a}$

$\qquad = (2-3)ab\sqrt{10a}$ $\qquad$ Distributive property

$\qquad = -ab\sqrt{10a}$

74. $\sqrt{99a^2b^2} - 5ab\sqrt{44} = \sqrt{9a^2b^2 \cdot 11} - 5ab\sqrt{4 \cdot 11}$

$\qquad = 3ab\sqrt{11} - 5ab \cdot 2\sqrt{11}$ $\qquad \sqrt{9a^2b^2} = 3ab$ and $\sqrt{4} = 2$

$\qquad = 3ab\sqrt{11} - 10ab\sqrt{11}$

$\qquad = (3ab - 10ab)\sqrt{11}$ $\qquad$ Distributive property

$\qquad = -7ab\sqrt{11}$

75. $\sqrt{3}\left(\sqrt{3} + 3\right) = \sqrt{3}\left(\sqrt{3}\right) + \sqrt{3}\,(3) = 3 + 3\sqrt{3}$

76. $\sqrt{5}\left(\sqrt{5} - 3\right) = \sqrt{5}\left(\sqrt{5}\right) - \sqrt{5}\,(3) = 5 - 3\sqrt{5}$

77. $4\sqrt{2}\left(\sqrt{3} + \sqrt{5}\right) = 4\sqrt{2}\left(\sqrt{3}\right) + 4\sqrt{2}\left(\sqrt{5}\right) = 4\sqrt{6} + 4\sqrt{10}$

78. $3\sqrt{7}\left(\sqrt{5} - \sqrt{2}\right) = 3\sqrt{7}\left(\sqrt{5}\right) - 3\sqrt{7}\left(\sqrt{2}\right) = 3\sqrt{35} - 3\sqrt{14}$

79. $(\sqrt{x} + 7)(\sqrt{x} - 7) = (\sqrt{x})^2 - 7^2 = x - 49$

80. $(\sqrt{x} + 4)(\sqrt{x} - 4) = (\sqrt{x})^2 - 4^2 = x - 16$

81. $(2\sqrt{5} - 4)(\sqrt{5} + 3) = 2\sqrt{5}(\sqrt{5}) + 2\sqrt{5}(3) + (-4)(\sqrt{5}) + (-4)(3)$

$\qquad\qquad\qquad\qquad$ **F** $\qquad$ **O** $\qquad$ **I** $\qquad$ **L**

$\qquad\qquad = 2 \cdot 5 + 6\sqrt{5} - 4\sqrt{5} - 12$

$\qquad\qquad = 10 + 2\sqrt{5} - 12$

$\qquad\qquad = 2\sqrt{5} - 2$

82. $(5\sqrt{7} - 8)(\sqrt{7} + 1) = 5\sqrt{7}(\sqrt{7}) + 5\sqrt{7}(1) - 8\sqrt{7} - 8(1)$

$\qquad\qquad\qquad\qquad$ **F** $\qquad$ **O** $\qquad$ **I** $\qquad$ **L**

$\qquad\qquad = 5(7) + 5\sqrt{7} - 8\sqrt{7} - 8$

$\qquad\qquad = 27 - 3\sqrt{7}$

83. $(\sqrt{x} + 5)^2 = (\sqrt{x})^2 + 2 \cdot 5\sqrt{x} + 5^2 = x + 10\sqrt{x} + 25$

84. $(\sqrt{x} - 2)^2 = (\sqrt{x})^2 - 2(\sqrt{x})(2) + 2^2 = x - 4\sqrt{x} + 4$

85. $\sqrt{x-3} = 3$

$\left(\sqrt{x-3}\right)^2 = 3^2$ Square both sides

$x - 3 = 9$

$x = 12$

Check: $\sqrt{12-3} \overset{?}{=} 3$

$\sqrt{9} \overset{?}{=} 3$

$3 = 3$ A true statement

The solution is 12.

86. $\sqrt{x+4} = 5$

$\left(\sqrt{x+4}\right)^2 = 5^2$ Square both sides

$x + 4 = 25$

$x = 21$

Check: $\sqrt{21+4} \overset{?}{=} 5$

$\sqrt{25} \overset{?}{=} 5$

$5 = 5$ A true statement

The solution is 21.

87. $\sqrt{3x-5} = 4$

$\left(\sqrt{3x-5}\right)^2 = 4^2$ Square both sides

$3x - 5 = 16$

$3x = 21$

$x = 7$

Check: $\sqrt{3 \cdot 7 - 5} \overset{?}{=} 4$

$\sqrt{21-5} \overset{?}{=} 4$

$\sqrt{16} \overset{?}{=} 4$

$4 = 4$ A true statement

The solution is 7.

88. $\sqrt{2x+3} = 5$

$\left(\sqrt{2x+3}\right)^2 = 5^2$ Square both sides

$2x + 3 = 25$

$2x = 22$

$x = 11$

Check: $\sqrt{2(11)+3} \overset{?}{=} 5$

$\sqrt{25} \overset{?}{=} 5$

$5 = 5$ A true statement

The solution is 11.

89. $5\sqrt{a} = 20$

$\left(5\sqrt{a}\right)^2 = 20^2$ Square both sides

$25a = 400$

$a = 16$

Check: $5\sqrt{16} \overset{?}{=} 20$

$5 \cdot 4 \overset{?}{=} 20$

$20 = 20$ A true statement

The solution is 16.

90. $7\sqrt{a} = 63$

$\sqrt{a} = 9$ Divide both sides by 7

$\left(\sqrt{a}\right)^2 = 9^2$ Square both sides

$a = 81$

Check: $7\sqrt{81} \overset{?}{=} 63$

$7(9) \overset{?}{=} 63$

$63 = 63$ A true statement

The solution is 81.

264

Chapter 7 Review

91. $\sqrt{3x-7}+6=2$

$\sqrt{3x-7}=-4$

$\left(\sqrt{3x-7}\right)^2=(-4)^2$

$3x-7=16$

$3x=23$

$x=\dfrac{23}{3}$

Check: $\sqrt{3\cdot\dfrac{23}{3}-7}+6\overset{?}{=}2$

$\sqrt{16}+6\overset{?}{=}2$

$4+6\overset{?}{=}2$

$10=2$ A false statement

No solution

92. $\sqrt{2x+1}+10=8$

$\sqrt{2x+1}=-2$

$\left(\sqrt{2x+1}\right)^2=(-2)^2$

$2x+1=4$

$2x=3$

$x=\dfrac{3}{2}$

Check: $\sqrt{2\left(\dfrac{3}{2}\right)+1}+10\overset{?}{=}8$

$\sqrt{4}+10\overset{?}{=}8$

$2+10\overset{?}{=}8$

$12=8$ A false statement

No solution

93. $\sqrt{7x+1}=x+1$

$\left(\sqrt{7x+1}\right)^2=(x+1)^2$ Square both sides

$7x+1=x^2+2x+1$

$0=x^2-5x$

$0=x(x-5)$

$x=0$ or $x-5=0$

$x=5$

Check: $\sqrt{7\cdot0+1}\overset{?}{=}0+1$ $\sqrt{7\cdot5+1}\overset{?}{=}5+1$

$\sqrt{1}\overset{?}{=}1$ $\sqrt{36}\overset{?}{=}6$

$1=1$ $6=6$

A true statement   A true statement

The solutions are 0 and 5.

94. $\sqrt{6x-2}=3x-5$

$\left(\sqrt{6x-2}\right)^2=(3x-5)^2$ Square both sides

$6x-2=9x^2-30x+25$

$0=9x^2-36x+27$ Standard form

$0=9(x-3)(x-1)$ Factor

$x-3=0$ or $x-1=0$

$x=3$ $x=1$

Check:

$\sqrt{6(3)-2}\overset{?}{=}3(3)-5$ $\sqrt{6(1)-2}\overset{?}{=}3(1)-5$

$\sqrt{16}\overset{?}{=}9-5$ $\sqrt{4}\overset{?}{=}-2$

$4=4$ $2=-2$

A true statement   A false statement

Possible solutions 3 and 1: only 3 checks.

95. $c=\sqrt{a^2+b^2}$; Let $a=1,\ b=\sqrt{2},\ c=x$

$x=\sqrt{1^2+\left(\sqrt{2}\right)^2}$

$x=\sqrt{1+2}$ $\left(\sqrt{2}\right)^2=2$

$x=\sqrt{3}$

96. $c=\sqrt{a^2+b^2}$; Let $a=8,\ b=8,\ c=x$

$x=\sqrt{8^2+8^2}$

$x=\sqrt{64+64}$

$x=\sqrt{128}$

$x=\sqrt{64\cdot2}$

$x=8\sqrt{2}$

97. $y = 4\sqrt[3]{x}$

x	y
8	$4\sqrt[3]{8} = 4 \cdot 2 = 8$
1	$4\sqrt[3]{1} = 4 \cdot 1 = 4$
0	$4\sqrt[3]{0} = 4 \cdot 0 = 0$
-1	$4\sqrt[3]{-1} = 4 \cdot -1 = -4$
-8	$4\sqrt[3]{-8} = 4 \cdot -2 = -8$

98. $y = 3\sqrt{x}$

x	y
4	$3\sqrt{4} = 3 \cdot 2 = 6$
1	$3\sqrt{1} = 3 \cdot 1 = 3$
0	$3\sqrt{0} = 3 \cdot 0 = 0$

See the graph in the back of the textbook.

99. $y = \sqrt{x} + 3$

x	y
9	$\sqrt{9} + 3 = 3 + 3 = 6$
4	$\sqrt{4} + 3 = 2 + 3 = 5$
1	$\sqrt{1} + 3 = 1 + 3 = 4$
0	$\sqrt{0} + 3 = 0 + 3 = 3$

100. $y = \sqrt[3]{x} + 2$

x	y
8	$\sqrt[3]{8} + 2 = 2 + 2 = 4$
1	$\sqrt[3]{1} + 2 = 1 + 2 = 3$
0	$\sqrt[3]{0} + 2 = 0 + 2 = 2$
-1	$\sqrt[3]{-1} + 2 = -1 + 2 = 1$
-8	$\sqrt[3]{-8} + 2 = -2 + 2 = 0$

See the graph in the back of the textbook.

CHAPTER 7 TEST

1. $\sqrt{16} = 4$ Because $(4)^2 = 16$

2. $-\sqrt{36} = -6$

Because -6 is the negative number we square to get 36.

3. The square roots of 49 and 7 and –7.

4. $\sqrt[3]{27} = 3$ Because $(3)^3 = 27$

5. $\sqrt[3]{-8} = -2$ Because $(-2)^3 = -8$

6. $-\sqrt[4]{81} = -3$ Because -3 is the negative number we raise to the fourth power to get 81.

7. $\sqrt{75} = \sqrt{25 \cdot 3} = \sqrt{25} \cdot \sqrt{3} = 5\sqrt{3}$

8. $\sqrt{32} = \sqrt{16 \cdot 2} = \sqrt{16} \cdot \sqrt{2} = 4\sqrt{2}$

9. $\sqrt{\dfrac{2}{3}} = \dfrac{\sqrt{2}}{\sqrt{3}}$ Property 2

$= \dfrac{\sqrt{2}}{\sqrt{3}} \cdot \dfrac{\sqrt{3}}{\sqrt{3}}$ Rationalize the denominator

$= \dfrac{\sqrt{6}}{3}$

10. $\dfrac{1}{\sqrt[3]{4}} = \dfrac{1}{\sqrt[3]{4}} \cdot \dfrac{\sqrt[3]{2}}{\sqrt[3]{2}}$ Rationalize the denominator

$= \dfrac{\sqrt[3]{2}}{\sqrt[3]{8}}$

$= \dfrac{\sqrt[3]{2}}{2}$

11. $3\sqrt{50x^2} = 3\sqrt{25x^2 \cdot 2}$

$= 3\sqrt{25x^2}\sqrt{2}$ Property 1

$= 3 \cdot 5x\sqrt{2}$ $\sqrt{25x^2} = 5x$

$= 15x\sqrt{2}$

12. $\sqrt{\dfrac{12x^2 y^3}{5}} = \dfrac{\sqrt{12x^2 y^3}}{\sqrt{5}}$ Property 2

$= \dfrac{\sqrt{4x^2 y^2 \cdot 3y}}{\sqrt{5}}$

$= \dfrac{\sqrt{4x^2 y^2}\sqrt{3y}}{\sqrt{5}}$ Property 1

$= \dfrac{2xy\sqrt{3y}}{\sqrt{5}}$ $\sqrt{4x^2 y^2} = 2xy$

$= \dfrac{2xy\sqrt{3y}}{\sqrt{5}} \cdot \dfrac{\sqrt{5}}{\sqrt{5}}$ Rationalize the denominator

$= \dfrac{2xy\sqrt{15y}}{5}$

13. $5\sqrt{12} - 2\sqrt{27} = 5\sqrt{4 \cdot 3} - 2\sqrt{9 \cdot 3}$

$= 5\sqrt{4}\sqrt{3} - 2\sqrt{9}\sqrt{3}$ Property 1

$= 5(2)\sqrt{3} - 2(3)\sqrt{3}$ $\sqrt{4} = 2$ and $\sqrt{9} = 3$

$= 10\sqrt{3} - 6\sqrt{3}$

$= 4\sqrt{3}$

14. $2x\sqrt{18} + 5\sqrt{2x^2} = 2x\sqrt{9 \cdot 2} + 5\sqrt{x^2 \cdot 2}$

$= 2x\sqrt{9}\sqrt{2} + 5\sqrt{x^2}\sqrt{2}$ Property 1

$= 2x(3)\sqrt{2} + 5x\sqrt{2}$ $\sqrt{9} = 3$ and $\sqrt{x^2} = x$

$= 6x\sqrt{2} + 5x\sqrt{2}$

$= 11x\sqrt{2}$

15. $\sqrt{3}\left(\sqrt{5}-2\right)=\sqrt{3}\sqrt{5}-\sqrt{3}(2)$

$\qquad =\sqrt{15}-2\sqrt{3}$

16. $\left(\sqrt{5}+7\right)\left(\sqrt{5}-8\right)=\sqrt{5}\sqrt{5}-8\sqrt{5}+7\sqrt{5}-7(8)$

$\qquad\qquad\qquad\quad$ **F** $\quad$ **O** $\quad$ **I** $\quad$ **L**

$\qquad\qquad =5-8\sqrt{5}+7\sqrt{5}-56$

$\qquad\qquad =-51-\sqrt{5}$

17. $\left(\sqrt{x}+6\right)\left(\sqrt{x}-6\right)=\left(\sqrt{x}\right)^{2}-6^{2}=x-36$

18. $\left(\sqrt{5}-\sqrt{3}\right)^{2}=\left(\sqrt{5}-\sqrt{3}\right)\left(\sqrt{5}-\sqrt{3}\right)$

$\qquad\qquad\quad =\left(\sqrt{5}\right)^{2}-2\left(\sqrt{5}\sqrt{3}\right)+\left(\sqrt{3}\right)^{2}$

$\qquad\qquad\quad =5-2\sqrt{15}+3$

$\qquad\qquad\quad =8-2\sqrt{15}$

19. $\dfrac{\sqrt{7}-\sqrt{3}}{\sqrt{7}+\sqrt{3}}=\dfrac{\sqrt{7}-\sqrt{3}}{\sqrt{7}+\sqrt{3}}\cdot\dfrac{\sqrt{7}-\sqrt{3}}{\sqrt{7}-\sqrt{3}}$ $\qquad$ Multiply by the conjugate $\sqrt{7}-\sqrt{3}$

$\qquad =\dfrac{\left(\sqrt{7}\right)^{2}-2\left(\sqrt{7}\sqrt{3}\right)+\left(\sqrt{3}\right)^{2}}{\left(\sqrt{7}\right)^{2}-\left(\sqrt{3}\right)^{2}}$

$\qquad =\dfrac{7-2\sqrt{21}+3}{7-3}$

$\qquad =\dfrac{10-2\sqrt{21}}{4}$

$\qquad =\dfrac{2\left(5-\sqrt{21}\right)}{4}$

$\qquad =\dfrac{5-\sqrt{21}}{2}$

20. $\dfrac{\sqrt{x}}{\sqrt{x}+5}=\dfrac{\sqrt{x}}{\sqrt{x}+5}\cdot\dfrac{\sqrt{x}-5}{\sqrt{x}-5}$ $\qquad$ Multiply by the conjugate $\sqrt{x}-5$

$\qquad =\dfrac{\sqrt{x}\sqrt{x}-5\sqrt{x}}{\left(\sqrt{x}\right)^{2}-25}$

$\qquad =\dfrac{x-5\sqrt{x}}{x-25}$

21. $\sqrt{2x+1}+2=7$

$\qquad \sqrt{2x+1}=5$

$\qquad \left(\sqrt{2x+1}\right)^{2}=5^{2}$ $\qquad$ Square both sides

$\qquad\qquad 2x+1=25$

$\qquad\qquad\quad 2x=24$

$\qquad\qquad\quad\ x=12$

22. $\sqrt{3x+1}+6=2$

$\qquad \sqrt{3x+1}=-4$

$\qquad \left(\sqrt{3x+1}\right)^{2}=(-4)^{2}$ $\qquad$ Square both sides

$\qquad\qquad 3x+1=16$

$\qquad\qquad\quad 3x=15$

$\qquad\qquad\quad\ x=5$

To check our solution when $x = 12$,

$$\sqrt{2(12)+1} + 2 \overset{?}{=} 7$$

$$\sqrt{24+1} + 2 \overset{?}{=} 7$$

$$\sqrt{25} + 2 \overset{?}{=} 7$$

$$5 + 2 \overset{?}{=} 7$$

$$7 = 7 \quad \text{A true statement}$$

The solution is 12.

To check our solution when $x = 5$,

$$\sqrt{3(5)+1} + 6 \overset{?}{=} 2$$

$$\sqrt{15+1} + 6 \overset{?}{=} 2$$

$$\sqrt{16} + 6 \overset{?}{=} 2$$

$$4 + 6 \overset{?}{=} 2$$

$$10 = 2 \quad \text{A false statement}$$

The solution is $\varnothing$.

23. $\sqrt{2x-3} = x - 3$

$$\left(\sqrt{2x-3}\right)^2 = (x-3)^2 \qquad \text{Square both sides}$$

$$2x - 3 = x^2 - 6x + 9$$

$$0 = x^2 - 8x + 12$$

$$0 = (x-6)(x-2) \quad \text{Factor}$$

$$0 = x - 6 \quad \text{or} \quad 0 = x - 2$$

$$6 = x \qquad\qquad 2 = x$$

To check our solution when $x = 6$,

$$\sqrt{2(6)-3} \overset{?}{=} 6 - 3$$

$$\sqrt{12-3} \overset{?}{=} 3$$

$$\sqrt{9} \overset{?}{=} 3$$

$$3 = 3 \quad \text{A true statement}$$

To check our solution $x = 2$,

$$\sqrt{2(2)-3} \overset{?}{=} 2 - 3$$

$$\sqrt{4-3} \overset{?}{=} -1$$

$$\sqrt{1} \overset{?}{=} -1$$

$$1 = -1 \quad \text{A false statement}$$

Possible solutions 2 and 6; only 6 checks.

24.   Let x = a number

$$x - 4 = 3\sqrt{x}$$

$$(x - 4)^2 = \left(3\sqrt{x}\right)^2 \qquad \text{Square both sides}$$

$$(x - 4)(x - 4) = \left(3\sqrt{x}\right)\left(3\sqrt{x}\right)$$

$$x^2 - 8x + 16 = 9x$$

$$x^2 - 17x + 16 = 0$$

$$(x - 16)(x - 1) = 0 \qquad \text{Factor}$$

$$x - 16 = 0 \quad \text{or} \quad x - 1 = 0$$

$$x = 16 \qquad\qquad x = 1$$

To check our solution when $x = 16$,

$$16 - 4 \overset{?}{=} 3\sqrt{16}$$

$$12 \overset{?}{=} 3(4)$$

$$12 = 12 \quad \text{A true statement}$$

To check our solution when $x = 1$,

$$1 - 4 \overset{?}{=} 3\sqrt{1}$$

$$-3 = 3 \quad \text{A false statement}$$

Possible solutions 1 and 16; only 16 checks.

25. $c = \sqrt{a^2 + b^2}$; $a = 1$, $b = \sqrt{5}$, $c = x$

$$x = \sqrt{1^2 + \left(\sqrt{5}\right)^2}$$

$$x = \sqrt{1 + 5}$$

$$x = \sqrt{6}$$

26. $y = \sqrt{x} - 2$

x	y
9	$\sqrt{9} - 2 = 3 - 2 = 1$
4	$\sqrt{4} - 2 = 2 - 2 = 0$
1	$\sqrt{1} - 2 = 1 - 2 = -1$
0	$\sqrt{0} - 2 = 0 - 2 = -2$

See the graph in the back of the textbook.

CHAPTER 8

SECTION 8.1

1. $x^2 = 9$

   $x = \pm\sqrt{9}$ Square root property

   $x = \pm 3$

   The two solutions are 3 and -3.

5. $y^2 = 8$

   $y = \pm\sqrt{8}$ Square root property

   $y = \pm\sqrt{4 \cdot 2}$

   $y = \pm\sqrt{4}\sqrt{2}$

   $y = \pm 2\sqrt{2}$

   The solutions are $2\sqrt{2}$ and $-2\sqrt{2}$.

9. $3a^2 = 54$

   $a^2 = 18$ Divide by 3

   $a = \pm\sqrt{18}$ Square root property

   $a = \pm\sqrt{9 \cdot 2}$

   $a = \pm\sqrt{9}\sqrt{2}$

   $a = \pm 3\sqrt{2}$

   The solutions are $3\sqrt{2}$ and $-3\sqrt{2}$.

13. $(x+1)^2 = 25$

   $(x+1) = \pm\sqrt{25}$ Square root property

   $x+1 = \pm 5$

   $x = -1 \pm 5$

   $x = -1+5$ or $x = -1-5$

   $x = 4$ $x = -6$

   Our solutions are 4 and -6.

17. $(y+1)^2 = 50$

   $y+1 = \pm\sqrt{50}$

   $y = -1 \pm \sqrt{50}$

   $y = -1 \pm \sqrt{25 \cdot 2}$

   $y = -1 \pm \sqrt{25}\sqrt{2}$

   $y = -1 \pm 5\sqrt{2}$

   The two solutions are $-1+5\sqrt{2}$ and $-1-5\sqrt{2}$.

21. $(4a-5)^2 = 36$

   $(4a-5) = \pm\sqrt{36}$ Square root property

   $4a-5 = \pm 6$

   $4a = 5 \pm 6$

   $a = \dfrac{5 \pm 6}{4}$

   $a = \dfrac{5+6}{4}$ or $a = \dfrac{5-6}{4}$

   $a = \dfrac{11}{4}$ $a = -\dfrac{1}{4}$

   Our solutions are $\dfrac{11}{4}$ and $-\dfrac{1}{4}$.

25. $(6x+2)^2 = 27$

   $(6x+2) = \pm\sqrt{27}$ Square root property

   $6x+2 = \pm\sqrt{9\cdot3}$

   $6x+2 = \pm\sqrt{9}\sqrt{3}$

   $6x+2 = \pm3\sqrt{3}$

   $6x = -2\pm3\sqrt{3}$

   $x = \dfrac{-2\pm3\sqrt{3}}{6}$

   $x = \dfrac{-2+3\sqrt{3}}{6}$ or $x = \dfrac{-2-3\sqrt{3}}{6}$

29. $(3x+6)^2 = 45$

   $\sqrt{(3x+6)^2} = \pm\sqrt{45}$ Square root property

   $3x+6 = \pm\sqrt{9\cdot5}$

   $3x+6 = \pm\sqrt{9}\sqrt{5}$

   $3x+6 = \pm3\sqrt{5}$

   $3x = -6\pm3\sqrt{5}$

   $x = \dfrac{-6\pm3\sqrt{5}}{3}$

   $x = \dfrac{-6+3\sqrt{5}}{3}$ or $x = \dfrac{-6-3\sqrt{5}}{3}$

   $x = \dfrac{3(-2+\sqrt{5})}{3}$ $x = \dfrac{3(-2-\sqrt{5})}{3}$

   $x = -2+\sqrt{5}$ $x = -2-\sqrt{5}$

33. $\left(x-\dfrac{2}{3}\right)^2 = \dfrac{25}{9}$

   $x-\dfrac{2}{3} = \pm\dfrac{5}{3}$ Square root property

   $x = \dfrac{2}{3}\pm\dfrac{5}{3}$

   $x = \dfrac{2}{3}+\dfrac{5}{3}$ or $x = \dfrac{2}{3}-\dfrac{5}{3}$

   $x = \dfrac{7}{3}$ $x = -\dfrac{3}{3}$

   $x = -1$

37. $\left(a-\dfrac{4}{5}\right)^2 = \dfrac{12}{25}$

   $a-\dfrac{4}{5} = \pm\dfrac{\sqrt{12}}{5}$ Square root property

   $a = \dfrac{4}{5}\pm\dfrac{2\sqrt{3}}{5}$ $\sqrt{12} = \sqrt{4\cdot3} = 2\sqrt{3}$

   $a = \dfrac{4\pm2\sqrt{3}}{5}$

   Our solutions are $\dfrac{4+2\sqrt{3}}{5}$ and $\dfrac{4-2\sqrt{3}}{5}$.

41. $x^2 - 2x + 1 = 9$

   $(x-1)^2 = 9$

   $x-1 = \pm3$ Square root property

   $x = 1\pm3$

   $x = 1+3$ or $x = 1-3$

   $x = 4$ $x = -2$

   Our solutions are -2 and 4.

45. When $x = -1+5\sqrt{2}$ the equation $(x+1)^2 = 50$ becomes

   $\left(-1+5\sqrt{2}+1\right)^2 \overset{?}{=} 50$

   $\left(5\sqrt{2}\right)^2 \overset{?}{=} 50$

   $\left(5\sqrt{2}\right)\left(5\sqrt{2}\right) \overset{?}{=} 50$

   $25(2) \overset{?}{=} 50$

   $50 = 50$ A true statement

49. Given $A = 100(1+r)^2$, solve for r

$$A = 100(1+r)^2 \quad \text{Original formula}$$

$$\frac{A}{100} = (1+r)^2 \qquad \text{Divide both sides by 100}$$

$$\pm\frac{\sqrt{A}}{10} = (1+r) \qquad \text{Square root property}$$

Since r represents interest rate, r will never be negative.

$$+\frac{\sqrt{A}}{10} = 1 + r$$

$$-1 + \frac{\sqrt{A}}{10} = r \qquad \text{Add } -1 \text{ to both sides}$$

53. Because the three sides of an equilateral triangle are equal, the height always divides

the base into two equal line segments. The figure below illustrates this fact. We find the

height by applying the Pythagorean Theorem.

$$6^2 = x^2 + 3^2$$

$$36 = x^2 + 9$$

$$x^2 = 36 - 9$$

$$x^2 = 27$$

$$x = \sqrt{27}$$

$$x = 3\sqrt{3} \text{ ft}$$

$$x \approx 5.20 \text{ feet}$$

6 ft   6 ft

x

3 ft   3 ft

(The front of the tent)

The question asks if a 5 foot 8 inch person could stand up inside the tent drawn above

in the figure. The tent is approximately 5.20 feet high, but a 5 foot 8 inch person is approximately 5.67 feet

tall (5 feet 8 inches = $5\frac{8}{12}$ feet ≈ 5.67 feet) so the answer is no.

57. $(x+3)^2 = (x+3)(x+3)$

$$= x^2 + 2(x)(3) + 3^2$$

$$= x^2 + 6x + 9$$

61. $x^2 - 12x + 36 = (x-6)^2$

Notice that the first and last terms are perfect squares $(x)^2$ and $(6)^2$.

Before going through the method for factoring trinomials by listing all

possible factors, we can check to see if it is a perfect square trinomial,

which it is.

65. $\sqrt[3]{8} = 2$ Because $2^3 = 8$

SECTION 8.2

1. Half of 6 is 3, the square of which is 9. If we add 9
    to the end, we have

    $$x^2 + 6x + 9 = (x+3)^2$$

5. Half of -8 is -4, the square of which is 16. If we add 16
    to the end, we have

    $$y^2 - 8y + 16 = (y-4)^2$$

9. Half of 16 is 8, the square of which is 64. If we add 64
    to the end, we have

    $$x^2 + 16x + 64 = (x+8)^2$$

13. Half of -7 is $-\dfrac{7}{2}$, the square of which is $\dfrac{49}{4}$. If we add $\dfrac{49}{4}$
    to the end, we have

    $$x^2 - 7x + \frac{49}{4} = \left(x - \frac{7}{2}\right)^2$$

17. Half of $-\dfrac{3}{2}$ is $-\dfrac{3}{4}$, the square of which is $\dfrac{9}{16}$. If we add $\dfrac{9}{16}$
    to the end, we have

    $$x^2 - \frac{3}{2}x + \frac{9}{16} = \left(x - \frac{3}{4}\right)^2$$

21. $x^2 - 6x = 16$

    $x^2 - 6x + 9 = 16 + 9$ Half of 6 squared is 9

    $(x-3)^2 = 25$

        $x - 3 = \pm 5$ Square root property

           $x = 3 \pm 5$ Add 3 to both sides

  $x = 3 + 5$ or $x = 3 - 5$

  $x = 8$ $x = -2$

The solutions are 8 and -2.

25. $x^2 - 10x = 0$

    $x^2 - 10x + 25 = 0 + 25$ Complete the square

    $(x-5)^2 = 25$

        $x - 5 = \pm 5$ Square root property

           $x = 5 \pm 5$ Add 5 to both sides

  $x = 5 + 5$ or $x = 5 - 5$

  $x = 10$ $x = 0$

The solutions are 10 and 0.

29. $x^2 + 4x - 3 = 0$

$\qquad x^2 + 4x = 3 \qquad$ Add 3 to both sides

$\qquad x^2 + 4x + \mathbf{4} = 3 + \mathbf{4} \qquad$ Complete the square

$\qquad (x+2)^2 = 7$

$\qquad\quad x + 2 = \pm\sqrt{7} \qquad$ Square root property

$\qquad\qquad x = -2 \pm \sqrt{7} \qquad$ Add -2 to both sides

$x = -2 + \sqrt{7} \quad$ or $\quad x = -2 - \sqrt{7}$

The solutions written in the shorter form are $-2 \pm \sqrt{7}$.

33. $\qquad\qquad a^2 = 7a + 8$

$\qquad\quad a^2 - 7a = 8 \qquad$ Add $-7a$ to both sides

$a^2 - 7a + \dfrac{\mathbf{49}}{\mathbf{4}} = 8 + \dfrac{\mathbf{49}}{\mathbf{4}} \qquad$ Complete the square

$\qquad \left(a - \dfrac{7}{2}\right)^2 = \dfrac{81}{4} \qquad\qquad 8 + \dfrac{49}{4} = \dfrac{32}{4} + \dfrac{49}{4} = \dfrac{81}{4}$

$\qquad\qquad a - \dfrac{7}{2} = \pm\dfrac{9}{2} \qquad$ Square root property

$\qquad\qquad\quad a = \dfrac{7}{2} \pm \dfrac{9}{2}$

$a = \dfrac{7}{2} + \dfrac{9}{2} \quad$ or $\quad a = \dfrac{7}{2} - \dfrac{9}{2}$

$a = \dfrac{16}{2} \qquad\qquad a = -\dfrac{2}{2}$

$a = 8 \qquad\qquad\quad a = -1$

The solutions are 8 and -1.

37. $2x^2 + 2x - 4 = 0$

$\qquad 2x^2 + 2x = 4 \qquad$ Add 4 to both sides

$\qquad\quad x^2 + x = 2 \qquad$ Divide by 2

$\qquad x^2 + x + \dfrac{\mathbf{1}}{\mathbf{4}} = 2 + \dfrac{\mathbf{1}}{\mathbf{4}} \qquad$ Complete the square

$\qquad \left(x + \dfrac{1}{2}\right)^2 = \dfrac{9}{4} \qquad\qquad 2 + \dfrac{1}{4} = \dfrac{8}{4} + \dfrac{1}{4} = \dfrac{9}{4}$

$\qquad\qquad x + \dfrac{1}{2} = \pm\dfrac{3}{2} \qquad$ Square root property

$\qquad\qquad\quad x = -\dfrac{1}{2} \pm \dfrac{3}{2} \qquad$ Add $-\dfrac{1}{2}$ to both sides

$$x = -\frac{1}{2} + \frac{3}{2} \quad \text{or} \quad x = -\frac{1}{2} - \frac{3}{2}$$

$$x = \frac{2}{2} \qquad\qquad x = -\frac{4}{2}$$

$$x = 1 \qquad\qquad x = -2$$

The solutions are 1 and -2.

41. $2x^2 - 2x = 1$

$$x^2 - x = \frac{1}{2} \qquad\qquad \text{Divide by 2}$$

$$x^2 - x + \frac{1}{4} = \frac{1}{2} + \frac{1}{4} \qquad \text{Complete the square}$$

$$\left(x - \frac{1}{2}\right)^2 = \frac{3}{4} \qquad \frac{1}{2} + \frac{1}{4} = \frac{2}{4} + \frac{1}{4} = \frac{3}{4}$$

$$x - \frac{1}{2} = \pm\frac{\sqrt{3}}{2}$$

$$x = \frac{1}{2} \pm \frac{\sqrt{3}}{2} \qquad \text{Add } \frac{1}{2} \text{ to both sides}$$

$$x = \frac{1}{2} + \frac{\sqrt{3}}{2} \quad \text{or} \quad x = \frac{1}{2} - \frac{\sqrt{3}}{2}$$

$$x = \frac{1 + \sqrt{3}}{2} \qquad\qquad x = \frac{1 - \sqrt{3}}{2}$$

The solutions written in shorter form are $\frac{1 \pm \sqrt{3}}{2}$.

45. $3y^2 - 9y = 2$

$$y^2 - 3y = \frac{2}{3} \qquad\qquad \text{Divide by 3}$$

$$y^2 - 3y + \frac{9}{4} = \frac{2}{3} + \frac{9}{4} \qquad\qquad \text{Complete the square}$$

$$\left(y - \frac{3}{2}\right)^2 = \frac{35}{12} \qquad\qquad \frac{2}{3} + \frac{9}{4} = \frac{8}{12} + \frac{27}{12} = \frac{35}{12}$$

$$y - \frac{3}{2} = \pm\frac{\sqrt{35}}{2\sqrt{3}} \qquad\qquad \frac{\sqrt{35}}{2\sqrt{3}} \cdot \frac{\sqrt{3}}{\sqrt{3}} = \frac{\sqrt{105}}{6}$$

$$y = \frac{3}{2} \pm \frac{\sqrt{105}}{6}$$

$$y = \frac{3}{2} + \frac{\sqrt{105}}{6} \quad \text{or} \quad y = \frac{3}{2} - \frac{\sqrt{105}}{6}$$

$$y = \frac{9}{6} + \frac{\sqrt{105}}{6} \qquad\qquad y = \frac{9}{6} - \frac{\sqrt{105}}{6}$$

$$y = \frac{9 + \sqrt{105}}{6} \qquad\qquad y = \frac{9 - \sqrt{105}}{6}$$

The solutions written in shorter form are $\frac{9 \pm \sqrt{105}}{6}$.

49. When $x = -1 + \sqrt{2}$ the equation $4x^2 + 8x - 4 = 0$ becomes

$$4\left(-1 + \sqrt{2}\right)^2 + 8\left(-1 + \sqrt{2}\right) - 4 = 0$$

$$4\left(3 - 2\sqrt{2}\right) + 8\left(-1 + \sqrt{2}\right) - 4 = 0$$

$$12 - 8\sqrt{2} - 8 + 8\sqrt{2} - 4 = 0 \quad \text{A true statement}$$

53. See the drawing in the back of the textbook.

57. When $a = 2$ the expression $2a$ becomes $2(2) = 4$.

61. When $a = 2$, $b = 4$, and $c = -3$ the expression $\sqrt{b^2 - 4ac}$ becomes $\sqrt{(4)^2 - 4(2)(-3)} = \sqrt{16 + 24}$

$$= \sqrt{40}$$

$$= \sqrt{4 \cdot 10}$$

$$= 2\sqrt{10}$$

65. $\sqrt{20x^2 y^3} = \sqrt{4x^2 y^2 \cdot 5y}$

$$= 2xy\sqrt{5y}$$

SECTION 8.3

1. For the equation $x^2 + 3x + 2 = 0$, when $a = 1$, $b = 3$, and $c = 2$,

$$x = \frac{-b \pm \sqrt{b^2 - 4ac}}{2a}, \quad \text{the quadratic formula becomes:}$$

$$= \frac{-3 \pm \sqrt{(3)^2 - 4(1)(2)}}{2(1)}$$

$$= \frac{-3 \pm \sqrt{9 - 8}}{2}$$

$$= \frac{-3 \pm 1}{2}$$

$$x = \frac{-3+1}{2} \quad \text{or} \quad x = \frac{-3-1}{2}$$

$$x = \frac{-2}{2} \qquad\qquad x = \frac{-4}{2}$$

$$x = -1 \qquad\qquad x = -2$$

The two solutions are -1 and -2.

5. For the equation $x^2 + 6x + 9 = 0$, when $a = 1$, $b = 6$, and $c = 9$,

$$x = \frac{-b \pm \sqrt{b^2 - 4ac}}{2a}, \quad \text{the quadratic formula becomes:}$$

$$= \frac{-6 \pm \sqrt{(6)^2 - 4(1)(9)}}{2(1)}$$

$$= \frac{-6 \pm \sqrt{36 - 36}}{2}$$

$$= \frac{-6}{2}$$

$$= -3$$

There is only one solution, -3.

9. For the equation $2x^2 + 5x + 3 = 0$, when $a = 2$, $b = 5$, and $c = 3$,

$$x = \frac{-b \pm \sqrt{b^2 - 4ac}}{2a}, \quad \text{the quadratic formula becomes:}$$

$$= \frac{-5 \pm \sqrt{(5)^2 - 4(2)(3)}}{2(2)}$$

$$= \frac{-5 \pm \sqrt{25 - 24}}{4}$$

$$= \frac{-5 \pm 1}{4}$$

$$x = \frac{-5+1}{4} \quad \text{or} \quad x = \frac{-5-1}{4}$$

$$x = \frac{-4}{4} \qquad\qquad x = \frac{-6}{4}$$

$$x = -1 \qquad\qquad x = -\frac{3}{2}$$

The two solutions are -1 and $-\frac{3}{2}$.

13. For the equation $x^2 - 2x + 1 = 0$, when $a = 1$, $b = -2$, and $c = 1$,

$$x = \frac{-b \pm \sqrt{b^2 - 4ac}}{2a}, \quad \text{the quadratic formula becomes:}$$

$$= \frac{-(-2) \pm \sqrt{(-2)^2 - 4(1)(1)}}{2(1)}$$

$$= \frac{2 \pm \sqrt{4 - 4}}{2}$$

$$= \frac{2}{2}$$

$$= 1$$

There is only one solution, 1.

17. For the equation $6x^2 - x - 2 = 0$, when $a = 6$, $b = -1$, and $c = -2$,

$$x = \frac{-b \pm \sqrt{b^2 - 4ac}}{2a}, \quad \text{the quadratic formula becomes:}$$

$$= \frac{-(-1) \pm \sqrt{(-1)^2 - 4(6)(-2)}}{2(6)}$$

$$= \frac{1 \pm \sqrt{1 + 48}}{12}$$

$$= \frac{1 \pm 7}{12}$$

$$x = \frac{1 + 7}{12} \quad \text{or} \quad x = \frac{1 - 7}{12}$$

$$x = \frac{8}{12} \qquad\qquad x = \frac{-6}{12}$$

$$x = \frac{2}{3} \qquad\qquad x = -\frac{1}{2}$$

The two solutions are $\frac{2}{3}$ and $-\frac{1}{2}$.

21. $(2x - 3)(x + 2) = 1$

    $2x^2 + x - 6 = 1$ Multiply using FOIL method

    $2x^2 + x - 7 = 0$ Standard form

      When $a = 2$, $b = 1$, and $c = -7$,

$$x = \frac{-b \pm \sqrt{b^2 - 4ac}}{2a}, \quad \text{the quadratic formula becomes:}$$

$$= \frac{-1 \pm \sqrt{(1)^2 - 4(2)(-7)}}{2(2)}$$

$$= \frac{-1 \pm \sqrt{1 + 56}}{4}$$

$$= \frac{-1 \pm \sqrt{57}}{4}$$

The two solutions are $\frac{-1+\sqrt{57}}{4}$ and $\frac{-1-\sqrt{57}}{4}$.

25. $\qquad 2x^2 = -6x + 7$

$\quad 2x^2 + 6x - 7 = 0 \qquad$ Standard form

$\qquad$ When $a = 2$, $b = 6$, and $c = -7$,

$$x = \frac{-b \pm \sqrt{b^2 - 4ac}}{2a}, \qquad \text{the quadratic formula becomes:}$$

$$= \frac{-6 \pm \sqrt{(6)^2 - 4(2)(-7)}}{2(2)}$$

$$= \frac{-6 \pm \sqrt{36 + 56}}{4}$$

$$= \frac{-6 \pm \sqrt{92}}{4}$$

$$= \frac{-6 \pm \sqrt{4 \cdot 23}}{4}$$

$$= \frac{-6 \pm 2\sqrt{23}}{4}$$

$$= \frac{2(-3 \pm \sqrt{23})}{4}$$

$$= \frac{-3 \pm \sqrt{23}}{2}$$

The two solutions are $\frac{-3+\sqrt{23}}{2}$ and $\frac{-3-\sqrt{23}}{2}$.

29. $\qquad 2x^2 - 5 = 2x$

$\quad 2x^2 - 2x - 5 = 0 \quad$ Standard form

$\qquad$ When $a = 2$, $b = -2$, and $c = -5$,

$$x = \frac{-b \pm \sqrt{b^2 - 4ac}}{2a}, \qquad \text{the quadratic formula becomes:}$$

$$= \frac{-(-2) \pm \sqrt{(-2)^2 - 4(2)(-5)}}{2(2)}$$

$$= \frac{2 \pm \sqrt{4 + 40}}{4}$$

$$= \frac{2 \pm \sqrt{4 \cdot 11}}{4}$$

$$= \frac{2 \pm 2\sqrt{11}}{4}$$

$$= \frac{2(1 \pm \sqrt{11})}{4}$$

$$= \frac{1 \pm \sqrt{11}}{2}$$

The two solutions are $\frac{1+\sqrt{11}}{2}$ and $\frac{1-\sqrt{11}}{2}$.

33. For the equation $3x^2 - 4x = 0$, when $a = 3$, $b = -4$, and $c = 0$,

$$x = \frac{-b \pm \sqrt{b^2 - 4ac}}{2a}, \quad \text{the quadratic formula becomes:}$$

$$= \frac{-(-4) \pm \sqrt{(-4)^2 - 4(3)(0)}}{2(3)}$$

$$= \frac{4 \pm \sqrt{16}}{6}$$

$$= \frac{4 \pm 4}{6}$$

$$x = \frac{4+4}{6} \quad \text{or} \quad x = \frac{4-4}{6}$$

$$x = \frac{8}{6} \qquad\qquad x = \frac{0}{6}$$

$$x = \frac{4}{3} \qquad\qquad x = 0$$

The two solutions are $\frac{4}{3}$ and 0.

37. $\left(2\sqrt{3}\right)\left(3\sqrt{5}\right) = (2 \cdot 3)\left(\sqrt{3} \cdot \sqrt{5}\right) = 6\sqrt{15}$

41. $\left(\sqrt{7} - \sqrt{2}\right)\left(\sqrt{7} + \sqrt{2}\right) = \left(\sqrt{7}\right)^2 - \left(\sqrt{2}\right)^2 \qquad (a-b)(a+b) = a^2 - b^2$

$$= 7 - 2$$

$$= 5$$

SECTION 8.4

1. $(3 - 2i) + 3i = 3 + (-2i + 3i)$ Associative property
    $\qquad\qquad\quad = 3 + i$ Combine similar terms

5. $(11 + 9i) - 9i = 11 + (9i - 9i)$ Associative property
    $\qquad\qquad\quad\; = 11$

9. $(5 + 7i) - (6 + 8i) = 5 + 7i - 6 - 8i$ Distributive property
    $\qquad\qquad\qquad = (5 - 6) + (7i - 8i)$ Commutative and Associative properties
    $\qquad\qquad\qquad = -1 - i$

13. $(6+i)-4i-(2-i)$
   $= 6+i-4i-2+i$ Distributive property
   $= (6-2)+(i-4i+i)$ Commutative and Associative properties
   $= 4-2i$

17. $(2+3i)-(6-2i)+(3-i)$
   $= 2+3i-6+2i+3-i$ Distributive property
   $= (2-6+3)+(3i+2i-i)$
   $= -1+4i$

21. $2i(8-7i) = 2i(8)-2i(7i)$ Distributive property
   $= 16i-14i^2$
   $= 16i-14(-1)$ $i^2 = -1$
   $= 14+16i$

25. $(2+i)(3-5i) = 2\cdot3+2(-5i)+i(3)+i(-5i)$ FOIL Method
   $= 6-10i+3i-5i^2$
   $= 6-7i-5(-1)$ $i^2 = -1$
   $= 11-7i$

29. $(2+i)(2-i) = (2)^2-(i)^2$
   $= 4-(-1)$
   $= 5$

33. $\dfrac{-3i}{2+3i}\left(\dfrac{2-3i}{2-3i}\right) = \dfrac{-6i+9i^2}{4-9i^2}$
   $= \dfrac{-6i-9}{4+9}$ $i^2 = -1$
   $= \dfrac{-6i-9}{13}$
   $= \dfrac{-9-6i}{13}$

37. $\dfrac{2+i}{2-i}\cdot\dfrac{2+i}{2+i} = \dfrac{4+2(2i)+i^2}{4-i^2}$
   $= \dfrac{4+4i-1}{4-(-1)}$ $i^2 = -1$
   $= \dfrac{3+4i}{5}$

41. $(x+3i)(x-3i) = x(x)+x(-3i)+3i(x)+3i(-3i)$
        F   **O**   **I**   **L**
   $= x^2-3ix+3ix-9i^2$
   $= x^2-9(-1)$ $i^2 = -1$
   $= x^2+9$

45. $(x-3)^2 = 25$

   $x - 3 = \pm 5$ Square root of each side

   $x = 3 \pm 5$ Add 3 to both sides

   $x = 3 + 5$ or $x = 3 - 5$

   $x = 8$ $x = -2$

   The solutions are 8 and -2.

49. $(x+3)^2 = 12$

   $x + 3 = \pm\sqrt{12}$ Square root of each side

   $x + 3 = \pm 2\sqrt{3}$ $\sqrt{12} = \sqrt{4 \cdot 3} = 2\sqrt{3}$

   $x = -3 \pm 2\sqrt{3}$

53. $\sqrt{\dfrac{8x^3 y^3}{3}} = \dfrac{\sqrt{8x^3 y^3}}{\sqrt{3}}$

   $= \dfrac{\sqrt{4x^2 y^2 \cdot 2y}}{\sqrt{3}}$

   $= \dfrac{2xy\sqrt{2y}}{\sqrt{3}}$

   $= \dfrac{2xy\sqrt{2y}}{\sqrt{3}} \cdot \dfrac{\sqrt{3}}{\sqrt{3}}$

   $= \dfrac{2xy\sqrt{6y}}{3}$

SECTION 8.5

1. $\sqrt{-16} = \sqrt{16(-1)} = \sqrt{16}\sqrt{-1} = 4i$

5. $\sqrt{-6} = \sqrt{6(-1)} = \sqrt{6}\sqrt{-1} = i\sqrt{6}$

9. $\sqrt{-32} = \sqrt{32(-1)} = \sqrt{32}\sqrt{-1} = 4i\sqrt{2}$

13. $\sqrt{-8} = \sqrt{8(-1)} = \sqrt{8}\sqrt{-1} = 2i\sqrt{2}$

17. $x^2 = 2x - 2$

$x^2 - 2x + 2 = 0$ Standard form

When $a = 1$, $b = -2$, and $c = 2$,

$$x = \frac{-b \pm \sqrt{b^2 - 4ac}}{2a}, \quad \text{the quadratic formula becomes:}$$

$$= \frac{-(-2) \pm \sqrt{(-2)^2 - 4(1)(2)}}{2(1)}$$

$$= \frac{2 \pm \sqrt{4 - 8}}{2}$$

$$= \frac{2 \pm \sqrt{-4}}{2}$$

$$= \frac{2 \pm 2i}{2}$$

$$= \frac{2(1 \pm i)}{2}$$

$$= 1 \pm i$$

The two solutions are $1 + i$ and $1 - i$.

21. $2x^2 + 5x = 12$

$2x^2 + 5x - 12 = 0$ Standard form

When $a = 2$, $b = 5$, and $c = -12$,

$$x = \frac{-b \pm \sqrt{b^2 - 4ac}}{2a}, \quad \text{the quadratic formula becomes:}$$

$$= \frac{-5 \pm \sqrt{(5)^2 - 4(2)(-12)}}{2(2)}$$

$$= \frac{-5 \pm \sqrt{25 + 96}}{4}$$

$$= \frac{-5 \pm \sqrt{121}}{4}$$

$$= \frac{-5 \pm 11}{4}$$

$x = \dfrac{-5 + 11}{4}$ or $x = \dfrac{-5 - 11}{4}$

$x = \dfrac{6}{4}$ $x = \dfrac{-16}{4}$

$x = \dfrac{3}{2}$ $x = -4$

The two solutions are $\frac{3}{2}$ and -4.

25. $\left(x+\dfrac{1}{2}\right)^2 = -\dfrac{9}{4}$

$\quad x+\dfrac{1}{2} = \pm\sqrt{-\dfrac{9}{4}} \qquad$ Square root of both sides

$\quad x+\dfrac{1}{2} = \pm\dfrac{3}{2}i \qquad \sqrt{-\dfrac{9}{4}} = \sqrt{\dfrac{9}{4}}\sqrt{-1} = \dfrac{3}{2}i$

$\quad\quad x = -\dfrac{1}{2} \pm \dfrac{3i}{2}$

$\quad\quad x = \dfrac{-1 \pm 3i}{2}$

The two solutions are $\dfrac{-1+3i}{2}$ and $\dfrac{-1-3i}{2}$.

29. $x^2 + x + 1 = 0$

When $a = 1$, $b = 1$, and $c = 1$,

$\quad x = \dfrac{-b \pm \sqrt{b^2 - 4ac}}{2a}, \qquad$ the quadratic formula becomes:

$\quad = \dfrac{-1 \pm \sqrt{(1)^2 - 4(1)(1)}}{2(1)}$

$\quad = \dfrac{-1 \pm \sqrt{1-4}}{2}$

$\quad = \dfrac{-1 \pm \sqrt{-3}}{2}$

$\quad = \dfrac{-1 \pm i\sqrt{3}}{2}$

The two solutions are $\dfrac{-1+i\sqrt{3}}{2}$ and $\dfrac{-1-i\sqrt{3}}{2}$.

33. $\dfrac{1}{2}x^2 + \dfrac{1}{3}x + \dfrac{1}{6} = 0$

    $3x^2 + 2x + 1 = 0$ Multiply both sides by LCD which is 6.

    When $a = 3$, $b = 2$, and $c = 1$,

$$x = \dfrac{-b \pm \sqrt{b^2 - 4ac}}{2a}, \quad \text{the quadratic formula becomes:}$$

$$= \dfrac{-2 \pm \sqrt{(2)^2 - 4(3)(1)}}{2(3)}$$

$$= \dfrac{-2 \pm \sqrt{4 - 12}}{6}$$

$$= \dfrac{-2 \pm \sqrt{-8}}{6}$$

$$= \dfrac{-2 \pm \sqrt{4(2)(-1)}}{6}$$

$$= \dfrac{-2 \pm 2i\sqrt{2}}{6}$$

$$= \dfrac{2\left(-1 \pm i\sqrt{2}\right)}{6}$$

$$= \dfrac{-1 \pm i\sqrt{2}}{3}$$

The two solutions are $\dfrac{-1 + i\sqrt{2}}{3}$ and $\dfrac{-1 - i\sqrt{2}}{3}$.

37. $(x + 2)(x - 3) = 5$

    $x^2 - x - 6 = 5$ FOIL Method

    $x^2 - x - 11 = 0$ Add -5 to both sides, standard form

    When $a = 1$, $b = -1$, and $c = -11$,

$$x = \dfrac{-b \pm \sqrt{b^2 - 4ac}}{2a}, \quad \text{the quadratic formula becomes:}$$

$$= \dfrac{-(-1) \pm \sqrt{(-1)^2 - 4(1)(-11)}}{2(1)}$$

$$= \dfrac{1 \pm \sqrt{1 + 44}}{2}$$

$$= \dfrac{1 \pm \sqrt{45}}{2}$$

$$= \dfrac{1 \pm 3\sqrt{5}}{2}$$

The two solutions are $\dfrac{1+3\sqrt{5}}{2}$ and $\dfrac{1-3\sqrt{5}}{2}$.

41. $(2x-2)(x-3)=9$

   $2x^2-8x+6=9$ FOIL Method

   $2x^2-8x-3=0$ Add -9 to both sides, standard form

   When $a=2$, $b=-8$, and $c=-3$,

   $$x=\frac{-b\pm\sqrt{b^2-4ac}}{2a}, \quad \text{the quadratic formula becomes:}$$

   $$=\frac{-(-8)\pm\sqrt{(-8)^2-4(2)(-3)}}{2(2)}$$

   $$=\frac{8\pm\sqrt{88}}{4}$$

   $$=\frac{8\pm2\sqrt{22}}{4}$$

   $$=\frac{2(4\pm\sqrt{22})}{4}$$

   $$=\frac{4\pm\sqrt{22}}{2}$$

   The two solutions are $\dfrac{4+\sqrt{22}}{2}$ and $\dfrac{4-\sqrt{22}}{2}$.

45. $3-7i$ because both solutions would be $3\pm7i$.

49. Graph $2x+4y=8$

x-Intercept		**y-Intercept**	
When	$y=0$	When	$x=0$
the equation	$2x+4y=8$	the equation	$2x+4y=8$
becomes	$2x+4(0)=8$	becomes	$2(0)+4y=8$
	$2x=8$		$4y=8$
	$x=4$		$y=2$

   The x-intercept is at $(4,0)$.   The y-intercept is at $(0,2)$.

   See the graph in the back of the textbook.

53. $\sqrt{24}-\sqrt{54}-\sqrt{150}=\sqrt{4\cdot6}-\sqrt{9\cdot6}-\sqrt{25\cdot6}$

   $$=2\sqrt{6}-3\sqrt{6}-5\sqrt{6}$$

   $$=-6\sqrt{6}$$

SECTION 8.6

1.

x	$y = x^2 - 4$	y
-3	$y = (-3)^2 - 4$	5
-2	$y = (-2)^2 - 4$	0
-1	$y = (-1)^2 - 4$	-3
0	$y = (0)^2 - 4$	-4
1	$y = (1)^2 - 4$	-3
2	$y = (2)^2 - 4$	0

See the graph in the back of the textbook.

5.

x	$y = (x+2)^2$	y
-4	$y = (-4+2)^2$	4
-2	$y = (-2+2)^2$	0
0	$y = (0+2)^2$	4

See the graph in the back of the textbook.

9.

x	$y = (x-5)^2$	y
3	$y = (3-5)^2$	4
4	$y = (4-5)^2$	1
5	$y = (5-5)^2$	0
6	$y = (6-5)^2$	1

See the graph in the back of the textbook.

13.

x	$y = (x+2)^2 - 3$	y
-4	$y = (-4+2)^2 - 3$	1
-3	$y = (-3+2)^2 - 3$	-2
-2	$y = (-2+2)^2 - 3$	-3
-1	$y = (-1+2)^2 - 3$	-2
0	$y = (0+2)^2 - 3$	1

See the graph in the back of the textbook.

17. $y = x^2 + 6x + 5$

$y = (x^2 + 6x + 9) + 5 - 9$ Completing the square on the first two terms

$y = (x + 3)^2 - 4$

x	$y = (x+3)^2 - 4$	y
0	$y = (0+3)^2 - 4$	5
-1	$y = (-1+3)^2 - 4$	0
-2	$y = (-2+3)^2 - 4$	-3
-3	$y = (-3+3)^2 - 4$	-4
-4	$y = (-4+3)^2 - 4$	-3
-5	$y = (-5+3)^2 - 4$	0

See the graph in the back of the textbook.

21.

x	$y = 4 - x^2$	y
-3	$y = 4 - (-3)^2$	-5
-2	$y = 4 - (-2)^2$	0
-1	$y = 4 - (-1)^2$	3
0	$y = 4 - (0)^2$	4
1	$y = 4 - (1)^2$	3
2	$y = 4 - (2)^2$	0
3	$y = 4 - (3)^2$	-5

See the graph in the back of the textbook.

25. To graph the line $y = x + 2$:

x-Intercept

When $y = 0$
the equation $y = x + 2$
becomes $0 = x + 2$
$-2 = x$

y-Intercept

When $x = 0$
the equation $y = x + 2$
becomes $y = 0 + 2$
$y = 2$

The x-intercept is at $(-2, 0)$. The y-intercept is at $(0, 2)$.

To graph the parabola $y = x^2$:

x	$y = x^2$	y
-2	$y = (-2)^2$	4
-1	$y = (-1)^2$	1
0	$y = (0)^2$	0
1	$y = (1)^2$	1
2	$y = (2)^2$	4

The two graphs intersect at $(-1, 1)$ and $(2, 4)$.

See the graph in the back of the textbook.

29. $\sqrt{x + 5} = 4$

$x + 5 = 16$ Square both sides

$x = 11$

Check: $\sqrt{11 + 5} \overset{?}{=} 4$

$\sqrt{16} \overset{?}{=} 4$

$4 = 4$ A true statement

The solution is 11.

33. $\sqrt{3a + 2} - 3 = 5$

$\sqrt{3a + 2} = 8$ Add 3 to both sides

$3a + 2 = 64$ Square both sides

$3a = 62$ Add -2 to each side

$a = \dfrac{62}{3}$ Divide each side by 3

Check: $\sqrt{3\left(\dfrac{62}{3}\right) + 2} - 3 \overset{?}{=} 5$

$\sqrt{62 + 2} - 3 \overset{?}{=} 5$

$\sqrt{64} - 3 \overset{?}{=} 5$

$8 - 3 \overset{?}{=} 5$

$5 = 5$ A true statement

The solution is $\dfrac{62}{3}$.

CHAPTER 8 REVIEW

1. $a^2 = 32$

    $a = \pm\sqrt{32}$ Square root property

    $a = \pm\sqrt{16 \cdot 2}$

    $a = \pm 4\sqrt{2}$

   The solutions are $-4\sqrt{2}$ and $4\sqrt{2}$.

2. $a^2 = 60$

    $a = \pm\sqrt{60}$ Square root property

    $a = \pm\sqrt{4 \cdot 15}$

    $a = \pm 2\sqrt{15}$

   The solutions are $2\sqrt{15}$ and $-2\sqrt{15}$.

3. $2x^2 = 32$

    $x^2 = 16$ Divide each side by 2

    $x = \pm\sqrt{16}$ Square root property

    $x = \pm 4$

   The solutions are 4 and -4.

4. $3x^2 = 27$

    $x^2 = 9$ Divide each side by 3

    $x = \pm\sqrt{9}$ Square root property

    $x = \pm 3$

   The solutions are 3 and -3.

5. $(x + 3)^2 = 36$

    $x + 3 = \pm 6$

      $x = -3 \pm 6$

  $x = -3 + 6$ or $x = -3 - 6$

  $x = 3$ $x = -9$

  The solutions are -9 and 3.

6. $(x - 2)^2 = 81$

    $x - 2 = \pm 9$ Square root property

      $x = 2 \pm 9$

  $x = 2 + 9$ or $x = 2 - 9$

  $x = 11$ $x = -7$

  The solutions are -7 and 11.

7. $(3x + 2)^2 = 16$

    $3x + 2 = \pm\sqrt{16}$ Square root property

    $3x + 2 = \pm 4$

       $3x = -2 \pm 4$

         $x = \dfrac{-2 \pm 4}{3}$

 $x = \dfrac{-2 + 4}{3}$ or $x = \dfrac{-2 - 4}{3}$

 $x = \dfrac{2}{3}$ $x = -\dfrac{6}{3}$

                   $x = -2$

  The solutions are -2 and $\dfrac{2}{3}$.

8. $(2x - 3)^2 = 49$

    $2x - 3 = \pm 7$ Square root property

       $2x = 3 \pm 7$

        $x = \dfrac{3 \pm 7}{2}$

 $x = \dfrac{3 + 7}{2}$ or $x = \dfrac{3 - 7}{2}$

 $x = \dfrac{10}{2}$ $x = \dfrac{-4}{2}$

 $x = 5$ $x = -2$

  The solutions are -2 and 5.

9. $(2x+5)^2 = 32$

$$2x+5 = \pm\sqrt{32}$$

$$2x+5 = \pm 4\sqrt{2} \qquad \sqrt{32} = \sqrt{16\cdot 2} = 4\sqrt{2}$$

$$2x = -5 \pm 4\sqrt{2}$$

$$x = \frac{-5 \pm 4\sqrt{2}}{2}$$

The solutions are $\dfrac{-5+4\sqrt{2}}{2}$ and $\dfrac{-5-4\sqrt{2}}{2}$.

10. $(3x-4)^2 = 27$

$$3x-4 = \pm\sqrt{27} \qquad \text{Square root property}$$

$$3x-4 = \pm\sqrt{9\cdot 3}$$

$$3x-4 = \pm 3\sqrt{3}$$

$$3x = 4 \pm 3\sqrt{3}$$

$$x = \frac{4 \pm 3\sqrt{3}}{3}$$

$$x = \frac{4+3\sqrt{3}}{3} \quad \text{or} \quad x = \frac{4-3\sqrt{3}}{3}$$

The solutions are $\dfrac{4+3\sqrt{3}}{3}$ and $\dfrac{4-3\sqrt{3}}{3}$.

11. $\left(x-\dfrac{2}{3}\right)^2 = -\dfrac{25}{9}$

$$x-\frac{2}{3} = \pm\sqrt{-\frac{25}{9}} \qquad \text{Square root property}$$

$$x-\frac{2}{3} = \pm\frac{5i}{3}$$

$$x = \frac{2}{3} \pm \frac{5i}{3}$$

$$x = \frac{2 \pm 5i}{3}$$

The solutions are $\dfrac{2+5i}{3}$ and $\dfrac{2-5i}{3}$.

12. $\left(x+\dfrac{1}{2}\right)^2 = -\dfrac{9}{2}$

$$x+\frac{1}{2} = \pm\sqrt{-\frac{9}{2}} \qquad \text{Square root property}$$

$$x+\frac{1}{2} = \pm\frac{3i}{\sqrt{2}}\cdot\frac{\sqrt{2}}{\sqrt{2}}$$

$$x+\frac{1}{2} = \pm\frac{3i\sqrt{2}}{2}$$

$$x = -\frac{1}{2} \pm \frac{3i\sqrt{2}}{2}$$

$$x = \frac{-1 \pm 3i\sqrt{2}}{2}$$

The solutions are $\dfrac{-1+3i\sqrt{2}}{2}$ and $\dfrac{-1-3i\sqrt{2}}{2}$.

13. $\qquad x^2 + 8x = 4$

$$x^2 + 8x + \frac{64}{4} = 4 + \frac{64}{4} \qquad \left(\frac{8}{2}\right)^2 = \frac{64}{4}$$

$$(x+4)^2 = 4 + 16 \qquad \frac{64}{4} = 16$$

$$x+4 = \pm\sqrt{20} \qquad \text{Square root property}$$

$$x+4 = \pm 2\sqrt{5} \qquad \sqrt{20} = \sqrt{4\cdot 5} = 2\sqrt{5}$$

$$x = -4 \pm 2\sqrt{5}$$

The solutions are $-4+2\sqrt{5}$ and $-4-2\sqrt{5}$.

14. $\qquad x^2 - 8x = 4$

$$x^2 - 8x + \frac{64}{4} = 4 + \frac{64}{4} \qquad \left(-\frac{8}{2}\right)^2 = \frac{64}{4}$$

$$(x-4)^2 = 4 + 16$$

$$x-4 = \pm\sqrt{20} \qquad \text{Square root property}$$

$$x-4 = \pm 2\sqrt{5} \qquad \sqrt{20} = \sqrt{4\cdot 5} = 2\sqrt{5}$$

$$x = 4 \pm 2\sqrt{5}$$

The solutions are $4+2\sqrt{5}$ and $4-2\sqrt{5}$.

15. $x^2 - 4x - 7 = 0$

$\qquad x^2 - 4x = 7 \qquad$ **Add 7 to both sides**

$x^2 - 4x + \dfrac{16}{4} = 7 + \dfrac{16}{4} \qquad \left(-\dfrac{4}{2}\right)^2 = \dfrac{16}{4}$

$\qquad (x-2)^2 = 11 \qquad\qquad \dfrac{16}{4} = 4$

$\qquad\qquad x - 2 = \pm\sqrt{11}$

$\qquad\qquad\qquad x = 2 \pm \sqrt{11}$

The solutions are $2 + \sqrt{11}$ and $2 - \sqrt{11}$.

16. $x^2 + 4x + 3 = 0$

$\qquad x^2 + 4x = -3 \qquad$ **Add -3 to both sides**

$x^2 + 4x + \dfrac{16}{4} = -3 + \dfrac{16}{4} \qquad \left(\dfrac{4}{2}\right)^2 = \dfrac{16}{4}$

$\qquad (x+2)^2 = -3 + 4 \qquad\qquad \dfrac{16}{4} = 4$

$\qquad (x+2)^2 = 1$

$\qquad\qquad x + 2 = \pm 1$

$\qquad\qquad\qquad x = -2 \pm 1$

The solutions are -3 and -1.

17. $\qquad a^2 = 9a + 3$

$\qquad a^2 - 9a = 3 \qquad$ **Add $-9a$ to both sides**

$a^2 - 9a + \dfrac{81}{4} = 3 + \dfrac{81}{4} \qquad \left(-\dfrac{9}{2}\right)^2 = \dfrac{81}{4}$

$\qquad \left(a - \dfrac{9}{2}\right)^2 = \dfrac{93}{4} \qquad\qquad 3 + \dfrac{81}{4} = \dfrac{12}{4} + \dfrac{81}{4} = \dfrac{93}{4}$

$\qquad\qquad a - \dfrac{9}{2} = \pm\sqrt{\dfrac{93}{4}}$

$\qquad\qquad a - \dfrac{9}{2} = \pm\dfrac{\sqrt{93}}{2}$

$\qquad\qquad\qquad a = \dfrac{9}{2} \pm \dfrac{\sqrt{93}}{2}$

$\qquad\qquad\qquad a = \dfrac{9 \pm \sqrt{93}}{2}$

The solutions are $\dfrac{9 + \sqrt{93}}{2}$ and $\dfrac{9 - \sqrt{93}}{2}$.

18. $\qquad a^2 = 5a + 6$

$\qquad a^2 - 5a = 6 \qquad$ **Add $-5a$ to both sides**

$a^2 - 5a + \dfrac{25}{4} = 6 + \dfrac{25}{4} \qquad \left(-\dfrac{5}{2}\right)^2 = \dfrac{25}{4}$

$\qquad \left(a - \dfrac{5}{2}\right)^2 = \dfrac{24}{4} + \dfrac{25}{4} \qquad\qquad 6 = \dfrac{24}{4}$

$\qquad \left(a - \dfrac{5}{2}\right)^2 = \dfrac{49}{4}$

$\qquad\qquad a - \dfrac{5}{2} = \pm\dfrac{7}{2}$

$\qquad\qquad\qquad a = \dfrac{5}{2} \pm \dfrac{7}{2}$

$$a = \frac{5}{2} + \frac{7}{2} \quad \text{or} \quad a = \frac{5}{2} - \frac{7}{2}$$

$$a = \frac{12}{2} \qquad\qquad a = -\frac{2}{2}$$

$$a = 6 \qquad\qquad a = -1$$

The solutions are -1 and 6.

19. $2x^2 + 4x - 6 = 0$

$$2x^2 + 4x = 6 \qquad\qquad \text{Add 6 to both sides}$$

$$x^2 + 2x = 3 \qquad\qquad \text{Divide each side by 2}$$

$$x^2 + 2x + \frac{4}{4} = 3 + \frac{4}{4} \qquad \left(\frac{2}{2}\right)^2 = \frac{4}{4}$$

$$(x+1)^2 = 4 \qquad\qquad \frac{4}{4} = 1$$

$$x + 1 = \pm 2$$

$$x = -1 \pm 2$$

$$x = -1 + 2 \quad \text{or} \quad x = -1 - 2$$

$$x = 1 \qquad\qquad x = -3$$

The solutions are -3 and 1.

20. $3x^2 - 6x - 2 = 0$

$$3x^2 - 6x = 2 \qquad\qquad \text{Add 2 to both sides}$$

$$x^2 - 2x = \frac{2}{3} \qquad\qquad \text{Divide each side by 3}$$

$$x^2 - 2x + \frac{4}{4} = \frac{2}{3} + \frac{4}{4} \qquad \left(-\frac{2}{2}\right)^2 = \frac{4}{4}$$

$$(x-1)^2 = \frac{5}{3} \qquad\qquad \frac{4}{4} = 1, \ \frac{2}{3} + \frac{3}{3} = \frac{5}{3}$$

$$x - 1 = \pm \frac{\sqrt{5}}{\sqrt{3}} \cdot \frac{\sqrt{3}}{\sqrt{3}}$$

$$x - 1 = \frac{\pm\sqrt{15}}{3}$$

$$x = 1 \pm \frac{\sqrt{15}}{3}$$

$$x = \frac{3 \pm \sqrt{15}}{3}$$

The solutions are $\frac{3 + \sqrt{15}}{3}$ and $\frac{3 - \sqrt{15}}{3}$.

21. $x^2 + 7x + 12 = 0$

Let $a = 1$, $b = 7$, $c = 12$, then

$$x = \frac{-b \pm \sqrt{b^2 - 4ac}}{2a}$$

$$= \frac{-7 \pm \sqrt{7^2 - 4(1)(12)}}{2(1)}$$

$$= \frac{-7 \pm \sqrt{49 - 48}}{2}$$

$$= \frac{-7 \pm 1}{2}$$

$x = \dfrac{-7 + 1}{2}$ or $x = \dfrac{-7 - 1}{2}$

$x = -\dfrac{6}{2}$ $\qquad$ $x = -\dfrac{8}{2}$

$x = -3$ $\qquad\quad$ $x = -4$

The two solutions are -4 and -3.

22. $x^2 - 8x + 16 = 0$

Let $a = 1$, $b = -8$, $c = 16$, then

$$x = \frac{-b \pm \sqrt{b^2 - 4ac}}{2a}$$

$$= \frac{-(-8) \pm \sqrt{(-8)^2 - 4(1)(16)}}{2(1)}$$

$$= \frac{8 \pm \sqrt{64 - 64}}{2}$$

$$= \frac{8}{2}$$

$x = 4$

23. $x^2 + 5x + 7 = 0$

Let $a = 1$, $b = 5$, $c = 7$, then

$$x = \frac{-b \pm \sqrt{b^2 - 4ac}}{2a}$$

$$= \frac{-5 \pm \sqrt{5^2 - 4(1)(7)}}{2(1)}$$

$$= \frac{-5 \pm \sqrt{25 - 28}}{2}$$

$$= \frac{-5 \pm \sqrt{-3}}{2}$$

$$x = \frac{-5 \pm i\sqrt{3}}{2}$$

The solutions are $\dfrac{-5 + i\sqrt{3}}{2}$ and $\dfrac{-5 - i\sqrt{3}}{2}$.

24. $x^2 + 4x - 2 = 0$

Let $a = 1$, $b = 4$, $c = -2$, then

$$x = \frac{-b \pm \sqrt{b^2 - 4ac}}{2a}$$

$$= \frac{-4 \pm \sqrt{4^2 - 4(1)(-2)}}{2(1)}$$

$$= \frac{-4 \pm \sqrt{16 + 8}}{2}$$

$$= \frac{-4 \pm \sqrt{24}}{2}$$

$$= \frac{-4 \pm 2\sqrt{6}}{2}$$

$$= \frac{2(-2 \pm \sqrt{6})}{2}$$

$$x = -2 \pm \sqrt{6}$$

The solutions are $-2 + \sqrt{6}$ and $-2 - \sqrt{6}$.

25.
$$2x^2 = -8x + 5$$

$2x^2 + 8x - 5 = 0$ Standard form

Let $a = 2$, $b = 8$, $c = -5$, then

$$x = \frac{-b \pm \sqrt{b^2 - 4ac}}{2a}$$

$$= \frac{-8 \pm \sqrt{8^2 - 4(2)(-5)}}{2(2)}$$

$$= \frac{-8 \pm \sqrt{64 + 40}}{4}$$

$$= \frac{-8 \pm \sqrt{104}}{4}$$

$$= \frac{-8 \pm 2\sqrt{26}}{4}$$

$$= \frac{2(-4 \pm \sqrt{26})}{4}$$

$$= \frac{-4 \pm \sqrt{26}}{2}$$

The two solutions are $\dfrac{-4 + \sqrt{26}}{2}$ and $\dfrac{-4 - \sqrt{26}}{2}$.

26.
$$3x^2 = -3x - 4$$

$3x^2 + 3x + 4 = 0$ Standard form

Let $a = 3$, $b = 3$, $c = 4$, then

$$x = \frac{-b \pm \sqrt{b^2 - 4ac}}{2a}$$

$$= \frac{-3 \pm \sqrt{3^2 - 4(3)(4)}}{2(3)}$$

$$= \frac{-3 \pm \sqrt{9 - 48}}{6}$$

$$= \frac{-3 \pm \sqrt{-39}}{6}$$

$$= \frac{-3 \pm i\sqrt{39}}{6}$$

The solutions are $\dfrac{-3 + i\sqrt{39}}{6}$ and $\dfrac{-3 - i\sqrt{39}}{6}$.

27.
$$\frac{1}{5}x^2 - \frac{1}{2}x = \frac{3}{10}$$

$\dfrac{1}{5}x^2 - \dfrac{1}{2}x - \dfrac{3}{10} = 0$ **Standard form**

$\quad 2x^2 - 5x - 3 = 0$ **Multiply each side by 10**

Let $a = 2$, $b = -5$, $c = -3$, then

$$x = \frac{-b \pm \sqrt{b^2 - 4ac}}{2a}$$

$$= \frac{-(-5) \pm \sqrt{(-5)^2 - 4(2)(-3)}}{2(2)}$$

$$= \frac{5 \pm \sqrt{25 + 24}}{4}$$

$$= \frac{5 \pm \sqrt{49}}{4}$$

$$= \frac{5 \pm 7}{4} \text{ this gives } x = \frac{5+7}{4} \text{ and } x = \frac{5-7}{4}$$

The solutions are $-\frac{1}{2}$ and 3.

28.
$$\frac{1}{2}x^2 - \frac{1}{3}x = -1$$

$\dfrac{1}{2}x^2 - \dfrac{1}{3}x + 1 = 0$ **Standard form**

$\quad 3x^2 - 2x + 6 = 0$ **Multiply both sides by 6**

Let $a = 3$, $b = -2$, $c = 6$, then

$$x = \frac{-b \pm \sqrt{b^2 - 4ac}}{2a}$$

$$= \frac{-(-2) \pm \sqrt{(-2)^2 - 4(3)(6)}}{2(3)}$$

$$= \frac{2 \pm \sqrt{-68}}{6}$$

$$= \frac{2 \pm 2i\sqrt{17}}{6}$$

$$= \frac{2(1 \pm i\sqrt{17})}{6} = \frac{1 \pm i\sqrt{17}}{3}$$

The solutions are $\dfrac{1 + i\sqrt{17}}{3}$ and $\dfrac{1 - i\sqrt{17}}{3}$.

29. $(2x+1)(2x-3)=-6$

$\quad 4x^2-4x-3=-6$ FOIL Method

$\quad 4x^2-4x+3=0$ Standard form

Let $a=4$, $b=-4$, $c=3$, then

$$x=\frac{-b\pm\sqrt{b^2-4ac}}{2a}$$

$$=\frac{-(-4)\pm\sqrt{(-4)^2-4(4)(3)}}{2(4)}$$

$$=\frac{4\pm\sqrt{16-48}}{8}$$

$$=\frac{4\pm\sqrt{-32}}{8}$$

$$=\frac{4\pm4i\sqrt{2}}{8}$$

$$=\frac{4\left(1\pm i\sqrt{2}\right)}{8}$$

$$=\frac{1\pm i\sqrt{2}}{2}$$

The solutions are $\dfrac{1+i\sqrt{2}}{2}$ and $\dfrac{1-i\sqrt{2}}{2}$.

30. $(2x+1)(2x-5)=-4$

$\quad 4x^2-8x-5=-4$ FOIL Method

$\quad 4x^2-8x-1=0$ Standard form

Let $a=4$, $b=-8$, $c=-1$, then

$$x=\frac{-b\pm\sqrt{b^2-4ac}}{2a}$$

$$=\frac{-(-8)\pm\sqrt{(-8)^2-4(4)(-1)}}{2(4)}$$

$$=\frac{8\pm\sqrt{64+16}}{8}$$

$$=\frac{8\pm\sqrt{80}}{8}$$

$$=\frac{8\pm4\sqrt{5}}{8}$$

$$=\frac{4\left(2\pm\sqrt{5}\right)}{8}$$

$$=\frac{2\pm\sqrt{5}}{2}$$

The solutions are $\dfrac{2+\sqrt{5}}{2}$ and $\dfrac{2-\sqrt{5}}{2}$.

31. $(4-3i)+5i=4+(-3i+5i)=4+2i$

32. $(2-5i)-3i=2+(-5i-3i)=2-8i$

33. $(5+6i)+(5-i)=(5+5)+(6i-i)=10+5i$

34. $(2+5i)+(3-7i)=(2+3)+(5i-7i)=5-2i$

35. $(3-2i)-(3-i)=3-2i-3+i$

$\qquad\qquad =(3-3)+(-2i+i)$

$\qquad\qquad =-i$

36. $(5-7i)-(6-2i)=5-7i-6+2i$

$\qquad\qquad =(5-6)+(-7i+2i)$

$\qquad\qquad =-1-5i$

37. $(3+i)-5i-(4-i)=3+i-5i-4+i$

$\qquad\qquad =(3-4)+(i-5i+i)$

$\qquad\qquad =-1-3i$

38. $(3+3i)-7i-(2+2i)=3+3i-7i-2-2i$

$\qquad\qquad =(3-2)+(3i-7i-2i)$

$\qquad\qquad =1-6i$

39. $(2-3i)-(5-2i)+(5-i)=2-3i-5+2i+5-i$

$\qquad\qquad =(2-5+5)+(-3i+2i-i)$

$\qquad\qquad =2-2i$

40. $(3+5i)+(4-3i)-(6-i) = 3+5i+4-3i-6+i$
$$= (3+4-6)+(5i-3i+i)$$
$$= 1+3i$$

41. $2(3-i) = 2(3)-2(i) = 6-2i$

42. $-6(4+2i) = (-6)(4)+(-6)(2i)$
$$= -24-12i$$

43. $4i(6-5i) = 4i(6)-4i(5i)$
$$= 24i-20i^2$$
$$= 24i-20(-1)$$
$$= 20+24i$$

44. $-3i(7+6i) = (-3i)(7)+(-3i)(6i)$
$$= -21i-18i^2$$
$$= 18-21i$$

45. $(2-i)(3+i) = 6+2i-3i-i^2$
$$\text{F O I L}$$
$$= 6-i+1 \qquad -i^2 = -(-1) = 1$$
$$= 7-i$$

46. $(3-4i)(5+i) = 15+3i-20i-4i^2$
$$= 15-17i+4 \qquad i^2 = -1$$
$$= 19-17i$$

47. $(3-i)^2 = 3^2+2(3)(-i)+(-i)^2$
$$= 9-6i-1 \qquad (-i)^2 = -1$$
$$= 8-6i$$

48. $(3+5i)^2 = 3^2+2(3)(5i)+(5i)^2$
$$= 9+30i+25i^2$$
$$= 9+30i-25 \qquad i^2 = -1$$
$$= -16+30i$$

49. $(4+i)(4-i) = 16-i^2$
$$= 16+1 \qquad -i^2 = -(-1) = 1$$
$$= 17$$

50. $(3+2i)(3-2i) = 9-4i^2$
$$= 9+4 \qquad i^2 = -1$$
$$= 13$$

51. $\dfrac{i}{3+i} = \left(\dfrac{i}{3+i}\right)\left(\dfrac{3-i}{3-i}\right)$
$$= \dfrac{3i-i^2}{9-i^2}$$
$$= \dfrac{3i+1}{9+1} \qquad -i^2 = -(-1)$$
$$= \dfrac{1+3i}{10}$$

52. $\dfrac{i}{2-i} = \left(\dfrac{i}{2-i}\right)\left(\dfrac{2+i}{2+i}\right)$
$$= \dfrac{2i+i^2}{4-i^2}$$
$$= \dfrac{2i-1}{4+1} \qquad i^2 = -1$$
$$= \dfrac{-1+2i}{5}$$

53. $\dfrac{5}{2+5i} = \left(\dfrac{5}{2+5i}\right)\left(\dfrac{2-5i}{2-5i}\right)$

$= \dfrac{10-25i}{4-25i^2}$

$= \dfrac{10-25i}{4+25} \quad i^2 = -1$

$= \dfrac{10-25i}{29}$

54. $\dfrac{3}{3-4i} = \left(\dfrac{3}{3-4i}\right)\left(\dfrac{3+4i}{3+4i}\right)$

$= \dfrac{9+12i}{9-16i^2}$

$= \dfrac{9+12i}{9+16} \quad i^2 = -1$

$= \dfrac{9+12i}{25}$

55. $\dfrac{2i}{4-i} = \left(\dfrac{2i}{4-i}\right)\left(\dfrac{4+i}{4+i}\right)$

$= \dfrac{8i+2i^2}{16-i^2}$

$= \dfrac{8i-2}{16+1} \quad i^2 = -1$

$= \dfrac{-2+8i}{17}$

56. $\dfrac{-3i}{3-2i} = \left(\dfrac{-3i}{3-2i}\right)\left(\dfrac{3+2i}{3+2i}\right)$

$= \dfrac{-9i-6i^2}{9-4i^2}$

$= \dfrac{-9i+6}{9+4} \quad i^2 = -1$

$= \dfrac{6-9i}{13}$

57. $\dfrac{3+i}{3-i} = \left(\dfrac{3+i}{3-i}\right)\left(\dfrac{3+i}{3+i}\right)$

$= \dfrac{9+3i+3i+i^2}{9-i^2}$

$= \dfrac{8+6i}{10}$

$= \dfrac{2(4+3i)}{10} \quad i^2 = -1$

$= \dfrac{4+3i}{5}$

58. $\dfrac{4-5i}{4+5i} = \left(\dfrac{4-5i}{4+5i}\right)\left(\dfrac{4-5i}{4-5i}\right)$

$= \dfrac{16+2(4)(-5i)+(5i)^2}{16-(5i)^2}$

$= \dfrac{16-40i-25}{16+25} \quad i^2 = -1$

$= \dfrac{-9-40i}{41}$

59. $\dfrac{2+3i}{3-5i} = \left(\dfrac{2+3i}{3-5i}\right)\left(\dfrac{3+5i}{3+5i}\right)$

$= \dfrac{6+19i+15i^2}{9-25i^2}$

$= \dfrac{6+19i-15}{9+25} \quad i^2 = -1$

$= \dfrac{-9+19i}{34}$

60. $\dfrac{3-2i}{5-i} = \left(\dfrac{3-2i}{5-i}\right)\left(\dfrac{5+i}{5+i}\right)$

$= \dfrac{15-7i-2i^2}{25-i^2}$

$= \dfrac{15-7i+2}{25+1} \quad i^2 = -1$

$= \dfrac{17-7i}{26}$

61. $\sqrt{-36} = \sqrt{36(-1)} = \sqrt{36}\sqrt{-1} = 6i$

62. $\sqrt{-144} = \sqrt{144(-1)} = \sqrt{144}\sqrt{-1} = 12i$

63. $\sqrt{-17} = \sqrt{17(-1)} = \sqrt{17}\sqrt{-1} = i\sqrt{17}$

64. $\sqrt{-31} = \sqrt{31(-1)} = \sqrt{31}\sqrt{-1} = i\sqrt{31}$

65. $\sqrt{-40} = \sqrt{4(-1)(10)} = \sqrt{4}\sqrt{-1}\sqrt{10} = 2i\sqrt{10}$

66. $\sqrt{-72} = \sqrt{36(-1)(2)} = \sqrt{36}\sqrt{-1}\sqrt{2} = 6i\sqrt{2}$

67. $\sqrt{-200} = \sqrt{100(-1)(2)} = \sqrt{100}\sqrt{-1}\sqrt{2} = 10i\sqrt{2}$

68. $\sqrt{-242} = \sqrt{121(-1)(2)} = \sqrt{121}\sqrt{-1}\sqrt{2} = 11i\sqrt{2}$

69. $y = x^2 + 2$

x	y
-2	$(-2)^2 + 2 = 6$
-1	$(-1)^2 + 2 = 3$
0	$0^2 + 2 = 2$
1	$1^2 + 2 = 3$
2	$2^2 + 2 = 6$

70. $y = x^2 - 2$

x	y
2	$2^2 - 2 = 4 - 2 = 2$
1	$1^2 - 2 = 1 - 2 = -1$
0	$0^2 - 2 = 0 - 2 = -2$
-1	$(-1)^2 - 2 = 1 - 2 = -1$
-2	$(-2)^2 - 2 = 4 - 2 = 2$

See the graph in the back of the textbook for problems 69-72.

71. $y = (x - 2)^2$

x	y
4	$(4-2)^2 = 2^2 = 4$
3	$(3-2)^2 = 1^2 = 1$
2	$(2-2)^2 = (0)^2 = 0$
1	$(1-2)^2 = (-1)^2 = 1$
0	$(0-2)^2 = (-2)^2 = 4$

72. $y = (x + 2)^2$

x	y
0	$(0+2)^2 = 2^2 = 4$
-1	$(-1+2)^2 = 1^2 = 1$
-2	$(-2+2)^2 = 0^2 = 0$
-3	$(-3+2)^2 = (-1)^2 = 1$
-4	$(-4+2)^2 = (-2)^2 = 4$

See the graph in the back of the textbook.

73. $y = (x+3)^2 - 2$

x	y
-5	$(-5+3)^2 - 2 = 2$
-4	$(-4+3)^2 - 2 = -1$
-3	$(-3+3)^2 - 2 = -2$
-2	$(-2+3)^2 - 2 = -1$
-1	$(-1+3)^2 - 2 = 2$

74. $y = (x-3)^2 - 2$

x	y
5	$(5-3)^2 - 2 = 2^2 - 2 = 2$
4	$(4-3)^2 - 2 = 1^2 - 2 = -1$
3	$(3-3)^2 - 2 = 0^2 - 2 = -2$
2	$(2-3)^2 - 2 = (-1)^2 - 2 = -1$
1	$(1-3)^2 - 2 = (-2)^2 - 2 = 2$

See the graph in the back of the textbook.

75. $y = x^2 + 4x + 7$

x	y
-1	$(-1)^2 + 4(-1) + 7 = 4$
-2	$(-2)^2 + 4(-2) + 7 = 3$
-3	$(-3)^2 + 4(-3) + 7 = 4$

76. $y = x^2 - 4x + 7$

x	y
3	$3^2 - 4(3) + 7 = 4$
2	$2^2 - 4(2) + 7 = 3$
1	$1^2 - 4(1) + 7 = 4$

See the graph in the back of the textbook.

CHAPTER 8 TEST

1. For the equation $x^2 - 7x - 8 = 0$, $a = 1$, $b = -7$, and $c = -8$:

$$x = \frac{-b \pm \sqrt{b^2 - 4ac}}{2a}$$

$$= \frac{-(-7) \pm \sqrt{(-7)^2 - 4(1)(-8)}}{2(1)}$$

$$= \frac{7 \pm \sqrt{49 + 32}}{2}$$

$$= \frac{7 \pm 9}{2}$$

$$x = \frac{7+9}{2} \quad \text{or} \quad x = \frac{7-9}{2}$$

$$x = \frac{16}{2} \qquad\qquad x = \frac{-2}{2}$$

$$x = 8 \qquad\qquad x = -1$$

The two solutions are -1 and 8.

2. $(x-3)^2 = 12$

$x^2 - 6x + 9 = 12$ F O I L Method

$x^2 - 6x - 3 = 0$ Add -12 to both sides

$a = 1, b = -6$ and $c = -3$

$$x = \frac{-b \pm \sqrt{b^2 - 4ac}}{2a}$$

$$= \frac{-(-6) \pm \sqrt{(-6)^2 - 4(1)(-3)}}{2(1)}$$

$$= \frac{6 \pm \sqrt{36 + 12}}{2}$$

$$= \frac{6 \pm \sqrt{48}}{2}$$

$$= \frac{6 \pm 4\sqrt{3}}{2}$$

$$= \frac{2(3 \pm 2\sqrt{3})}{2}$$

$$= 3 \pm 2\sqrt{3}$$

The solutions are $3 + 2\sqrt{3}$ and $3 - 2\sqrt{3}$.

3. $\left(x - \dfrac{5}{2}\right)^2 = -\dfrac{75}{4}$

$x - \dfrac{5}{2} = \pm\sqrt{-\dfrac{75}{4}}$

$x - \dfrac{5}{2} = \pm\dfrac{5i\sqrt{3}}{2}$ $\sqrt{-\dfrac{75}{4}} = \dfrac{\sqrt{25(-1)(3)}}{\sqrt{4}} = \dfrac{5i\sqrt{3}}{2}$

$x = \dfrac{5}{2} \pm \dfrac{5i\sqrt{3}}{2}$

$x = \dfrac{5 \pm 5i\sqrt{3}}{2}$

The solutions are $\dfrac{5 + 5i\sqrt{3}}{2}$ and $\dfrac{5 - 5i\sqrt{3}}{2}$.

4. $\dfrac{1}{3}x^2 = \dfrac{1}{2}x - \dfrac{5}{6}$

$6\left(\dfrac{1}{3}x^2\right) = 6\left(\dfrac{1}{2}x\right) - 6\left(\dfrac{5}{6}\right)$ LCD $= 6$

$2x^2 = 3x - 5$

$2x^2 - 3x + 5 = 0$ Standard form

$a = 2$, $b = -3$ and $c = 5$

$$x = \frac{-b \pm \sqrt{b^2 - 4ac}}{2a}$$

$$= \frac{-(-3) \pm \sqrt{(-3)^2 - 4(2)(5)}}{2(2)}$$

$$= \frac{3 \pm \sqrt{9 - 40}}{4}$$

$$x = \frac{3 \pm \sqrt{-31}}{4}$$

$$x = \frac{3 \pm i\sqrt{31}}{4}$$

The solutions are $\dfrac{3 + i\sqrt{31}}{4}$ and $\dfrac{3 - i\sqrt{31}}{4}$.

5. $\qquad 3x^2 = -2x + 1$

$3x^2 + 2x - 1 = 0$

$a = 3, b = 2,$ and $c = -1$

$$x = \frac{-b \pm \sqrt{b^2 - 4ac}}{2a}$$

$$= \frac{-(2) \pm \sqrt{(2)^2 - 4(3)(-1)}}{2(3)}$$

$$= \frac{-2 \pm \sqrt{4 + 12}}{6}$$

$$= \frac{-2 \pm 4}{6}$$

$$= \frac{2(-1 \pm 2)}{6} \quad \text{this gives } x = \frac{-2 + 4}{6} \text{ and } x = \frac{-2 - 4}{6}$$

The two solutions are $\frac{1}{3}$ and -1.

6. $(x + 2)(x - 1) = 6$

$\qquad x^2 + x - 2 = 6 \qquad$ FOIL Method

$\qquad x^2 + x - 8 = 0$

$a = 1, b = 1$ and $c = -8$

$$x = \frac{-b \pm \sqrt{b^2 - 4ac}}{2a}$$

$$= \frac{-1 \pm \sqrt{(1)^2 - 4(1)(-8)}}{2(1)}$$

$$= \frac{-1 \pm \sqrt{1 + 32}}{2}$$

$$= \frac{-1 \pm \sqrt{33}}{2}$$

The solutions are $\dfrac{-1 + \sqrt{33}}{2}$ and $\dfrac{-1 - \sqrt{33}}{2}$.

7. $9x^2 + 12x + 4 = 0$

$a = 9$, $b = 12$ and $c = 4$

$$x = \frac{-b \pm \sqrt{b^2 - 4ac}}{2a}$$

$$= \frac{-12 \pm \sqrt{(12)^2 - 4(9)(4)}}{2(9)}$$

$$= \frac{-12 \pm \sqrt{144 - 144}}{18}$$

$$= \frac{-12}{18}$$

$$= -\frac{2}{3}$$

The solution is $-\frac{2}{3}$.

8. $x^2 - 6x - 6 = 0$

$\qquad x^2 - 6x = 6 \qquad$ Add 6 to both sides

$\qquad x^2 - 6x + \mathbf{9} = 6 + \mathbf{9} \qquad$ Complete the square

$\qquad (x - 3)^2 = 15$

$\qquad x - 3 = \pm\sqrt{15}$

$\qquad x = 3 \pm \sqrt{15}$

9. $\sqrt{-9} = \sqrt{9(-1)} = \sqrt{9}\,\sqrt{-1} = 3i$

10. $\sqrt{-121} = \sqrt{121(-1)} = \sqrt{121}\,\sqrt{-1} = 11i$

11. $\sqrt{-72} = \sqrt{72(-1)} = \sqrt{36}\,\sqrt{-1}\,\sqrt{2} = 6i\sqrt{2}$

12. $\sqrt{-18} = \sqrt{18(-1)} = \sqrt{9}\,\sqrt{-1}\,\sqrt{2} = 3i\sqrt{2}$

13. $(3i + 1) + (2 + 5i) = (1 + 2) + (3i + 5i) = 3 + 8i$

14. $(6 - 2i) - (7 - 4i) = 6 - 2i - 7 + 4i$

$\qquad\qquad = (6 - 7) + (-2i + 4i)$

$\qquad\qquad = -1 + 2i$

15. $(2+i)(2-i) = (2)^2 - (i)^2$

          $= 4 - (-1) = 5$

    Remember $(a+bi)(a-bi) = (a)^2 - (bi)^2$

16. $(3+2i)(1+i) = 3(1) + 3(i) + 2i(1) + 2i(i)$

               F   **O**   **I**   **L**

           $= 3 + 3i + 2i + 2i^2$

           $= 3 + 5i + 2(-1)$

           $= 1 + 5i$

17. $\dfrac{i}{3-i} = \dfrac{i}{3-i}\left(\dfrac{3+i}{3+i}\right)$

    $= \dfrac{3i+i^2}{9-i^2}$

    $= \dfrac{3i-1}{9-(-1)}$

    $= \dfrac{-1+3i}{10}$

18. $\dfrac{2+i}{2-i} = \dfrac{2+i}{2-i}\left(\dfrac{2+i}{2+i}\right)$

    $= \dfrac{4 + 2(2i) + i^2}{4-i^2}$

    $= \dfrac{4 + 4i - 1}{4-(-1)}$

    $= \dfrac{3+4i}{5}$

19.

x	$y = x^2 - 4$	y
-3	$y = (-3)^2 - 4$	5
-2	$y = (-2)^2 - 4$	0
-1	$y = (-1)^2 - 4$	-3
0	$y = (0)^2 - 4$	-4
1	$y = (1)^2 - 4$	-3
2	$y = (2)^2 - 4$	0

20.

x	$y = (x-4)^2$	y
2	$y = (2-4)^2$	4
3	$y = (3-4)^2$	1
4	$y = (4-4)^2$	0
5	$y = (5-4)^2$	1
6	$y = (6-4)^2$	4

See the graph in the back of the textbook.

21.

x	$y = (x+3)^2 - 4$	y
-6	$y = (-6+3)^2 - 4$	5
-5	$y = (-5+3)^2 - 4$	0
-3	$y = (-3+3)^2 - 4$	-4
-1	$y = (-1+3)^2 - 4$	0
0	$y = (0+3)^2 - 4$	5

22.

x	$y = x^2 - 6x + 11$	y
1	$y = 1^2 - 6(1) + 11$	6
3	$y = 3^2 - 6(3) + 11$	2
5	$y = 5^2 - 6(5) + 11$	6

See the graph in the back of the textbook.

CHAPTER 9

SECTION 9.1

1. $x < -1$ or $x > 5$

   First we graph each inequality separately:

   $x < -1$

   $x > 5$

   Then, since the original two inequalities are connected by the word "or" , we graph everything
   that is on either graph above:

   $x < -1$ or $x > 5$

   See the graph of the solution in the back of the textbook.

5. $x \le 6$ and $x > -1$

   First we graph each inequality separately:

   $x \le 6$

   $x > -1$

   See the graph of the solution in the back of the textbook.

   Then, since the original two inequalities are connected by the word "and", we graph all points
   that are common to the graphs above:

   $x \le 6$ and $x > -1$

9. The graph in the back of the textbook shows all points that are both greater than or equal to -2
   and less than or equal to 4.

13. The graph in the back of the textbook shows all the points that are between -1 and 3 on the number line.
    That means the graph is all points that are both greater than -1 and less than 3.

17. $3x - 1 < 5$ or $5x - 5 > 10$
    $3x < 6$ $5x > 15$
    $x < 2$ $x > 3$

21. $11x < 22$ or $12x > 36$
    $x < 2$ $x > 3$

See the graph of the solution in the back of the textbook.

25. $2x - 3 < 8$ and $3x + 1 > -10$

     $2x < 11$ $x > -11$

       $x < \dfrac{11}{2}$ $x > -\dfrac{11}{3}$

See the graph of the solution in the back of the textbook.

29. $-1 \leq \quad x - 5 \quad \leq 2$

     $4 \leq \quad\quad x \quad\quad \leq 7$

33. $-3 < \quad 2x + 1 \quad < 5$

     $-4 < \quad\quad 2x \quad\quad < 4$ Add -1 to each member

     $-2 < \quad\quad x \quad\quad < 2$ Multiply through by $\dfrac{1}{2}$

See the graph of the solution in the back of the textbook.

37. $-7 < \quad 2x + 3 \quad < 11$

     $-10 < \quad\quad 2x \quad\quad < 8$ Add -3 to each member

      $-5 < \quad\quad x \quad\quad < 4$ Multiply through by $\dfrac{1}{2}$

See the graph of the solution in the back of the textbook.

41. $10 < \quad x + 5 \quad < 20$

      $5 < \quad\quad x \quad\quad < 15$ Add -5 to each member

45. Let width $= x$, length $= x + 4$. Remember $P = 2L + 2W$.

       $20 < \quad\quad\quad P \quad\quad\quad < 30$

       $20 < \quad\quad 2L + 2W \quad\quad < 30$

       $20 < \quad 2(x + 4) + 2x \quad < 30$

       $20 < \quad\; 2x + 8 + 2x \quad\; < 30$ Distributive property

       $20 < \quad\quad\; 4x + 8 \quad\quad\; < 30$ Simplify

       $12 < \quad\quad\quad 4x \quad\quad\quad < 22$ Add -8 to each member

        $3 < \quad\quad\quad x \quad\quad\quad < \dfrac{11}{2}$ Multiply through by $\dfrac{1}{4}$

The width is between 3 inches and $\dfrac{11}{2}$ inches.

49. $-3 - 4(-2) = -3 + 8$ Multiply

              $= 5$

53. $5-2\left[-3(5-7)-8\right]=5-2\left[-3(-2)+-8\right]$ Add

$\qquad = 5-2\left[6+-8\right]$ Multiply

$\qquad = 5-2\left[-2\right]$ Add

$\qquad = 5+4$ Multiply

$\qquad = 9$

57. $\dfrac{1}{2}(4x-6)=\dfrac{1}{2}(4x)-\dfrac{1}{2}(6)$ Distributive property

$\qquad = 2x-3$ Multiply

61. $7-3(2x-4)-8=7-6x+12-8$ Distributive property

$\qquad = -6x+11$

65. $8-2(x+7)=2$

$\quad 8-2x-14=2$ Distributive property

$\qquad -2x-6=2$ Simplify

$\qquad -2x=8$ Add 6 to both sides

$\qquad x=-4$ Multiply each side by $-\dfrac{1}{2}$

69. $3-2x>5$

$\quad -2x>2$ Add -3 to both sides

$\quad x<-1$ Multiply each side by $-\dfrac{1}{2}$ and reverse
the direction of the inequality symbol.

See the graph in the back of the textbook.

73.   Width $= x$

Length $= 3x+5$.   Remember $P=2L+2W$

Perimeter $= 26$ inches

$\qquad P=2L+2W$

$\qquad 26=2(3x+5)+2x$

$\qquad 26=6x+10+2x$ Distributive property

$\qquad 26=8x+10$ Simplify

$\qquad 16=8x$ Add -10 to both sides

$\qquad 2=x$ Multiply by $\dfrac{1}{8}$

Width is 2 inches, length is 11 inches.

SECTION 9.2

1. $2x - 3y < 6$

Step 1: Find the intercepts for the boundary.

x-Intercept	**y-Intercept**
When $y = 0$	When $x = 0$
$2x - 3(0) = 6$	$2(0) - 3y = 6$
$2x = 6$	$-3y = 6$
$x = 3$	$y = -2$

Step 2: Since the inequality symbol is $<$, we graph the boundary as a broken line with x-intercept 3 and y-intercept -2.

Step 3: Using $(0, 0)$ in $2x - 3y < 6$, we have

$$0 - 0 < 6$$

$$0 < 6 \quad \text{A true statement}$$

Since $(0, 0)$ lies above the boundary, the solution set also lies above the boundary.

See the graph in the back of the textbook.

5. $x - y \leq 2$

Step 1: Find the intercepts for the boundary.

x-Intercept	**y-Intercept**
When $y = 0$	When $x = 0$
$x - 0 = 2$	$0 - y = 2$
$x = 2$	$-y = 2$
	$y = -2$

Step 2: Since the original inequality symbol is $\leq$, we graph a solid line with x-intercept 2 and y-intercept -2.

Step 3: Using $(0, 0)$ in $x - y \leq 2$, we have

$$0 - 0 \leq 2$$

$$0 \leq 2 \quad \text{A true statement}$$

Since $(0, 0)$ lies above the boundary, the solution set also lies above the boundary.

See the graph in the back of the textbook.

9. $5x - y \leq 5$

Step 1: Find the intercepts for the boundary.

x-Intercept	**y-Intercept**
When $y = 0$	When $x = 0$
$5x - 0 = 5$	$5(0) - y = 5$
$5x = 5$	$-y = 5$
$x = 1$	$y = -5$

Step 2: Since the inequality symbol is $\leq$ we have a solid line with
x-intercept 1 and y-intercept -5 for our boundary.

Step 3: Using $(0, 0)$ in $5x - y \leq 5$, we have

$$0 - 0 \leq 5$$

$$0 \leq 5 \quad \text{A true statement}$$

Since $(0, 0)$ lies above the boundary, the solution set also lies above the boundary.

See the graph in the back of the textbook.

13. $x \geq 1$

Step 1: $x = 1$ \quad The boundary is $x = 1$, which is a vertical line.

Step 2: The symbol is $\geq$, therefore the boundary is a solid line.

Step 3: Using $(0, 0)$ in $x \geq 1$, we have

$$0 \geq 1 \quad \text{A false statement}$$

Since $(0, 0)$ is to the left of the boundary the solution set lies to the right of the boundary.

See the graph in the back of the textbook.

17. $y < 2$

Step 1: $y = 2$ \quad The boundary is $y = 2$, which is a horizontal line.

Step 2: The symbol is $<$, therefore we have a broken line for the boundary.

Step 3: Using $(0, 0)$ in $y < 2$, we have

$$0 < 2 \quad \text{A true statement}$$

Since $(0, 0)$ lies below the boundary, the solution set also lies below the boundary.

See the graph in the back of the textbook.

21. $y \leq 3x - 1$

Step 1: The boundary line is $y = 3x - 1$. The slope is 3 and the y-intercept is -1.

Step 2: The symbol is $\leq$, therefore we have a solid line for the boundary.

Step 3: Using $(0, 0)$ in $y \leq 3x - 1$, we have

$$0 \leq 3(0) - 1$$

$$0 \leq -1 \quad \text{A false statement}$$

The solution set lies below the boundary. Since $(0, 0)$ lies above the boundary, the solution set lies below the boundary.

See the graph in the back of the textbook.

25. $y \leq -\frac{1}{2}x + 2$

Step 1: The boundary is $y = -\frac{1}{2}x + 2$, which is a line with slope $-\frac{1}{2}$ and y-intercept 2.

Step 2: The symbol is $\leq$, so the boundary is a solid line.

Step 3: Using (0, 0) in $y \le -\frac{1}{2}x + 2$, we have

$$0 \le -\frac{1}{2}(0) + 2$$

$$0 \le 2 \qquad \text{A true statement}$$

Since (0, 0) lies below the boundary, the solution set also lies below the boundary.

See the graph in the back of the textbook.

29. $y = \frac{1}{2}x + 3$ when $x = -2$ $y = \frac{1}{2}x + 3$ when $x = 0$ $y = \frac{1}{2}x + 3$ when $x = 2$

$y = \frac{1}{2}(-2) + 3$ $y = \frac{1}{2}(0) + 3$ $y = \frac{1}{2}(2) + 3$

$y = -1 + 3$ $y = 3$ $y = 1 + 3$

$y = 2$ $y = 4$

The ordered pairs are $(-2, 2)$, $(0, 3)$ and $(2, 4)$.

33. Let $(x_1, y_1) = (-1, 2)$ and $(x_2, y_2) = (5, -4)$

$$m = \frac{y_2 - y_1}{x_2 - x_1}$$

$$m = \frac{-4 - 2}{5 - (-1)} = \frac{-6}{6} = -1$$

37. $3x + 2y = 1$ No change $\rightarrow$ $3x + 2y = 1$

     $2x + y = 3$ Multiply by $-2 \rightarrow$ $-4x - 2y = -6$

                                               $-x = -5$

                                                 $x = 5$

Substituting $x = 5$ into

$2x + y = 3$ becomes

$2(5) + y = 3$

$10 + y = 3$

$y = -7$

The ordered pair is $(5, -7)$.

41. Step 1: We know that there is $1200 to invest at 8% and 10%. The total interest rate from both accounts for the year is $104. We do not know how much is invested in each account.

Step 2: Let x equal the amount in the 8% account and y equal the amount in the 10% account. If the two accounts have $1200 this translates to:

$$x + y = 1200$$

To find the total interest, we multiply the amount invested by the interest rate and add.

$$0.08x + 0.10y = 104$$

Step 3: The two equations are:

$$x + y = 1200$$
$$0.08x + 0.10y = 104$$

Step 4: To solve for y we will rewrite the first equation in terms of x.

$$x + y = 1200$$
$$y = 1200 - x$$

Substituting $y = 1200 - x$ into the second equation gives us

$$0.08x + 0.10(1200 - x) = 104$$
$$0.08x + 120 - 0.10x = 104$$
$$-0.02x + 120 = 104$$
$$-0.02x = -16$$
$$x = \$800$$
$$y = 1200 - x = \$400$$

Step 5: $800 is invested at 8% and $400 at 10%.

Step 6: $800 + $400 = $1,200

The interest on $800 at 8% is 0.08(800) = 64

~~The interest on $400 at 10% is 0.10(400) = 40~~

The total interest   = 104

SECTION 9.3

1. $(x+4)(x-4) = x^2 - 16$

5.
$$
\begin{array}{r}
x^2 - 2x + 4 \\
x + 2 \\
\hline
x^3 - 2x^2 + 4x \\
2x^2 - 4x + 8 \\
\hline
x^3 \qquad\qquad + 8
\end{array}
$$

9.
$$
\begin{array}{r}
x^2 - 4x + 16 \\
x + 4 \\
\hline
x^3 - 4x^2 + 16x \\
4x^2 - 16x + 64 \\
\hline
x^3 \qquad\qquad + 64
\end{array}
$$

13.
$$
\begin{array}{r}
x^2 - xy + y^2 \\
x + y \\
\hline
x^3 - x^2y + xy^2 \\
x^2y - xy^2 + y^3 \\
\hline
x^3 \qquad\qquad + y^3
\end{array}
$$

17. $x^3 + 7^3 = (x+7)(x^2 - 7x + 49)$

21. $x^3 + 8 = x^3 + 2^3$
$$= (x+2)(x^2 - 2x + 4)$$

25. $x^3 + 1 = x^3 + 1^3$

$\quad\quad = (x+1)(x^2 - x + 1)$

29. $27x^3 - 8y^3 = (3x)^3 - (2y)^3$

$\quad\quad\quad\quad = (3x - 2y)(9x^2 + 6xy + 4y^2)$

33. $x^3 + 64y^3 = x^3 + (4y)^3$

$\quad\quad\quad = (x + 4y)(x^2 - 4xy + 16y^2)$

37. $\dfrac{x^4}{x^{-3}} = x^{4-(-3)} = x^{4+3} = x^7$

41. $20ab^2 - 16ab^2 + 6ab^2 = (20 - 16 + 6)ab^2$

$\quad\quad\quad\quad\quad\quad\quad = 10ab^2$

45. $2x^2(3x^2 + 3x - 1) = 2x^2(3x^2) + 2x^2(3x) + 2x^2(-1)$

$\quad\quad\quad\quad\quad\quad = 6x^4 + 6x^3 - 2x^2$

49. $(2a^2 + 7)(2a^2 - 7) = (2a^2)^2 - 7^2$

$\quad\quad\quad\quad\quad\quad = 4a^4 - 49$

53. $3ax - 2a + 15x - 10 = a(3x - 2) + 5(3x - 2)$

$\quad\quad\quad\quad\quad\quad\quad = (a + 5)(3x - 2)$

57. $2x^2 + xy - 21y^2$

Possible Factors	Middle Term when Multiplied
$(2x - 3y)(x + 7y)$	$11xy$
$(2x - 7y)(x + 3y)$	$-xy$
$(2x - y)(x + 21y)$	$41xy$
$(2x - 21y)(x + y)$	$-19xy$

$2x^2 + xy - 21y^2 = (2x + 7y)(x - 3y)$

61. $\quad\quad\quad 8x^2 = -2x + 15$

$\quad 8x^2 + 2x - 15 = 0$ $\quad\quad$ Standard form

$\quad (4x - 5)(2x + 3) = 0$ $\quad\quad$ Factor the left side

$\quad 4x - 5 = 0 \quad$ or $\quad 2x + 3 = 0$

$\quad\quad 4x = 5 \quad\quad\quad\quad 2x = -3$

$\quad\quad x = \dfrac{5}{4} \quad\quad\quad\quad x = -\dfrac{3}{2}$

The set of answers are $-\frac{3}{2}$ and $\frac{5}{4}$.

SECTION 9.4

1. When $y = 10$ and $x = 5$, $y = kx$ becomes
$$10 = k(5)$$
$$2 = k$$
Therefore, $y = 2x$. Since $x = 4$, we have
$$y = 2(4)$$
$$y = 8$$

9. When $y = 75$ and $x = 5$, $y = kx^2$ becomes
$$75 = k(5)^2$$
$$75 = 25k$$
$$3 = k$$
Therefore, $y = 3x^2$. Since $x = 1$, we have
$$y = 3(1)^2$$
$$y = 3$$

17. When $y = 5$ and $x = 3$, $y = \dfrac{k}{x}$ becomes
$$5 = \dfrac{k}{3}$$
$$15 = k$$
Therefore, $y = \dfrac{15}{x}$. Since $y = 15$, we have
$$15 = \dfrac{15}{x}$$
$$15x = 15$$
$$x = 1$$

5. When $y = -24$ and $x = 4$, $y = kx$ becomes
$$-24 = k(4)$$
$$-6 = k$$
Therefore, $y = -6x$. Since $y = -30$, we have
$$-30 = -6x$$
$$5 = x$$

13. When $y = 5$ and $x = 2$, $y = \dfrac{k}{x}$ becomes
$$5 = \dfrac{k}{2}$$
$$10 = k$$
Therefore, $y = \dfrac{10}{x}$. Since $x = 5$, we have
$$y = \dfrac{10}{5}$$
$$y = 2$$

21. When $y = 4$ and $x = 5$, $y = \dfrac{k}{x^2}$ becomes
$$4 = \dfrac{k}{5^2}$$
$$4 = \dfrac{k}{25}$$
Therefore, $y = \dfrac{100}{x^2}$. Since $x = 2$, we have
$$y = \dfrac{100}{2^2}$$
$$y = \dfrac{100}{4}$$
$$y = 25$$

25. When $t = 42$ lbs and $d = 2$ inches, $t = kd$ becomes
$$42 = k(2)$$
$$21 = k$$
Therefore, $t = 21d$. Since $d = 4$, we have
$$t = 21(4)$$
$$t = 84$$
The tension is 84 lbs.

29. When $h = 20$ hours and $M = \$157$, $M = kh$ becomes

$$157 = k(20)$$
$$7.85 = k$$

Therefore, $M = 7.85h$. Since $h = 30$ hours, we have

$$M = 7.85(30)$$
$$M = \$235.50$$

She makes $235.50.

33. When $I = 30$ amps and $R = 2$ ohms, $I = \dfrac{k}{R}$ becomes

$$30 = \frac{k}{2}$$
$$60 = k$$

Therefore, $I = \dfrac{60}{R}$. Since $R = 5$ ohms, we have

$$I = \frac{60}{5}$$
$$I = 12$$

The current is 12 amps.

37. $\dfrac{x^2 - 25}{x + 4} \cdot \dfrac{2x + 8}{x^2 - 9x + 20} = \dfrac{(x+5)(x-5) \cdot 2(x+4)}{(x+4)(x-5)(x-4)} = \dfrac{2(x+5)}{x-4}$

41. $\dfrac{1 - \frac{25}{x^2}}{1 - \frac{8}{x} + \frac{15}{x^2}} = \dfrac{x^2\left(1 - \frac{25}{x^2}\right)}{x^2\left(1 - \frac{8}{x} + \frac{15}{x^2}\right)}$

$= \dfrac{x^2 - 25}{x^2 - 8x + 15}$

$= \dfrac{(x+5)(x-5)}{(x-5)(x-3)}$

$= \dfrac{x+5}{x-3}$

45. We know a pool can be filled in 8 hours and can drain in 12 hours. They want to know how long it will take to fill if you leave the inlet pipe and drain open.

Let $x =$ amount of time to fill the pool with both pipes open. In one hour,

Amount of water let in by inlet pipe		Amount of water let out by drain		Total amount of water in pool
$\frac{1}{8}$	$-$	$\frac{1}{12}$	$=$	$\frac{1}{x}$

$$24x\left(\frac{1}{8}\right) \quad - \quad 24x\left(\frac{1}{12}\right) \quad = \quad 24x\left(\frac{1}{x}\right) \quad \text{Find LCD } 24x$$

$$3x \quad - \quad 2x \quad = \quad 24$$

$$x \quad = \quad 24 \text{ hours}$$

It will take 24 hours to fill the pool.

$$\frac{1}{8} - \frac{1}{12} \overset{?}{=} \frac{1}{24} \qquad \text{LCD} = 24$$

$$\frac{3}{24} - \frac{2}{24} \overset{?}{=} \frac{1}{24}$$

$$\frac{1}{24} = \frac{1}{24}$$

SECTION 9.5

1. Domain $= \{1, 3, 5\}$, Range $= \{2, 4, 6\}$

5. Domain $= \{1, 2, 3, 4\}$, Range $= \{1, 2, 3, 4\}$

9. Domain $= \{0, 2, 4\}$, Range $= \left\{2, \frac{2}{3}, -\frac{2}{3}\right\}$

13. $y = \dfrac{1}{x - 2}$

$x = 2$ makes the denominator zero:

Domain $= \{x \mid x \text{ is real}, \ x \neq 2\}$

17. $y = \dfrac{4}{(x+1)(x+2)}$

$x = -1$ and $x = -2$ makes the denominator zero:

Domain $= \{x \mid x \text{ is real}, \ x \neq -1, \ x \neq -2\}$

21. $y = \dfrac{3}{x^2 - 9} = \dfrac{3}{(x+3)(x-3)}$

$x = 3$ and -3 makes the denominator zero:

Domain $= \{x \mid x \text{ is real}, \ x \neq 3, \ x \neq -3\}$

25. $y = \dfrac{x}{2x^2 - 5x - 3} = \dfrac{x}{(2x+1)(x-3)}$

$x = -\dfrac{1}{2}$ and $x = 3$ makes the denominator zero:

Domain $= \left\{x \mid x \text{ is real}, \ x \neq -\dfrac{1}{2}, \ x \neq 3\right\}$

29. $f(x) = x - 5$

$f(2) = 2 - 5 = -3$

$f(0) = 0 - 5 = -5$

$f(-1) = -1 - 5 = -6$

$f(a) = a - 5$

33. $f(x) = x^2 + 1$

$f(2) = 2^2 + 1 = 5$

$f(3) = 3^2 + 1 = 10$

$f(4) = 4^2 + 1 = 17$

$f(a) = a^2 + 1$

37. $f(x) = x^2 + 2x$

$f(3) = 3^2 + 2(3) = 9 + 6 = 15$

$f(-2) = (-2)^2 + 2(-2) = 4 - 4 = 0$

$f(a) = a^2 + 2a$

$f(a - 2) = (a - 2)^2 + 2(a - 2)$

$= a^2 - 4a + 4 + 2a - 4$

$= a^2 - 2a$

41. $\sqrt{49} = 7$ because $7^2 = 49$

45. $\sqrt{\dfrac{2}{5}} = \dfrac{\sqrt{2}}{\sqrt{5}} \cdot \dfrac{\sqrt{5}}{\sqrt{5}}$

$= \dfrac{\sqrt{10}}{5}$

49. $(\sqrt{6} + 2)(\sqrt{6} - 5) = (\sqrt{6})^2 + \sqrt{6}(-5) + 2\sqrt{6} + 2(-5)$

                 F   **O**   **I**   **L**

$= 6 - 5\sqrt{6} + 2\sqrt{6} - 10$

$= -4 - 3\sqrt{6}$

53. $\sqrt{2x - 5} = 3$

$2x - 5 = 9$ Square both sides

$2x = 14$

$x = 7$

Check: $\sqrt{2(7) - 5} \overset{?}{=} 3$

$\sqrt{9} \overset{?}{=} 3$

$3 = 3$

The solution is 7.

SECTION 9.6

1. $4^{1/2} = \sqrt{4} = 2$ The exponent of $\dfrac{1}{2}$ indicates that we are to find the square root of 4.

5. $27^{1/3} = \sqrt[3]{27} = 3$ The exponent of $\dfrac{1}{3}$ indicates that we are to find the cube root of 27.

9. $81^{1/4} = \sqrt[4]{81} = 3$ The exponent of $\dfrac{1}{4}$ indicates that we are to find the fourth root of 81.

13. $8^{2/3} = (8^{1/3})^2$ Separate exponents

$= (\sqrt[3]{8})^2$ Write as cube root

$= 2^2$ $\sqrt[3]{8} = 2$

$= 4$ $2^2 = 4$

17. $16^{3/4} = \left(16^{1/4}\right)^3$ Separate exponents

     $= \left(\sqrt[4]{16}\right)^3$ Write as fourth root

     $= 2^3$ $\sqrt[4]{16} = 2$

     $= 8$ $2^3 = 8$

21. $4^{3/2} = \left(4^{1/2}\right)^3$ Separate exponents

     $= \left(\sqrt{4}\right)^3$ Write as square root

     $= 2^3$ $\sqrt{4} = 2$

     $= 8$ $2^3 = 8$

25. $(-32)^{1/5} = \sqrt[5]{-32}$ Write as a fifth root

       $= -2$ $(-2)^5 = -32$

29. $16^{3/4} + 27^{2/3} = \left(16^{1/4}\right)^3 + \left(27^{1/3}\right)^2$ Separate exponents

           $= \left(\sqrt[4]{16}\right)^3 + \left(\sqrt[3]{27}\right)^2$ Write as roots

           $= 2^3 + 3^2$ $\sqrt[4]{16} = 2, \ \sqrt[3]{27} = 3$

             $= 8 + 9$ $2^3 = 8, \ \ 3^2 = 9$

             $= 17$

33. $x^{1/4} \cdot x^{3/4} = x^{1/4 + 3/4}$

          $= x^1$

          $= x$

37. $\dfrac{a^{3/5}}{a^{1/5}} = a^{3/5 - 1/5}$

       $= a^{2/5}$

41. $\left(9a^4 b^2\right)^{1/2} = 9^{1/2} \cdot a^{4(1/2)} \cdot b^{2(1/2)}$

           $= \sqrt{9} \cdot a^2 \cdot b^1$

           $= 3a^2 b$

45. $25^{-1/2} = \dfrac{1}{25^{1/2}}$

       $= \dfrac{1}{\sqrt{25}}$

       $= \dfrac{1}{5}$

49. $27^{-2/3} = \dfrac{1}{27^{2/3}}$

$= \dfrac{1}{\left(27^{1/3}\right)^2}$

$= \dfrac{1}{\left(\sqrt[3]{27}\right)^2}$

$= \dfrac{1}{3^2}$

$= \dfrac{1}{9}$

53. $\qquad 3x^2 = 4x + 2$

$3x^2 - 4x - 2 = 0$

Letting $a = 3$, $b = -4$ and $c = -2$ in

$x = \dfrac{-b \pm \sqrt{b^2 - 4ac}}{2a}$, we have:

$= \dfrac{-(-4) \pm \sqrt{(-4)^2 - 4(3)(-2)}}{2(3)}$

$= \dfrac{4 \pm \sqrt{16 + 24}}{6}$

$= \dfrac{4 \pm \sqrt{40}}{6}$

$= \dfrac{4 \pm 2\sqrt{10}}{6}$

$= \dfrac{2\left(2 \pm \sqrt{10}\right)}{6}$

$x = \dfrac{2 \pm \sqrt{10}}{3}$

The solutions are $\dfrac{2 + \sqrt{10}}{3}$ and $\dfrac{2 - \sqrt{10}}{3}$.

57. $(4 + 3i) - (2 - 5i) = 4 + 3i - 2 + 5i$

$= 4 - 2 + 3i + 5i$

$= 2 + 8i$

CHAPTER 9 REVIEW

1 - 4. See the graph in the back of the textbook.

5. $-5x \geq 25$ or $2x - 3 \geq 9$
$x \leq -5$ $2x \geq 12$
$x \geq 6$

See the graph in the back of the textbook.

6. $-3x \geq -6$ or $2x - 7 \geq 7$
$x \leq 2$ $2x \geq 14$
$x \geq 7$

See the graph in the back of the textbook.

7. $-1 < 2x + 1 < 9$
$-2 < 2x < 8$
$\dfrac{-2}{2} < \dfrac{2x}{2} < \dfrac{8}{2}$
$-1 < x < 4$

See the graph in the back of the textbook.

8. $-3 < 2x + 3 < 5$
$-6 < 2x < 2$
$\dfrac{-6}{2} < \dfrac{2x}{2} < \dfrac{2}{2}$
$-3 < x < 1$

See the graph in the back of the textbook.

9. $-1 \leq 3x + 5 \leq 8$
$-6 \leq 3x \leq 3$
$\dfrac{-6}{3} \leq \dfrac{3x}{3} \leq \dfrac{3}{3}$
$-2 \leq x \leq 1$

See the graph in the back of the textbook.

10. $-5 \leq 4x + 3 \leq 11$
$-8 \leq 4x \leq 8$
$\dfrac{-8}{4} \leq \dfrac{4x}{4} \leq \dfrac{8}{4}$
$-2 \leq x \leq 2$

See the graph in the back of the textbook.

11. $x - y < 3$

$x - y = 3$, let $x = 0$ $x - y = 3$, let $y = 0$
$0 - y = 3$ $x - 0 = 3$
   $y = -3$ (y-intercept) $x = 3$ (x-intercept)

The line is broken because of $<$. Try $(0, 0)$.

$x - y < 3$
$0 - 0 < 3$
   $0 < 3$ true

Since $(0, 0)$ lies above the boundary, the solution also lies above the boundary.

See the graph in the back of the textbook.

12. $2x + 3y \geq 6$

$2x + 3y = 6$, let $x = 0$ $2x + 3y = 6$, let $y = 0$

$2(0) + 3y = 6$ $2x + 3(0) = 6$

   $3y = 6$ $2x = 6$

    $y = 2$ $(y\text{-intercept})$ $x = 3$ $(x\text{-intercept})$

The line is solid because of the $\geq$. Try $(0, 0)$.

  $2x + 3y \geq 6$

$2(0) + 3(0) \geq 6$

    $0 \geq 6$ false

Since $(0, 0)$ lies below the boundary, the solution lies above the boundary.

See the graph in the back of the textbook.

13. $x \geq -3$ y is all real numbers and the line is solid. The solution lies to the right of the boundary.

See the graph in the back of the textbook.

14. $y < 2$ x is all real numbers and the line is broken. The solution lies below the boundary.

See the graph in the back of the textbook.

15. $y \leq -2x + 3$

$y = -2x + 3$ let $x = 0$ $y = -2x + 3$, let $y = 0$

$y = -2(0) + 3$ $0 = -2x + 3$

$y = 3$ $(y\text{-intercept})$ $-3 = -2x$

          $\dfrac{3}{2} = x$ $(x\text{-intercept})$

The line is solid because of $\leq$. Try $(0, 0)$.

  $y \leq -2x + 3$

$0 \leq -2(0) + 3$

$0 \leq 3$ true

Since $(0, 0)$ lies below the boundary, the solution also lies below the boundary.

See the graph in the back of the textbook.

16. $y > 3x - 4$

    $y = 3x - 4,$ let $x = 0,$ $y = 3x - 4,$ let $y = 0$

    $y = 3(0) - 4$ $0 = 3x - 4$

    $y = -4$ (y-intercept) $4 = 3x$

                                $\dfrac{4}{3} = x$ (x-intercept)

The line is broken because of $>$. Try $(0, 0)$.

    $y > 3x - 4$

    $0 > 0 - 4$

    $0 > -4$ true

Since $(0, 0)$ lies above the boundary, the solution also lies above the boundary.

See the graph in the back of the textbook.

17. $a^3 - 7^3 = (a - 7)(a^2 + 7a + 49)$

18. $a^3 + 6^3 = (a + 6)(a^2 - 6a + 36)$

19. $x^3 - 8 = x^3 - 2^3 = (x - 2)(x^2 + 2x + 4)$

20. $x^3 + 8 = x^3 + 2^3 = (x + 2)(x^2 - 2x + 4)$

21. $27x^3 + y^3 = (3x)^3 + y^3 = (3x + y)(9x^2 - 3xy + y^2)$

22. $27x^3 - y^3 = (3x)^3 - y^3 = (3x - y)(9x^2 + 3xy + y^2)$

23. $64x^3 + 125 = (4x)^3 + 5^3 = (4x + 5)(16x^2 - 20x + 25)$

24. $8x^3 - 27 = (2x)^3 - 3^3 = (2x - 3)(4x^2 + 6x + 9)$

25.   When $y = -20$ and $x = 4,$ $y = kx$ becomes

                     $-20 = k(4)$

                       $-5 = k$

     Therefore, $y = -5x.$ Since $x = 7,$ we have

                     $y = -5(7)$

                     $y = -35$

26.   When $y = 45$ and $x = 3,$ $y = kx^2$ becomes

                       $45 = k(3)^2$

                       $45 = 9k$

                       $5 = k$

     Therefore, $y = 5x^2.$ Since $x = 5,$ we have

                     $y = 5(5)^2$

                     $y = 125$

27.   When $y = 3$ and $x = 2$, $y = \dfrac{k}{x}$ becomes

$$3 = \frac{k}{2}$$

$$6 = k$$

Therefore, $y = \dfrac{6}{x}$. Since $x = 12$, we have

$$y = \frac{6}{12}$$

$$y = \frac{1}{2}$$

28.   When $y = 8$ and $x = 36$, $y = \dfrac{k}{\sqrt{x}}$ becomes

$$8 = \frac{k}{\sqrt{36}}$$

$$8 = \frac{k}{6}$$

$$48 = k$$

Therefore, $y = \dfrac{48}{\sqrt{x}}$. Since $x = 144$, we have

$$y = \frac{48}{\sqrt{144}}$$

$$y = \frac{48}{12}$$

$$y = 4$$

29.   When $V = 5$ and $T = 200$, $V = kT$ becomes

$$5 = k(200)$$

$$\frac{5}{200} = k$$

$$\frac{1}{40} = k$$

Therefore, $V = \dfrac{1}{40}T$. Since $T = 360$, we have

$$V = \frac{1}{40}(360)$$

$$V = 9$$

30.   When $I = 50$ and $R = 6$ ohms, $I = \dfrac{k}{R}$ becomes

$$50 = \frac{k}{6}$$

$$300 = k$$

Therefore, $I = \dfrac{300}{R}$. Since $R = 10$, we have

$$I = \frac{300}{10}$$

$$I = 30 \text{ amps}$$

31. Domain = $\{2, -3, 4\}$,   Range = $\{1, 5, -2\}$

32. Domain = $\{-3, 1, -5\}$, Range = $\{-2, -4\}$

33. $y = 3x - 1$ for $x = 2$ $y = 3x - 1$ for $x = 4$ $y = 3x - 1$ for $x = 6$
   $y = 3(2) - 1$ $y = 3(4) - 1$ $y = 3(6) - 1$
   $y = 5$ $y = 12 - 1$ $y = 18 - 1$
                                  $y = 11$ $y = 17$
   Domain = $\{2, 4, 6\}$ and Range = $\{5, 11, 17\}$

34. $y = 2x + 3$ for $y = -1$ $y = 2x + 3$ for $y = 1$ $y = 2x + 3$ for $y = 3$
   $-1 = 2x + 3$ $1 = 2x + 3$ $3 = 2x + 3$
   $-4 = 2x$ $-2 = 2x$ $0 = 2x$
   $-2 = x$ $-1 = x$ $0 = x$
   Domain = $\{-2, -1, 0\}$ and Range = $\{-1, 1, 3\}$

35. $y = \dfrac{1}{(x+3)(x+7)}$

   $x = -3$ and $x = -7$ makes the denominator zero:

   Domain = $\{x \mid x$ is real, $x \neq -3, x \neq -7\}$.

36. $y = \dfrac{1}{(x-2)(x+5)}$

   $x = 2$ and $x = -5$ makes the denominator zero:

   Domain = $\{x \mid x$ is real, $x \neq 2, x \neq -5\}$.

37. $y = \dfrac{5}{x^2 - 8x + 15} = \dfrac{5}{(x-3)(x-5)}$

   $x = 3$ and $x = 5$ makes the denominator zero:

   Domain = $\{x \mid x$ is real, $x \neq 3, x \neq 5\}$.

38. $y = \dfrac{2x}{x^2 - 3x - 28} = \dfrac{2x}{(x+4)(x-7)}$

   $x = -4$ and $x = 7$ makes the denominator zero:

   Domain = $\{x \mid x$ is real, $x \neq -4, x \neq 7\}$.

39. $f(x) = -2x$
   $f(-3) = -2(-3) = 6$
   $f(0) = -2(0) = 0$
   $f(1) = -2(1) = -2$

40. $f(x) = -2x - 1$
   $f(-2) = -2(-2) - 1 = 3$
   $f(1) = -2(1) - 1 = -3$
   $f(3) = -2(3) - 1 = -7$

41. $f(x) = 2x^2 + 1$
   $f(-2) = 2(-2)^2 + 1 = 2(4) + 1 = 9$
   $f(2) = 2(2)^2 + 1 = 2(4) + 1 = 9$
   $f(a) = 2a^2 + 1$

42. $f(x) = -3x^2 + 2x + 1$
   $f(-1) = -3(-1)^2 + 2(-1) + 1 = -3 - 2 + 1 = -4$
   $f(3) = -3(3)^2 + 2(3) + 1 = -27 + 6 + 1 = -20$
   $f(c) = -3c^2 + 2c + 1$

43. $f(x) = \dfrac{1}{x}$

$f(-2) = -\dfrac{1}{2}$

$f(4) = \dfrac{1}{4}$

$f\left(\dfrac{1}{2}\right) = \dfrac{1}{\frac{1}{2}} = 1 \cdot \dfrac{2}{1} = 2$

$f\left(\dfrac{2}{3}\right) = \dfrac{1}{\frac{2}{3}} = 1 \cdot \dfrac{3}{2} = \dfrac{3}{2}$

44. $f(x) = \sqrt{x}$

$f(36) = \sqrt{36} = 6$

$f(81) = \sqrt{81} = 9$

$f(b) = \sqrt{b}$

45. $f(x) = x^3 + 1$

$f(-2) = (-2)^3 + 1 = -8 + 1 = -7$

$f(2) = (2)^3 + 1 = 8 + 1 = 9$

$f(a) = a^3 + 1$

46. $f(x) = x^3 - 3$

$f(-1) = (-1)^3 - 3 = -4$

$f(3) = 3^3 - 3 = 24$

$f(b) = b^3 - 3$

47. $36^{1/2} = \sqrt{36} = 6$

48. $100^{1/2} = \sqrt{100} = 10$

49. $27^{2/3} = \left(27^{1/3}\right)^2 = \left(\sqrt[3]{27}\right)^2 = 3^2 = 9$

50. $8^{4/3} = \left(8^{1/3}\right)^4 = \left(\sqrt[3]{8}\right)^4 = 2^4 = 16$

51. $(-8)^{2/3} = \left[(-8)^{1/3}\right]^2 = \left(\sqrt[3]{-8}\right)^2 = (-2)^2 = 4$

52. $(-32)^{2/5} = \left[(-32)^{1/5}\right]^2 = \left(\sqrt[5]{-32}\right)^2 = (-2)^2 = 4$

53. $4^{5/2} + 27^{2/3}$

$= \left(4^{1/2}\right)^5 + \left(27^{1/3}\right)^2$

$= \left(\sqrt{4}\right)^5 + \left(\sqrt[3]{27}\right)^2$

$= 2^5 + 3^2$

$= 32 + 9$

$= 41$

54. $(-8)^{1/3} + (8)^{2/3}$

$= \sqrt[3]{-8} + \left(\sqrt[3]{8}\right)^2$

$= -2 + (2)^2$

$= 2$

55. $\left(16y^{20}\right)^{1/4}$

$= 16^{1/4}\left(y^{20}\right)^{1/4}$

$= \sqrt[4]{16}\left(y^5\right)$

$= 2y^5$

56. $\left(27y^{15}\right)^{1/3}$

$= (27)^{1/3}\left(y^{15}\right)^{1/3}$

$= \sqrt[3]{27}\left(y^5\right)$

$= 3y^5$

57. $27^{-1/3}$

$$= \frac{1}{27^{1/3}}$$

$$= \frac{1}{\sqrt[3]{27}}$$

$$= \frac{1}{3}$$

58. $32^{-2/5}$

$$= \frac{1}{32^{2/5}}$$

$$= \frac{1}{\left(\sqrt[5]{32}\right)^2}$$

$$= \frac{1}{2^2}$$

$$= \frac{1}{4}$$

CHAPTER 9 TEST

1. $x < 2$ or $x > 3$

2. $x \geq -2$ and $x \leq 3$

The graphs of the solutions are in the back of the textbook.

3. $3 - 4x \geq -5$ or $2x \geq 10$

$ 3 - 4x \geq -5 \qquad 2x \geq 10$

$ -4x \geq -8 \qquad x \geq 5$

$ x \leq 2$

4. $-7 < 2x - 1 < 9$

$ -6 < 2x \quad < 10$ Add 1 to each member

$ -3 < x \quad < 5$ Divide each by 2

5. $y < -x + 4$

    Step 1: $y = -x + 4$

        Graph the line with slope $= -1$ and y-intercept $= 4$.

    Step 2: The symbol is $<$, therefore we have a broken line.

    Step 3: Using $(0, 0)$ in $y < -x + 4$, we have

$$0 < -0 + 4$$

$$0 < 4 \quad \text{A true statement}$$

    Since $(0, 0)$ lies below the boundary, the solution set also lies below the boundary.

    The graph is in the back of the textbook.

6. $3x - 4y \geq 12$

    Step 1: $3x - 4y = 12$

x-Intercept	**y-Intercept**
$y = 0$	$x = 0$
$3x - 4(0) = 12$	$3(0) - 4y = 12$
$3x = 12$	$-4y = 12$
$x = 4$	$y = -3$
x-intercept at $(4, 0)$	y-intercept at $(0, -3)$

Step 2: The symbol is $\geq$, therefore we have a solid line.

Step 3: Using $(0, 0)$ in $3x - 4y \geq 12$, we have

$$3(0) - 4(0) \geq 12$$

$$0 \geq 12 \quad \text{A false statement}$$

Since $(0, 0)$ lies above the boundary, the solution set lies below the boundary.

The graph is in the back of the textbook.

7. $x^3 - 7^3 = (x - 7)(x^2 + 7x + 49)$

8. $x^3 + 27 = x^3 + 3^3 = (x + 3)(x^2 - 3x + 9)$

9. $(2x)^3 + (5y)^3 = (2x + 5y)(4x^2 - 10xy + 25y^2)$

10. $8x^3 - 27y^3 = (2x)^3 - (3y)^3 = (2x - 3y)(4x^2 + 6xy + 9y^2)$

11. $y = kx^2 \quad$ when $x = 3$ and $y = 36$

$36 = k3^2$

$36 = 9k$

$4 = k \quad$ therefore

$y = 4x^2 \quad$ substitute $x = 5$

$y = 4(5)^2$

$y = 4(25)$

$y = 100$

12. $y = \dfrac{k}{x} \quad$ when $x = 3$ and $y = 6$

$6 = \dfrac{k}{3}$

$18 = k \quad$ therefore

$y = \dfrac{18}{x} \quad$ substitute $x = 9$

$y = \dfrac{18}{9}$

$y = 2$

13. Domain = $\{1, 3, 5\}$ and Range = $\{2, 4, 6\}$

14. $y = \dfrac{5}{x^2 - 25} = \dfrac{5}{(x + 5)(x - 5)}$

$x = 5$ and $x = -5$ makes the denominator zero:

Domain = $\{x | x \text{ is real}, x \neq 5, x \neq -5\}$.

15. $f(x) = 3x^2 - 1$

$f(3) = 3(3)^2 - 1$

$\quad = 3(9) - 1$

$\quad = 27 - 1$

$\quad = 26$

16. $f(x) = 3x^2 - 1$

$f(-2) = 3(-2)^2 - 1$

$\quad = 3(4) - 1$

$\quad = 12 - 1$

$\quad = 11$

17. $g(x) = 5x + 2$
$\quad g(0) = 5(0) + 2$
$\quad\quad\quad = 2$

18. $g(x) = 5x + 2$
$\quad f(-5) = 5(-5) + 2$
$\quad\quad\quad = -25 + 2$
$\quad\quad\quad = -23$

19. $25^{1/2} = \sqrt{25} = 5$

20. $8^{2/3} = \left(8^{1/3}\right)^2$ Separate exponents

$\quad\quad = \left(\sqrt[3]{8}\right)^2$ Write as cube root

$\quad\quad = 2^2$ $\sqrt[3]{8} = 2$

$\quad\quad = 4$ $2^2 = 4$

21. $x^{3/4} x^{1/4} = x^{3/4 + 1/4}$
$\quad\quad\quad\quad = x^1$
$\quad\quad\quad\quad = x$

22. $\left(27x^6 y^9\right)^{1/3} = 27^{1/3} x^{6 \cdot 1/3} y^{9 \cdot 1/3}$
$\quad\quad\quad\quad\quad = \sqrt[3]{27} x^2 y^3$
$\quad\quad\quad\quad\quad = 3x^2 y^3$